CW00796346

Chronicles *of* Drug Discovery

Volume 3

Chronicles *of* Drug Discovery

Volume 3

EDITED BY

Daniel Lednicer

National Cancer Institute, National Institutes of Health

American Chemical Society, Washington, DC 1993

Library of Congress Cataloging-in-Publication Data

(Revised for vol. 3)

Chronicles of drug discovery.
"A Wiley-Interscience publication."
Vol. 3 published by American Chemical Society, Washington, DC.
Includes bibliographical references and index.
1. Drugs—Research—History. 2. Drugs—History. I. Bindra, Jasjit S. II. Lednicer, Daniel, 1929- . [DNLM: 1. Drugs—Personal narratives. 2. Research—Personal narratives. 3. Technology, Pharmaceutical—Personal narratives. QV 55 C553]

RS122.C48
615'.1'09047 81-11471

ISBN 0-471-06516-1 (v. 1)
ISBN 0-471-89135-5 (v. 2)
ISBN 0-8412-2523-0 (clothbound) — ISBN 0-8412-2733-0 (paperback) (v. 3)

The paper used in this publication meets the minimum requirements of American National Standard for Information Sciences—Permanence of Paper for Printed Library Materials, ANSI Z39.48-1984.

PRINTED IN THE UNITED STATES OF AMERICA

1993 ACS Books Advisory Board

Contents

About the Editor

Daniel Lednicer worked in a wide variety of therapeutic areas for close to 20 years as a bench chemist for the Upjohn Company before he moved on to research management at a number of other pharmaceutical laboratories. For the past several years he has been with the Departmental Therapeutics Program at the National Cancer Institute (National Institutes of Health), where he oversees outside contracts that deal with various aspects of cancer drug development.

He is the author of numerous publications in medicinal chemistry, ranging from research papers to books on which he has served as editor or coeditor. He is coauthor of the four volumes of *The Organic Chemistry of Drug Synthesis,* published by John Wiley & Sons, Inc.

Contributors

Elden H. Banitt
3M Pharmaceuticals
3M Center 270–2S–06
St. Paul, MN 55144–1000

Allen Barnett
Schering-Plough Research Institute
2015 Galloping Hill Road
Kenilworth, NJ 07033–0539

John Bradshaw
Glaxo Group Research Limited
Park Road
Ware, Hertfordshire SG12 0DP
United Kingdom

C. M. Cimarusti
Bristol-Myers Squibb
Pharmaceutical Research Institute
One Squibb Drive
P.O. Box 191
New Brunswick, NJ 08903–0191

Paul W. Collins
Department of Medicinal Chemistry
G. D. Searle and Company
4901 Searle Parkway
Skokie, IL 60077

William A. Denny
Cancer Research Laboratory
Auckland Medical School
University of Auckland
Private Bag 92019
Auckland, New Zealand

R. Deraedt
23, allée Jean-Baptiste Clément
93320 Pavillons-Sous-Bois
France

Paul W. Erhardt
Berlex Laboratories
100 East Hanover Avenue
Cedar Knolls, NJ 07927

Michael J. Green
Schering-Plough Research Institute
2015 Galloping Hill Road
Kenilworth, NJ 07033–0539

Hirozumi Inoue
Organic Chemistry Research
Laboratory
Tanabe Seiyaku Company
2–50 Kawagishi 2–chome
Toda–shi, Saitama 335
Japan

Taku Nagao
Department of Toxicology and
Pharmacology
Faculty of Pharmaceutical Sciences
University of Tokyo
3–1 Hongo 7–chome
Bunkyo–ku, Tokyo 113
Japan

Arthur A. Patchett
Merck Research Laboratories
Merck & Co. Inc.
P.O. Box 2000
Rahway, NJ 07065–0900

D. Philibert
Recherche Sante
Departement Endocrinologie
Roussel UCLAF
102, 111 Route de Noisy
F–93230 Romainville
France

Jack R. Schmid
3M Pharmaceuticals
3M Center 270–2S–06
St. Paul, MN 55144–1000

Hartmann Stähelin
Hardstrasse 80
CH–4052 Basel
Switzerland

G. Teutsch
Recherche Sante
Departement Endocrinologie
Roussel UCLAF
102, 111 Route de Noisy
F–93230 Romainville
France

Julien P. H. Verheyden
Syntex Discovery Research
Institute of Bio-Organic Chemistry
3401 Hillview Avenue
P.O. Box 10850
Palo Alto, CA 94303

Albert von Wartburg
In der Au 33
CH–4125 Riehen
Switzerland

Monroe E. Wall
Research Triangle Institute
P.O. Box 12194
3040 Cornwallis Road
Research Triangle Park,
NC 27709–2194

Preface

THE EFFECTS OF CHEMICAL COMPOUNDS ON PHYSIOLOGICAL FUNCTION has been the focus of human interest for many centuries. This interest has led to continuing fascination with the effects of such compounds (commonly referred to as drugs if their use is fairly widespread), which are known either as a means of curing disease or as the cause of societal ills. The process of discovering new drugs that offer therapeutic potential is still far from a rote exercise. Though they are not the subject of this book, the same observation probably applies to drugs of abuse.

Drug discovery often involves a good measure of luck and serendipity. The process is sufficiently unpredictable so that pursuit of a new drug can involve the excitement of the chase. The present volume of *Chronicles of Drug Discovery* contains accounts of the discovery of 14 drugs that were fairly recently approved for sale or are in the late stages of development. The stories are told by the people most directly involved in the initial discovery stages. Because the majority of drugs are synthetic compounds, more often than not those individuals are chemists. Authors have been urged to depart from the formalities of the scientific literature; they have been encouraged instead to present an informal narrative account of what actually happened, warts and all.

The first volume of *Chronicles of Drug Discovery* was published about 10 years ago. The intervening decade saw a series of major advances in medicinal chemistry and its associated fields that promise to fairly revolutionize the process of drug discovery. As an example, the mode of interaction of the quinolone antibiotics with DNA gyrase has been elucidated to the molecular level. Consequently, it is often possible to seriously factor detailed descriptions of drug receptor interactions into drug design. Those developments may be expected to remove much of the empiricism that marked earlier medicinal chemistry research.

However, over the past decade these advances have not appreciably shortened the time lapse between the laboratory discovery of a new therapeutic entity and its application to human medicine. Neglecting for the moment some notable exceptions such as zidovudine, we see that the development period has not varied much from an average of about 10 years. Most of the chapters in this volume consequently describe drugs whose development was initiated not too long after the publication of Volume 1. Stories describing the discovery of drugs made possible by the newer developments in medicinal chemistry will have to wait for a later volume.

The accounts that follow very aptly illustrate the varying sources of inspiration for the discovery of new drugs. At least several of those stories involve targeted synthesis aimed at optimizing the activity of a promising substance uncovered by a screening program. The role of the biologist is highlighted by the drugs whose discovery hinged on the availability of some new screen. More than one drug in the collection was made possible by either the prior discovery of a new organic reaction or the exploitation of a new intermediate, reagent, or starting material. Several more entities came from programs guided by the precepts of rational drug design. That such very disparate strategies have all succeeded perhaps serves to emphasize the statement that drug discovery is still far from a cut and dried predictable exercise. This lack of predictability generates both excitement in practitioners and despair for those who preside over balance sheets and bottom lines.

DANIEL LEDNICER
Rockville, MD 20852

May 1992

Mifepristone

G. Teutsch, R. Deraedt, and D. Philibert
Roussel UCLAF

The report in 1981–1982 on the antiglucocorticoid and antiprogestational activities of the steroid RU 486 (now mifepristone, **1**) (*1–3*) is considered a landmark in endocrine research. Among the five classic steroid–hormone classes—estrogens, androgens, progestins, glucocorticoids, and mineralocorticoids—antihormones had been found for three of them (estrogens, androgens, and mineralocorticoids), whether of steroidal or nonsteroidal structure (Figure 1). A report on antiglucocorticoids (*4*) met with little success, as the compounds, 17β-carboxamides like **2**, displayed very weak or no in vivo activity. Antiprogestins, an actively sought species because of their

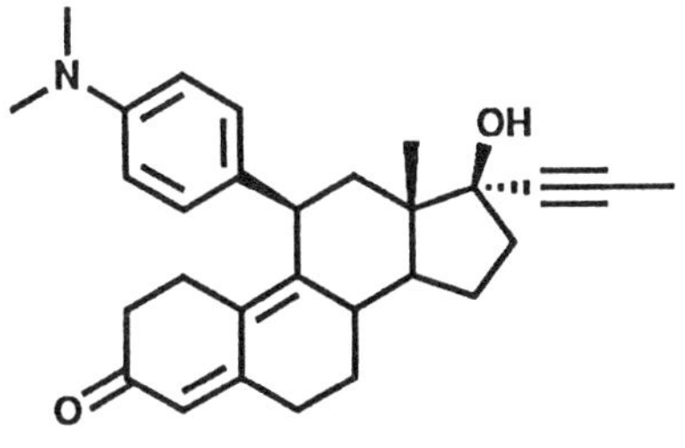

1

2523–0/93/0001$11.75/0

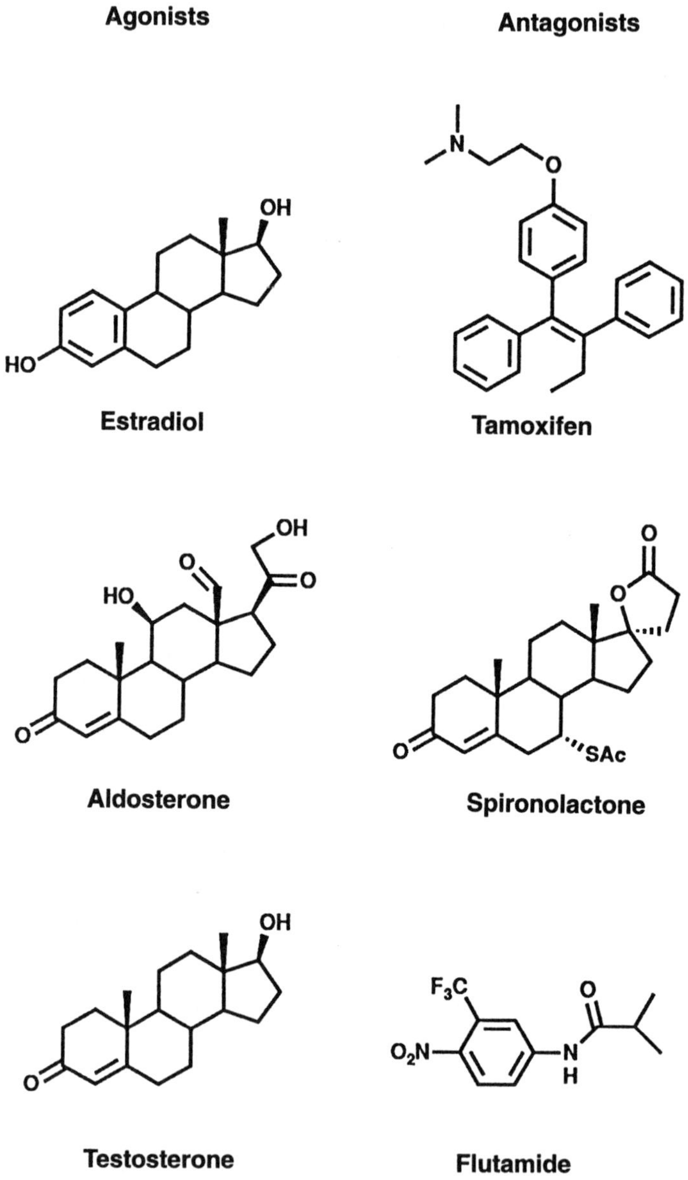

Agonists | **Antagonists**

Estradiol | **Tamoxifen**

Aldosterone | **Spironolactone**

Testosterone | **Flutamide**

Figure 1. Antihormones known prior to 1981 for three steroid–hormone classes.

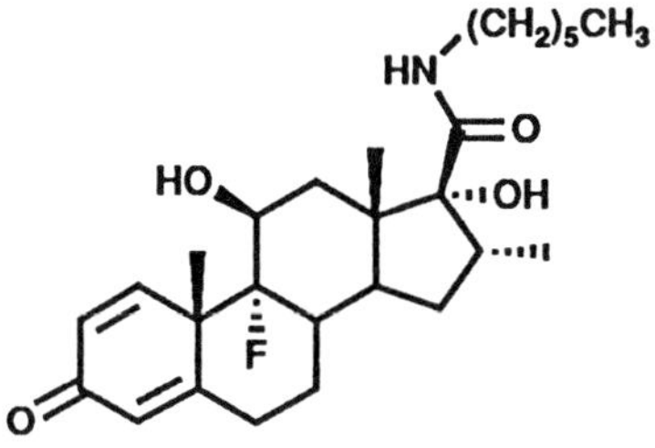

2

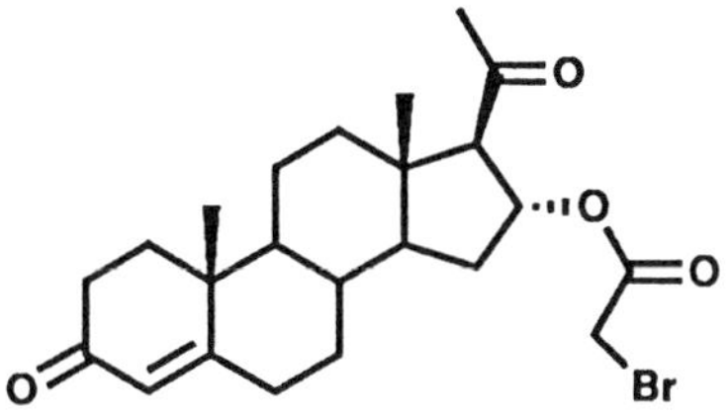

3

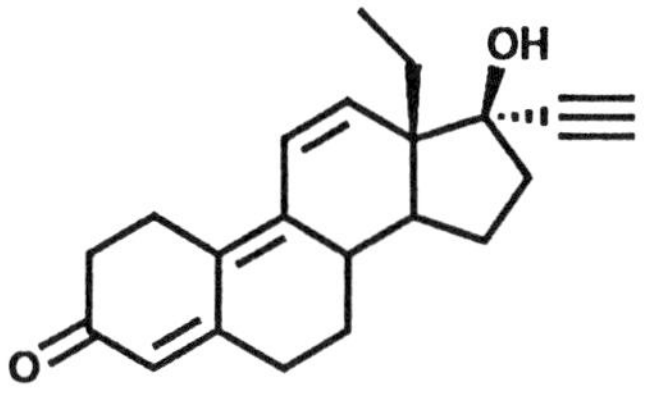

4: RU 2323

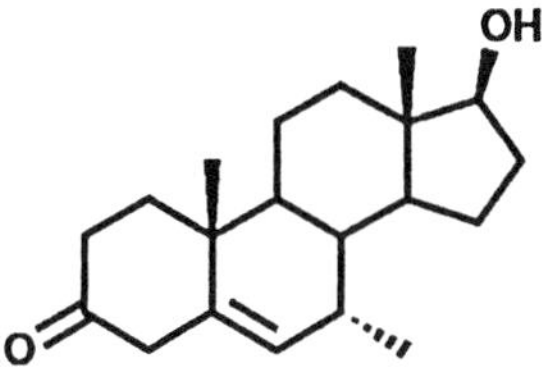

5: RMI 12936

predictable impact on reproductive function in females (*5–7*), were completely missing, with the exception of a few alleged antiprogestins like 16α-bromoacetoxy progesterone (**3**) (*8*), R 2323 (**4**) (*9*), or RMI 12936 (**5**) (*10, 11*) whose antiprogestational qualities have never been confirmed. In those days, there was some confusion about antigestational, antiprogestational, and antideciduogenic activities, and almost anything could be called an antiprogestin. The synthesis of RU 486 helped clarify the semantics and simultaneously appeared to solve the two last problems in steroid endocrinology.

To date, close to 1,000 research papers have reported on mifepristone as a biochemical or pharmacological tool. The present report is a personal account, by the first author, of the tribulations that led from the simple idea of innovation in steroid structure to innovation in steroid biology.

Historical Setting

I joined Roussel UCLAF in the fall of 1970. My first contact, a few weeks earlier, with Robert Bucourt, who had been prominent in applying the torsional angle concept to conformational analysis (*12*), was decisive in my choosing Roussel UCLAF, essentially known for its steroid chemistry. Our early collaborative work involved the enantioselective total synthesis of corticoids via a Johnson-type biomimetic polyene cyclization. We never published anything about this work because it did not stand up to our expectations, even though some interesting chemistry came out of it. But soon a personal interest first expressed during my postdoctoral work with Elliot Shapiro at Schering-Plough Corporation in 1968–1969 emerged again: synthesizing biologically active compounds is more fun than making hydrocortisone, which is well-known to everybody. Moreover, I wondered how it was that Roussel UCLAF, the leading producer of bulk corticosteroids, did not market any original compound of its own apart from the recently introduced desoximetasone (*13, 14*). Robert Bucourt gave some historical reasons and suggested further that it was too late to search for novel corticoids. I had heard similar comments at Schering, while trying to shift the 6-azido-Δ6 modification from the progestin-antiandrogen series, where we first applied it, to the corticosteroid series (*15*). I believe that my interest had contributed to the resumption of steroid antiinflammatory research by Schering (*16*). Perhaps something similar could happen in Romainville.

Roussel UCLAF had numerous assets for designing original compounds, both by the availability of synthetic intermediates from the bile acid route and by the existence of valuable compounds from the total synthesis of 19-norsteroids. To a youthful chemist, Roussel UCLAF's assets were a vast treasure chamber waiting to be explored. It seemed to me then that the synthetic chemists were confining their work to minor modifications around the steroid nucleus. With hindsight, I now know that this was not a fair evaluation of the situation, and that there were many truly innovative structures within Roussel UCLAF's collection of chemical compounds. But the lessons I learned from Elliot Shapiro were still vivid: science should not be stopped by received opinion. This state of mind was at the time quite out of place in Romainville. Indeed, without the openmindedness and personal sympathy of Robert Bucourt, this story would no doubt have stopped there. In fact, most chemists who had been active in 19-nor total synthesis had been converted to nonsteroidal chemistry, although a few were still making derivatives for affinity chromatography at the request of E. E. Baulieu, a longtime consultant with the company, with the objective of purifying the estrogen receptor (*17*).

In the meantime, I had been granted my first assistant, Christian Richard, and with our combined force it seemed reasonable that we could, in addition to the total synthesis project, have a try at making an "active compound." Among the possible candidates that could lead to original corticoids were the epoxides of the general formula **6**. The epoxides had been synthesized a few years earlier by Lucien Nédélec (*18*), who, together with Jean-Claude Gasc, had used these steroids in a formal total synthesis of cortisone (*19*). The epoxides came to my attention when Jean-Claude Gasc, working at the bench next to me, began experimenting with a new reagent, hexafluoracetone-peroxyhydrate, that had been described in *Tetrahedron Letters* for the Bayer-Villiger reaction (*20*). Gasc wanted to determine if the new reagent could be used for epoxidation as well. Indeed, the epoxides of interest were made with some difficulty and poor regio- and stereoselectivity by treatment with peroxy acids (*18*). The new reagent gave highly improved results, and the 5α,10α-epoxides became more readily available (*21*). However, no scientist was available to use them, and there were no more accepted steroid projects. Lucien Nédélec had already been transferred to a different laboratory, and Gasc was assigned to a total synthesis of vitamin B_{12}.

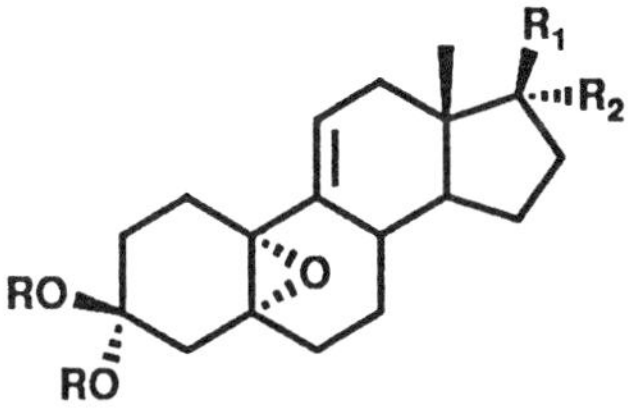

RO RO R1 R2 O

6

OAc O OH OH O

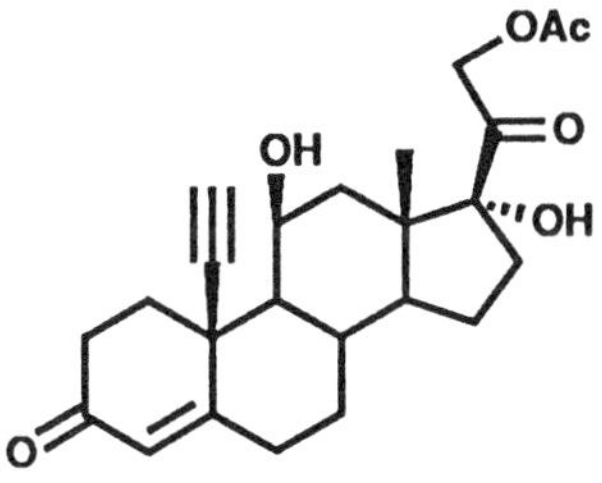

7

When I proposed to Robert Bucourt that we use this type of epoxide to prepare the 10-ethynyl analogue of hydrocortisone, he consented, but warned that it should not take too much time from the corticoid total synthesis. In any event, Bucourt's consent was not a go-ahead signal for a new project. Bucourt had to be very diplomatic to reconcile the directives of the research director, Gérard Nominé, with the enthusiasms of his subordinate. Nevertheless, we succeeded in making the desired compound **7** (*22*) without neglecting the main project. Although the hydrocortisone analogue had no improved activity over the parent compound, it did not deter me from occasionally attempting an epoxide-ring opening with various nucleophiles such as azide, thiocyanate, and mercaptides (Figure 2). But for a 2-year period the company was interested in peptide synthesis. Following a 2-month stay at the Saint Antoine Hospital in Paris, in the peptide laboratory of G. Milhaud, we engaged in the solid-phase synthesis of luteinizing hormone-releasing hormone (LHRH) (Merrifield technique), which was eventually commercialized in France as a diagnostic tool. But peptide projects were not born under a good omen, and after we had synthesized somatostatin as well as some analogues, and were engaged in the task of preparing human calcitonin, we were informed that "peptides never would make useful drugs". Peptide chemistry was abandoned as quickly as it had been started.

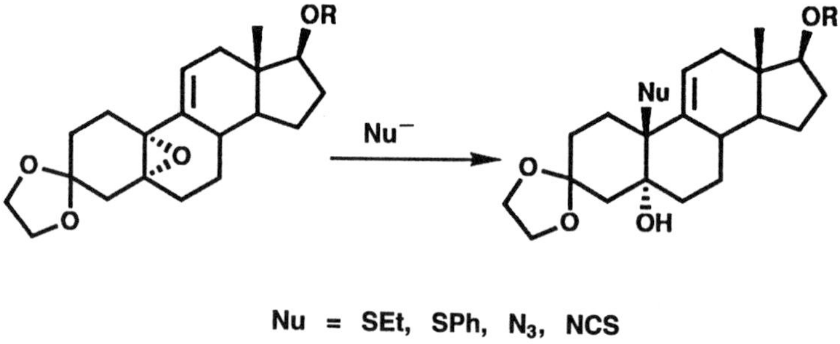

Nu = SEt, SPh, N_3, NCS

Figure 2. Experimental epoxide-ring openings of the 10-ethynyl analogue of hydrocortisone.

Meanwhile, our small team of two people had been expanded by the arrival of Germain Costerousse, who had previously worked on the total synthesis of vinca alkaloids (23). One of his first tasks was to complete the corticoid total synthesis that had given us some trouble. The initial synthetic scheme failed at the decisive cyclization step—a result we would now easily predict—and we embarked on several far-fetched alternatives until we finally realized that the synthesis could not be achieved. Nevertheless, Costerousse had had the opportunity to tackle some interesting chemistry, which to his disappointment we never published.

A Research Project on Glucocorticoids

In the meanwhile, management's position on new corticosteroids progressed, perhaps influenced by an impressive report on this class of compounds by the two people who made up our basic pharmacochemistry unit, Jean-Cyr Gaignault and Christian Marchandeau. The decision to initiate research on new corticosteroids was made at the end of January 1975. I had my first true medicinal chemistry project on steroids. Entitled "New steroids with antiallergic, topical or anti-stress activities", it became known as Project 249. The poorly focused title suggested that we were targeting both agonists and antagonists, as well as various routes of administration, but the main thrust was directed toward agonists. One of the possible chemical modifications that we mentioned in the original project was replacement of the angular 10-methyl group by other substituents. The way was apparently open for profound modifications of the steroid substitution pattern.

Although the first compounds to be synthesized, like RU 24643 (**8**), were close analogues of dichlorisone (**9**) (*24*) and displayed remarkable local antiinflammatory activity, we soon engaged in substitutions, which have a somewhat closer though still distant relationship to RU 486. The original idea arose from a series of papers emanating from Schering AG on fluocortin butyl (**10**), which was effective as a topical antiinflammatory agent while being devoid of systemic activity (*25, 26*). The question that arose in my mind was simple: What other chemical arrangement could mimic this system? A well-known example of a bioisosteric relationship was close at hand with the progestational steroids: the 17β-acetyl side chain of progesterone can be replaced by the 17α-ethynyl-17β-hydroxy system, preserving essentially the biological activity. Why not apply this substitution to the classic corticosteroid side chain? This analogy led to the postulate that Schering's carboxylic esters should be mimicked by 17β-hydroxy-17α-propiolates (Figure 3).

The daringness of the analogy did not deter us from making the compound, which is very easily accessible. Indeed, to our surprise the first molecule to be synthesized (**11**) presented the expected antiinflammatory activity when applied topically in the mouse croton oil edema test, with an EC_{50} (effective concentration that induces 50% of the maximal response) of 1 mg/mL (the EC_{50} of cortisol is 2.5 mg/mL). The corresponding acid (**12**) was inactive, a result in agreement with the hypothesis. We then wondered whether the ester contributed significantly to the activity. To answer this question we investigated the corresponding 17α-ethynyl (**13**) and 17α-propynyl (**14**) compounds. The first one proved to be inactive, but the second, the propynyl analogue, displayed not only local activity (four times that of cortisol) but also systemic activity (granuloma and thymus weight inhibition) when given orally. This in vivo activity was supported by a substantial affinity of **14** for the rat glucocorticoid receptor, which reached 70% that of dexamethasone, while cortisol bound only to the extent of 30% (Table 1).

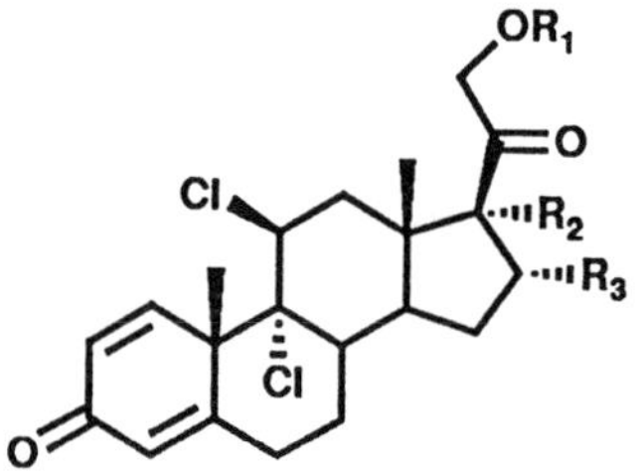

8: R_1 = (ester of dicyclohexylmethanol, –C(=O)–O–)

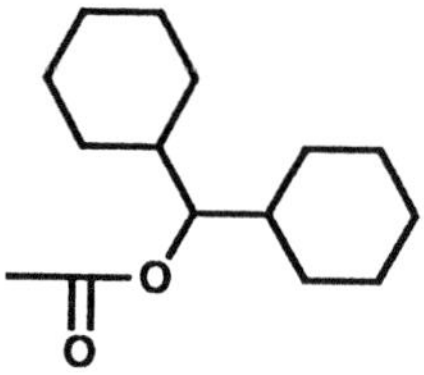

R_2 = H

R_3 = Me

9: Dichlorisone

R_1 = R_3 = H

R_2 = OH

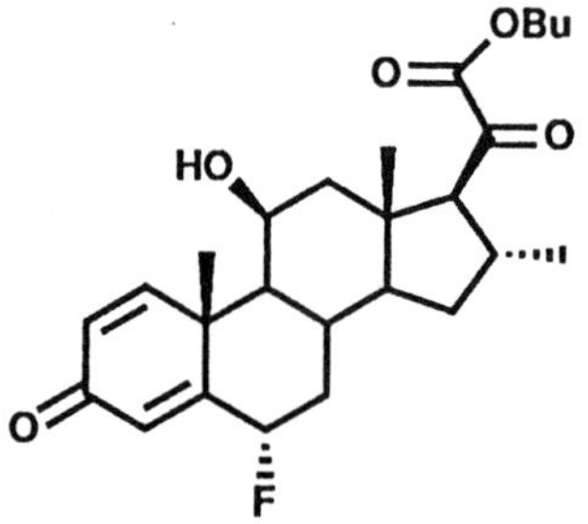

10: Fluocortin butyl

A common feature of these 17α-propynyl analogues of corticoids was their high specificity in binding to the glucocorticoid receptor while being essentially devoid of any affinity for the mineralocorticoid receptor (*27, 28*). The pure glucocorticoid RU 28362 (**15**) has since been used extensively by the scientific community as a tool for biochemical and pharmacological investigations.

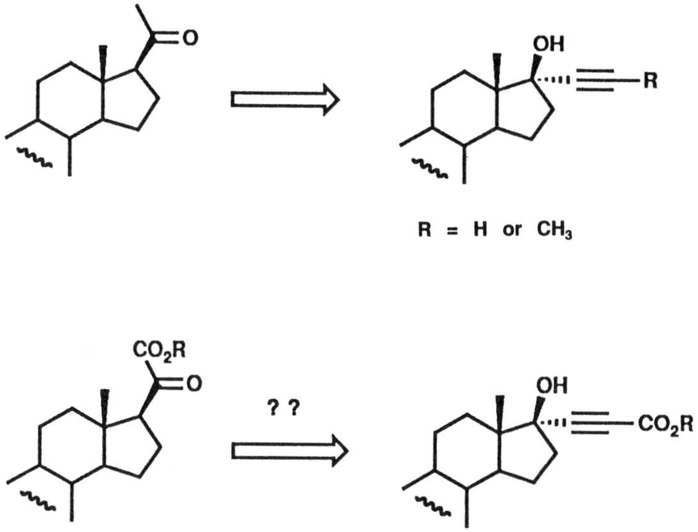

R = H or CH_3

CO_2R

? ?

Figure 3. Top: Bioisosteric relationship for progestins. Bottom: Potential application to corticosteroid side chain of 17β-hydroxy-17α-propiolates by Roussel UCLAF.

11β-Substitution

Meanwhile, my ongoing interest in use of my favorite epoxide had not subsided. From time to time I would come back to it and have another try at a new type of substitution. The results with the 10β-ethynyl steroids, although not of transcending value, were not discouraging, and I decided to synthesize some 10β-phenyl steroids to see if the epoxide-opening reaction could be achieved. When, by the end of January 1975, I reacted phenyl magnesium bromide with epoxide **16** and observed the transformation by thin-layer chromatography analysis, I was confident that the reaction had worked. But from the NMR spectrum it became clear that what I had considered a pure compound was in fact a 4:1 mixture of two structures: the dienol **17** and the minor compound, possibly **18**, in which the phenyl group is attached to position 11. A close look at molecular models confirmed that the large shielding seen on the angular 18-methyl group could only be explained by an 11β-configuration for the phenyl group. This unexpected result prompted me to repeat the reaction with diphenyl copper lithium

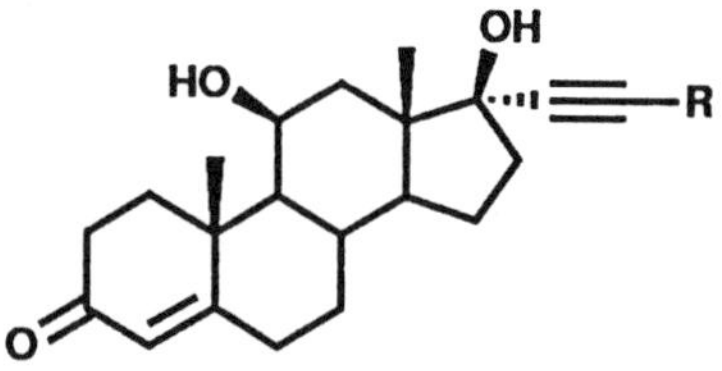

11: RU 24450, R = CO_2Me

12: RU 25101, R = CO_2H

13: RU 200, R = H

14: RU 25458, R = Me

Table 1. Biological Activities of 17α-Alkynyl 4-Androstenones

Compound	*RBA-GR*[a]	*Granuloma*[b]	*Thymolysis*[b]	*Edema*[c]
11	14	>50	IN 50	1
12	1.5	>50	>50	IN 1
13	5	IN 50	IN 50	IN 1
14	68	50	13	0.6
Cortisol	31	15	10	2.5

NOTE: IN denotes inactive at the dose indicated.
[a] RBA-GR is relative binding affinity for rat thymus glucocorticoid receptor measured after 4 h of incubation at 0 °C (value for dexamethasone is 100).
[b] Values are ED_{50} (in milligrams per kilogram) in rats.
[c] Values are ED_{50} (in milligrams per milliliter) in mice.

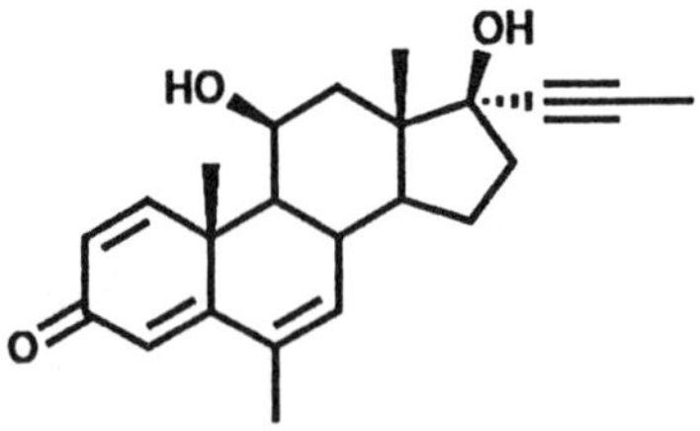

15: RU 28362

instead of the Grignard reagent, in the hope of shifting the reaction pathway entirely to the 11β-substituted compound. It worked. Later I learned that conjugate epoxide openings with copper reagents had previously been performed on butadiene monoepoxide (*29, 30*) and cyclohexadiene monoepoxide (*31, 32*). Whereas the reaction was reported to be entirely regiospecific in the first instance, in the second the authors had obtained a mixture of regioisomers. In retrospect, it is clear that the total regiocontrol we observed could not have been predicted with certainty from a hypothesis based on the reported data alone. This is especially true when we consider that the bulky substituent was introduced at a position expected to exhibit strong 1,3-diaxial interaction with the angular 18-methyl group.

16 —PhMgBr→ **17 (major)** + **18 (minor)**

Stimulated by these results, I checked the generality of the reaction by using other cuprates such as methyl and ethyl. I also demonstrated that the same transformation could be achieved by using Grignard reagents in the presence of a catalytic amount of copper salts. Thus, for example, with benzyl magnesium chloride, the benzyl group went entirely to the 10β-position, but in the presence of cuprous chloride, this group was introduced at 11β (Scheme 1) (*33*).

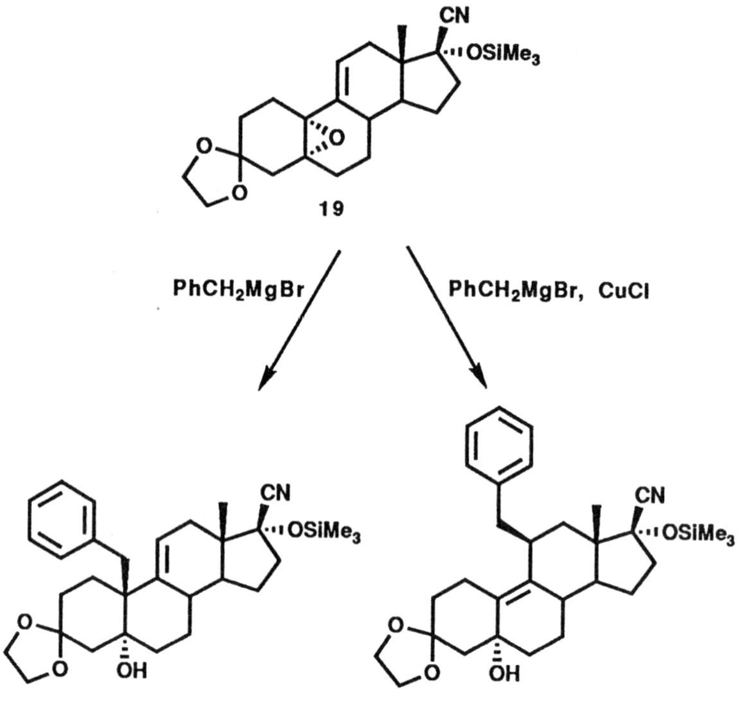

Scheme 1

First Attempts at Biological Evaluation

By April 1975 I was convinced that we had discovered an original and potentially exciting class of steroids. The first compounds had been made from available epoxides and belonged either to the estrane series, structurally related to 19-nortestosterone, or to the 19-norpregnane series. Some were 17-benzoates, a feature that can be beneficial for testing the compounds in vivo but rather troublesome for in vitro studies such as receptor-binding experiments. Others had the 17α-hydroxyprogesterone-type substitution known to be deleterious to progesterone-receptor binding. When I attempted to contact Roussel UCLAF researchers who had studied the biology of steroid sex hormones, I learned that the group was no longer operational, and I was further informed that "it is well known that steroids lose their activity when substituents are introduced at the 11β-position".

I wondered why my informants held this belief. There were few reports on 11β-substituted steroids in the literature at that time, and only three relating to 11β-alkyl steroids (*34–36*). In fact, all three were on 11β-methyl, and only one paper reported biological activity, which indeed was increased as compared to the unsubstituted analogue (*36*). In addition, a report in *Steroids* by the Glaxo group had concluded that introduction of an 11β-chlorine in the 19-nor series consistently improved the progestational activity (*37*).

There had been some in-house experience with 11β-substitution, although from my chemist's point of view I thought it not especially relevant to our problem: 11β-methoxy-ethynylestradiol (**22**) (R 2858) had improved estrogenic activity over the parent compound (*38, 39*). The improved activity was partly attributed to its reduced binding to plasma proteins, whereas the corresponding 11α-analogue (**23**) had some "antiestrogenic" activity, later termed "impeded agonist" activity (*40, 41*). Further examples included 11β-alkoxy, thioalkyl, and azido estradienones of general formula **24** (*42*), which had been found to be weak progestins.

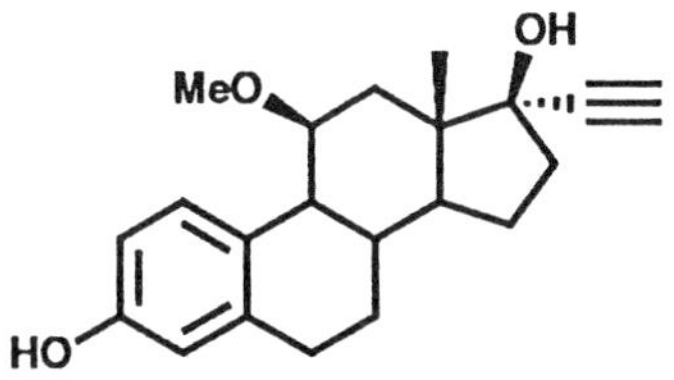

22: R 2858

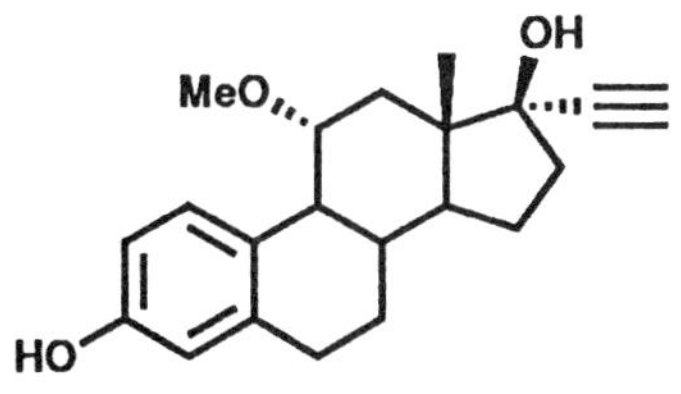

23: RU 16117

Fortunately for the future of mifepristone, I had made the acquaintance of Daniel Philibert. Philibert was trained as a physicist and had come to Roussel UCLAF in 1967 to work with Jean-Pierre Raynaud on uterine contractility, a project he soon abandoned to take part, with the same group, in the elaboration of the receptor screening. The objective was to adapt the recent basic results from the pioneering work of Jensen, Gorski, Baulieu and Milgrom, O'Malley and others on receptor binding (see refs. 43–45 for reviews) to an efficient tool able to screen the hundreds to thousands of steroids available at Roussel UCLAF (*46*). Philibert agreed to measure the binding affinities of the handful of compounds I had to offer him. The

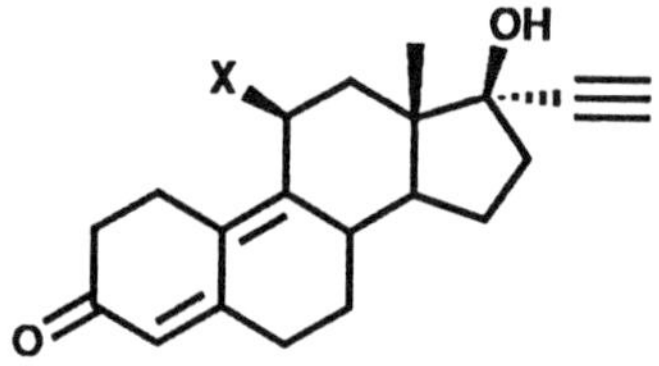

24: X = OR, SMe, N_3, SCN

results with respect to progesterone–receptor binding were neither spectacular nor discouraging (Table 2). I approached Robert Bucourt to initiate a full research project based on this novel class of compounds, arguing that by choosing the right substituents both in the 11β-position and the D-ring we might find something completely new.

Within a few days, however, we were informed that the company was no longer interested in sex hormones, and that starting a project along these lines was out of the question. In defense of management's position, it should

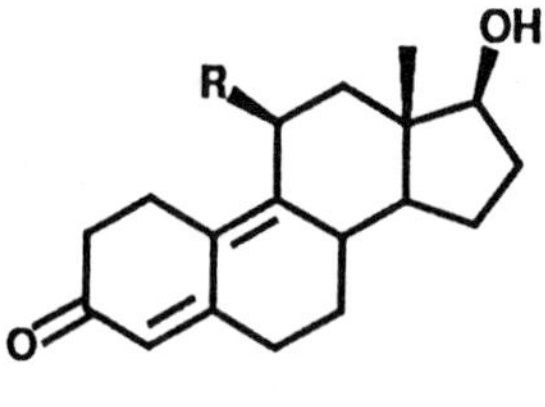

25

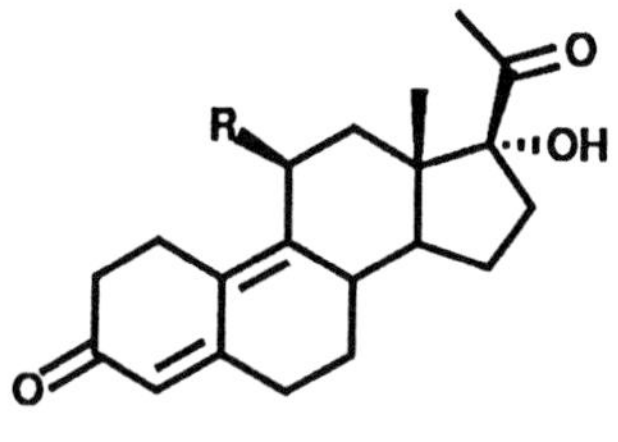

26

Table 2. Relative Binding Affinities of Various Compounds for the Rabbit Progesterone Receptor

Compound	*Code*	*R*	*RBA (2H)*[a]
25a	R 3118	H	17
25b	RU 23647	Et	7
25c	RU 23650	Ph	1.5
26a	R 2721	H	5
26b	RU 23655	Ph	0.4
26c	RU 23656	Et	7

[a] RBA (2H) denotes relative binding affinity after 2 h of incubation. Value for progesterone is 100.

be remembered that 1974 was the year in which the development of R 2323 (**4**) was compromised. This compound had been slated to become a "once-a-week" contraceptive (*9, 47*). But as clinical testing enrolled more and more volunteers, the initially very good Pearl index decreased to unacceptable levels, placing the compound into the category of a "too weak" contraceptive. No one wanted to risk another failure with sex-hormone steroids.

Substitutions on the Steroid Nucleus

The grim future for 11β-substituted steroids lightened somewhat a few weeks later when Fernand Labrie, a scientist at Laval University, in Quebec, approached Roussel UCLAF to inquire about a research position for one of his co-workers anxious to get acquainted with steroid chemistry. I was delighted to have another assistant, but asked that he work on 11β-substitution. Because the researcher would be paid by a grant from the Canadian Research Council and not by Roussel UCLAF, the proposal appeared fair, and was accepted within hours. So, when in September 1975 Alain Bélanger arrived in Paris, en route to nearby Romainville, the main guidelines for a working program were waiting for him. My aim was to investigate the influence on biological activity of 11β-substitution, in a first instance on receptor binding, in the three classes of sex hormones: progestins, estrogens, and androgens. The proposal to Alain was for him to work through the estrane series, to synthesize all the target compounds (**30**–**32**) derived from a single intermediate, shown in Scheme 2. The choice was logical. It was known that 19-norsteroids were active in the sexual sphere but not, so far, on other targets. From the chemical point of view we now know that protecting the 17-keto group was unnecessary. However, the presence of a silylated cyanohydrin had been shown to improve the stereoselectivity of the epoxide-formation step. Jean-Pierre Raynaud and Daniel Philibert assured us that the compounds would be tested quickly for their receptor-binding affinity, even though they were not originating in an official project.

Alain Bélanger worked hard, and when he left, at the end of June 1976, the synthesis of about 35 compounds had been achieved. An eclectic set of substituents had been introduced in the β-position, ranging from the classic alkyl groups (Et, *n*-propyl, i-propyl, *t*-butyl, *n*-decyl) to aromatics and heteroaromatics (substituted phenyl, benzyl, and thienyl) as well as alkenyls (vinyl, isopropenyl, and allyl). We tried *t*-butyl only to show that the reaction would be limited by bulk, but to our surprise it gave a quantitative yield. Unfortunately, a few substituents that included a nitrogen atom, such

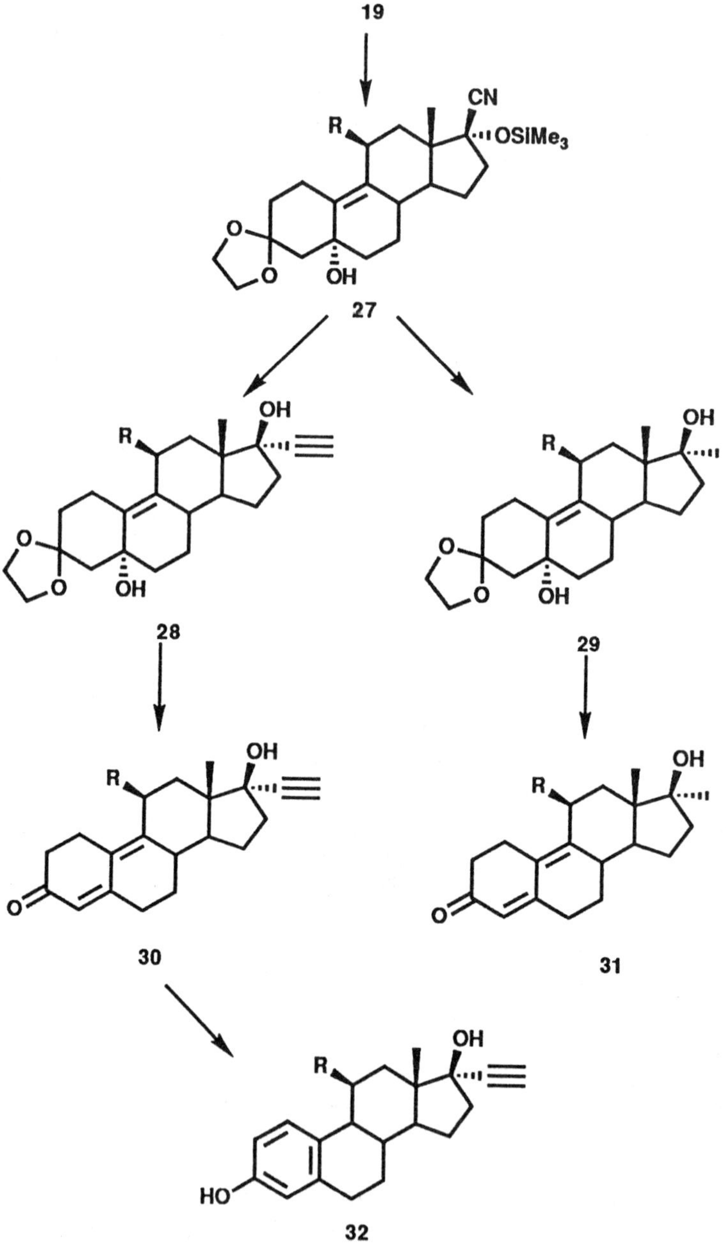

19

CN
R
$OSiMe_3$
O
O
OH
27

OH
R
O
O
OH
28

OH
R
O
O
OH
29

OH
R
O
30

OH
R
O
31

OH
R
HO
32

Scheme 2

as pyrrole, pyridines, and a thiazolidine, could not be introduced during our first attempt, and we had to return to them later. Nevertheless, we were convinced of the appropriateness of our research direction.

By the time we got back the results from the binding experiments, Bélanger had already returned to Quebec. We remained in contact with him, and he was informed of the exciting conclusions of his work, some of which we subsequently published in *Steroids* (48) (Tables 3–5).

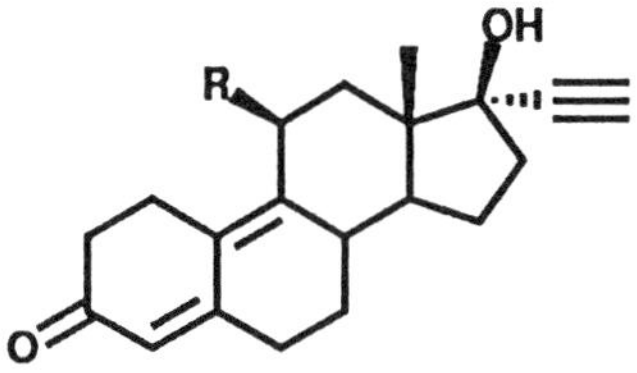

30

Table 3. Relative Binding Affinities for the Progesterone (PR) and Glucocorticoid (GR) Receptors

		PR[a]		*GR*[b]	
Compound	*R*	*2 h*	*24 h*	*4 h*	*24 h*
30a	H	43	34	6	2
30b	Ethyl	40	25	15	n.d.
30c	*n*-Propyl	18	12	35	25
30d	i-Propyl	2.3	1.3	5	6.5
30e	*n*-Decyl	3.6	n.d.	0.1	n.d.
30f	Vinyl	170	535	65	40
30g	i-Propenyl	75	95	10	8
30h	Allyl	50	45	8	n.d.
30i	Phenyl	20	65	60	20
30j	*p*-OMe Ph	130	335	100	70
30k	*o*-OMe Ph	1	0.7	10	3.5
30l	*p*-F Ph	38	36	100	55
30m	Thienyl	70	85	120	40
30n	Benzyl	0	0	0	n.d.

[a] Relative binding affinities for progesterone receptor determined after 2 and 24 h of incubation at 0 °C in rabbit uterus preparation. Value for progesterone is 100. n.d. denotes no data.
[b] Relative binding affinities for glucocorticoid receptor determined after 4 and 24 h of incubation at 0 °C in rat liver preparation. Value for dexamethasone is 100. n.d. denotes no data.

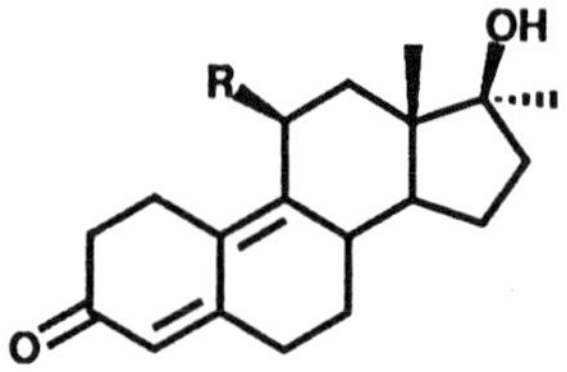

31

Table 4. Relative Binding Affinities of Various Compounds for the Androgen (AR), Glucocorticoid (GR), and Progesterone (PR) Receptors

Compound	*R*	*Code*	*AR*[a]	*GR*[b]	*PR*[c]
31a	H	R 3467	92	6	71
31b	Benzyl	RU 25054	2.3	4.6	1.0
31c	i-Propyl	RU 25070	0.7	6.2	0.3
31d	*n*-Propyl	RU 25071	17.6	28.5	2.3
31e	Vinyl	RU 25333	46	31	23
31f	Allyl	RU 25334	18	6	12
31g	*n*-Decyl	RU 25337	0.2	1.4	9
31h	Thienyl	RU 25339	46	78	34
31i	Isopropenyl	RU 25340	78	20	15
31j	*p*-MeO phenyl	RU 25341	7	56	66

[a] Relative binding affinity for androgen receptor after 2 h of incubation at 0 °C in a rat prostate preparation. Value for testosterone is 100.
[b] Relative binding affinity for glucocorticoid receptor after 4 h of incubation at 0 °C in a rat liver preparation. Value for dexamethasone is 100.
[c] Relative binding affinity for progesterone receptor after 2 h of incubation at 0 °C in a rabbit uterus preparation. Value for progesterone is 100.

Focusing on the sex-hormone receptors, we found that estradienes of general formula **30** with unsaturated 11β-substituents such as vinyl, isopropenyl, phenyl, or para-substituted phenyl rings, displayed an unexpectedly high affinity for the progesterone receptor. However, when the double bond was displaced from the 11β-carbon by just one carbon unit, the corresponding compound exhibited a greatly reduced affinity for the receptor (compare vinyl and allyl or phenyl and benzyl). In the androgen–receptor-directed series (**31**) no similar remarkable feature could be distinguished. (We exclude for the time being the results on the glucocorticoid receptor.)

Our novel substitutions on the steroid nucleus permitted us to draw two main conclusions:

32

Table 5. Relative Binding Affinities of Various Compounds for the Estrogen (ER) and Progesterone (PR) Receptors

		ER[a]		PR[b]	
Compound	*R*	*2 h, 0 °C*	*5 h, 25 °C*	*2 h*	*24 h*
32a	H	110	245	21	10
32b	*n*-Propyl	95	735	7	3
32c	i-Propyl	85	710	8	5
32d	*n*-Decyl	0	n.d.	0	n.d.
32e	Vinyl	90	415	60	77
32f	Allyl	93	555	17	10
32g	*p*-OMe Ph	60	66	34	48
32h	*o*-OMe Ph	15	8	0.7	1
32i	Thienyl	75	90	22	17
32j	Benzyl	77	53	0.4	0.1

[a] Relative binding affinity for estrogen receptor after 2 h of incubation at 0 °C and 5 h of incubation at 25 °C in a mouse uterus preparation. Value for estradiol is 100.
[b] Relative binding affinity for progesterone receptor after 2 and 24 h of incubation at 0 °C in the rabbit uterus preparation. Value for progesterone is 100.

1. The presence of an unexpected large hydrophobic pocket in the progesterone receptor, able to accommodate substituents like *p*-methoxyphenyl.
2. The existence of a remarkable specificity in the binding interaction, as demonstrated by the difference in affinity between an unsaturated substituent (vinyl = 535) and the corresponding saturated substituent (ethyl = 25), a difference of great significance if we would find an explanation for it.

Concerning the estrogen receptor, it was found that most 11β-substituted derivatives of ethynyl estradiol retained a substantial affinity. For some substituents (*n*-propyl, isopropyl, vinyl, and allyl), this affinity considerably exceeded the value for the unsubstituted reference compound when measured after 5 h of incubation at 25 °C, instead of after 2 h of incubation at 0 °C. This difference indicated a more slowly dissociating ligand–receptor complex. Combined with the significant progesterone–receptor binding of some of these compounds (i.e., **32e**, Table 5), these results gave us some thoughts about the possibility of designing drugs with some desirable ratio of progestational and estrogenic activities, a realization of the "progoestrone" dream (*49, 50*). However, subsequent experiments showed that in the presence of estrogenic activity, the progestational effect was not detectable.

In Vivo Testing

At the end of 1976 I presented our results at the annual Scientific Colloquium, which was attended by most Roussel UCLAF scientists as well as by R&D management. My main point was to demonstrate that original chemistry could be at the origin of innovative drugs, and that according to the unusual binding profile, the new class of compounds would be expected to have some unusual biological properties. All that remained was to test the compounds in vivo. To illustrate the novelty of this class of compounds, I showed the photograph of a cow with horns topped by yellow balloons, symbolizing the new substitution and asking the question about the unknown properties of this kind of animal (Figure 4). The talk generated considerable discussion, but the compounds were not tested, for the good reason that the endocrine pharmacology laboratory had been more or less deactivated after the decision to stop development of Gestrinone (R 2323). On February 18, 1977, I presented the same topic again, with the same conclusions, at the Hoechst Pharmacolloquium in Frankfurt; again, the response was silence.

What I was not aware of in 1976 was a discovery made recently in the laboratory notebook of one of the endocrine pharmacologists who has now retired from the company: most compounds had indeed been tested in vivo, and some of them had even displayed abortive activity. If this finding could appear of little originality for compounds with an estradiol-like structure, it should have raised a serious question with RU 25056 (**30j**, Table 3), the 11β-*p*-methoxyphenyl-substituted estradiene derivative. Why this important clue, which I needed so badly, would not have been communicated to the chemist remains a mystery. It may be that the unhappy experience with R 2323, which had also shown abortive activity, was still very vivid.

Figure 4. Balloons query the unknown properties of a new animal.

Although the idea that some of these compounds could antagonize progesterone had crossed my mind, it seemed too fantastic to express without hard supporting evidence. Curiously, the idea of finding progesterone antagonists had been around for a moment (*51*), and, under the impetus of Jean-Pierre Raynaud, screening of nonsteroidal compounds had been initiated with the purpose of detecting "inhibition of uptake of progesterone." The project, inspired by a report on the nonsteroidal antiandrogen flutamide (*52*), was waiting for a definitive go-ahead, but was cleanly killed by our new research director, a CNS specialist. Oddly, nobody in the company seems to have thought about steroids as antiprogestins, and when Robert Bucourt, who was also interested in the in vivo testing of the 11β-substituted series, asked me whether they could possibly have antiprogestational activity, I answered in the affirmative, with the secret concern that they might not. Indeed, one of the 11β-vinyl derivatives, RU 25253 (**33**), had finally been tested in the Clauberg assay and found to be a potent progestin (*53*). The idea appeared fleetingly in a single report, then vanished again, along with the ephemeral antiprogesterone project. Everything seemed to end there, and sex hormones became completely outlawed.

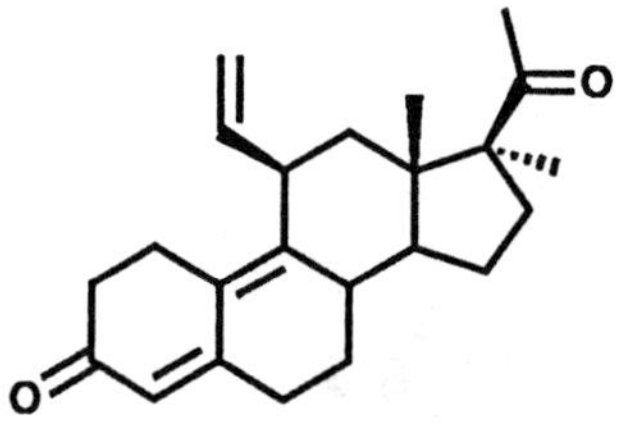

33: RU 25253

Some Hints of Antiglucocorticoid Activity

Fortunately, there was another unexplored facet that appeared just as exciting: the affinity of some of these compounds for the glucocorticoid receptor. The first results were obtained using the hepatic receptor of the rat. Later, Philibert switched to the thymic-receptor preparation, which has the advantage of containing fewer metabolizing enzymes. Tables 3 and 4 summarize these first results, which were especially striking for the 11β-vinyl, *p*-methoxyphenyl, *p*-fluorophenyl, and thienyl derivatives. The fact that we had an ongoing program on glucocorticoids led us to the in vivo testing of one of these compounds for possible antiinflammatory activity. The thienyl derivative RU 25055 (**30m**) was chosen because it was available in sufficient amounts and because it presented the highest binding (this was true at the shorter incubation time which corresponded to the first available results). At this point, Philibert was no longer engaged actively with these compounds, and the biologist with whom I interacted for the glucocorticoid project was Roger Deraedt. Our cooperation had started with Project 249, in the beginning of 1975, and by the time the binding results of RU 25055 became available, we had done considerable work together, mainly on the 9α,11β-dihalopregnane series (*54*), and also on the previously mentioned 17α-alkynyl derivative RU 24450 (**11**). However, RU 25055 turned out to be inactive in vivo as an antiinflammatory compound in the cotton granuloma and thymus involution assays (at a dosage of 5 mg/kg). We were disappointed, and without much further thought we figured that, in addition to binding, some other features would be needed, such as the presence of a D-ring substitution, which more closely resembled what we were used to for glucocorticoids. The two compounds RU 25375 (**34**) and RU 25402 (**35**) were synthesized in a short time by Christian Richard in my laboratory. The first steroid had a side chain that resembled the classic corticoid side chain except for the 17α-methyl (it had been chosen because a convenient intermediate was available); the second had the 17α-propiolic substituent that had conferred antiinflammatory activity on RU 24450. Unfortunately, both compounds were inactive at a dosage of 50 mg/kg, and their affinity for the glucocorticoid receptor was quite weak (40% and 5% of dexamethasone's, respectively).

Elsewhere at Roussel UCLAF another group was independently working on glucocorticoids. The project, headed by Jean-Cyr Gaignault and including Christian Marchandeau, Daniel Philibert, and Jean-Pierre Raynaud, had been proposed in August 1974 to set up a simple, specific in vitro screening for the determination of glucocorticoid or antiglucocorticoid activities. These tests were to be carried out in the laboratory of Philippe Meyer in Paris for the mouse thymocytes and in Fernand Labrie's facility in Quebec for the adrenocorticotropin (ACTH)-producing rat pituitary cells. The work, which involved a number of reference compounds on the thymocyte model, was published in *Molecular Pharmacology* in 1977 (*55*). Although the classic corticoids predictably displayed agonistic activity, potent progestins or androgens were classified as antagonists. Subsequently, relations were established with Ebel's group, and especially with Gisèle Beck, in Strasbourg, who worked on a model involving the induction of tyrosine amino transferase (TAT) in rat hepatoma tissue cells (HTC) by glucocorticoids. I had been unaware of this work until Christian Marchandeau, who participated in both projects, proposed sending along with a series of classic steroids, some of our bizarre compounds. We chose those that bound most tightly to the glucocorticoid receptor, including RU 25055 (**30m**) and RU 25593 (**30l**) (Table 3). The results were very interesting: not only did the compounds not induce TAT, but they antagonized the effect of dexamethasone (*56, 57*). For the first time I had the impression of standing face to face with an antiglucorticoid. Others in the group, however, were of the opinion that the compounds appeared as glucocorticoid antagonists because they also bound to the progesterone receptor. Progesterone itself acted as an antiglucocorticoid in this model (*58*). Still another hard-lived idea was that an agonist could be distinguished from an antagonist by its receptor-binding profile: agonists would have similar or higher affinity at longer incubation times, whereas antagonists would dissociate rapidly from the receptor, and hence the affinity would be much lower at the longer incubation times (*59–61*). This latter situation seemed to apply to RU 25055. Nevertheless, I remained convinced that in this case it must be different: The chemical structure was so much different, why should not the mode of action be different? Roger Deraedt, too, was excited and engaged in major work to demonstrate the antiglucorticoid activity in vivo.

The results were not very encouraging, because no clear-cut antagonism could be demonstrated in rats on the cotton granuloma test, on the thymus involution assay, or on TAT induction in vivo, irrespective of the glucocorticoid to be opposed (dexamethasone, prednisolone, or cortisol) (Table 6). The very few effects seen were limited to an apparently dose-dependent rise in corticosterone levels in the intact rat following subcutaneous injection of 1 to 100 mg/kg, and a moderate hyperglycemia in mice at 50 and 100 mg/kg. The latter effect, contrary to the first one, would have been expected for an agonist but was not reproducible. Could it have been the consequence of a rise in endogenous corticosterone production? In any case, there was no clear-cut answer to our question.

As a chemist, I sought a solution by designing a more potent analogue. I believed this could be achieved by increasing the binding affinity for the glucocorticoid receptor. That is why we synthesized RU 28289 (**36**) during November 1978, and in sufficient amounts (5 g) for in vivo testing. Our previous work on glucocorticoids had shown that in the 11β,17β-dihydroxy-4-androsten-3-one series, a switch from 17α-ethynyl to 17α-propynyl substitution would increase the receptor binding activity for the glucocorticoid receptor from 5% to 68%. A similar increase could be expected when going from RU 25055 to RU 28289. It turned out exactly that way (Table 7). Not only did RU 28289 have a higher affinity than RU 25055, but this affinity was maintained at the longer incubation time. Despite these

Table 6. Search for Antiglucocorticoid Activity of RU 25055

Test	*Dose RU 25055*	*Result*
Induction of hepatic TAT		
By dex., 0.01 mg/kg SC	1 mg/kg SC	No antagonism
By ACTH, 0.005 mg/kg ?	1 mg/kg SC	No antagonism
	10 mg/kg SC	No antagonism
Neoglycogenesis by hydrocortisone		
1mg/kg PO	10, 20 mg/kg PO	No antagonism
Thymolysis by dex.		
0.05 mg/kg PO	0.5, 5, 10 mg/kg PO	Slight potentiation?
0.05 mg locally in pellet	1 mg in pellet	No antagonism
Cotton granuloma		
0.05 mg/kg dex. PO	0.5, 5, 10 mg/kg PO	No antagonism
0.05 mg/kg dex. in pellet	1 mg in pellet	No antagonism
Croton oil edema		
0.1 mg/mL dex.	5 mg/mL	No antagonism
Inhibition of RES by prednisolone		
10 mg/kg PO	20 mg/kg SC	No antagonism

NOTE: TAT, (rat) hepatoma tissue cells; dex., dexamethasone; ACTH, adrenocorticotropin; PO, per os; SC, subcutaneously; RES, reticuloendothelial system.

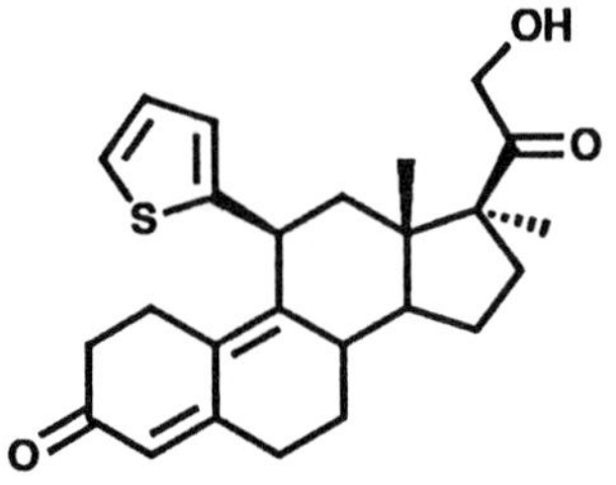

34: RU 25375

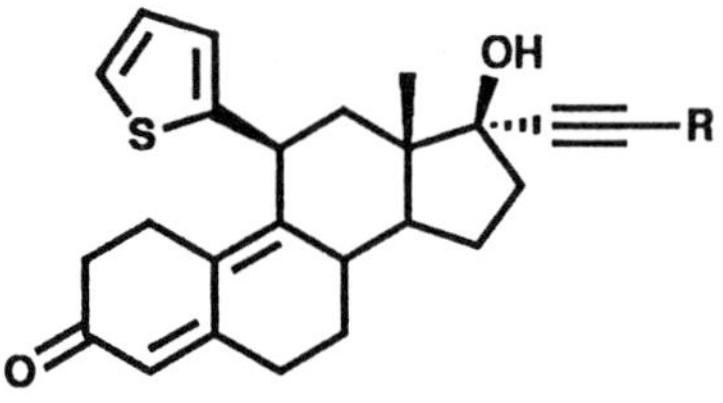

30m: R = H, RU 25055

35: R = CO_2Me, RU 25402

36: R = Me, RU 28289

encouraging results, in vivo testing of the compound was not immediately initiated. The pharmacologists were busy with RU 25055, and we waited well over a year to learn that the antiglucocorticoid activity was not significantly improved. In retrospect, our hypothesis appears quite daring, because it was recently shown that in the 17β-carboxamide series, increased affinity for the glucocorticoid receptor results in a shift from antagonistic to agonistic activity (62).

A Revival of Steroid Research

In the late 1970s, the structure of the research center had changed, under the leadership of J. R. Boissier. Robert Bucourt had left Romainville to head the Roussel Research Center in Swindon, Great Britain, whereas J. B. Taylor had come to Romainville from Swindon. Daniel Philibert had left Jean-Pierre Raynaud's group to join Roger Deraedt, and soon Raynaud would leave research. The glucocorticoid project had been completed. The only steroids we worked on were the ones expected to be devoid of any hormonal activity and intended to be submitted to general pharmacological screening. I called these new compounds the "parabiliary steroids" because they were derived from bile acids. Compound **37** is a typical example. In the midst of this intense activity, the request to convene a brainstorming group on steroids reached us. Jean Mathieu, who was in charge of the small prospective department, organized a few meetings between March and October 1979.

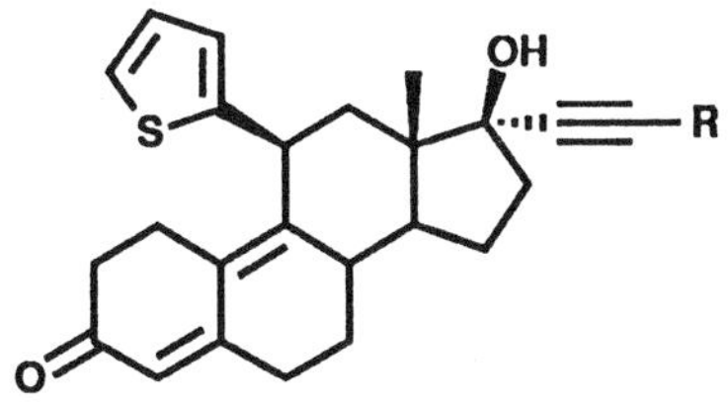

S OH R O

Table 7. Influence of 17α-Substitution on Relative Binding Affinity (RBA) for the Rat Thymic Glucocorticoid Receptor

		RBA[a]	
Code	*R*	*4 h*	*24 h*
RU 25055	H	135	55
RU 28289	Me	270	300

[a] Relative binding affinity determined after 4 h and 24 h of incubation at 0 °C in a rat thymus preparation. Value for dexamethasone is 100.

The group included Sir Derek Barton, Nobel laureate in organic chemistry; Etienne Emile Baulieu, a top steroid–receptor researcher; Roger Deraedt; Daniel Philibert; and myself. The task assigned to this group was to propose half a dozen projects involving steroids, but excluding sex hormones. Although we did not know the origin of the prosteroid trend, it was probably stimulated by Baulieu's influence on his friend Edouard Sakiz, president of Roussel UCLAF.

When we met for the first session in March 1979, at the Roussel UCLAF headquarters in Paris, Deraedt, Philibert, and I were well-prepared to defend

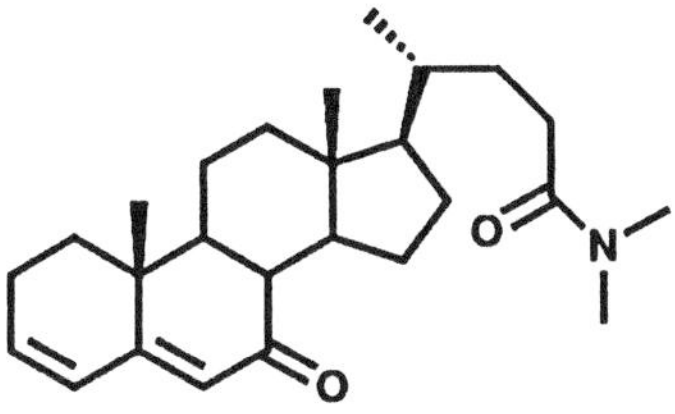

O N O

37

an antiglucocorticoid project, based on our preliminary results with RU 25055, and Deraedt presented a summary of our experience with this compound. Although clear-cut evidence of in vivo activity of the compound was not available, we were confident that the design of an antiglucocorticoid was within reach. Our strongest argument rested on Philibert's finding that, when administered to animals, the compound effectively occupied the glucocorticoid receptors while being devoid of any agonistic activity. The discussion turned mainly on the therapeutic utility of antiglucocorticoids and the biological tests most adapted to the detection of activity. Etienne Baulieu queried the permissive effects of glucocorticoids in a physiological context, a still unsolved problem that did not seem amenable, then, to a simple screening procedure.

A number of other potential projects were discussed, including inhibition of cholesterol biosynthesis, steroids involved in the CNS, triterpene antibiotics, and pure glucocorticoids.

After hearing our report on 11β-substituted 19-norsteroids as antiglucocorticoids, Baulieu asked whether the introduction of the tamoxifen side chain in position 11β of estradiol would lead to an antiestrogen. (From a recent conversation with him, it turns out that he does not remember this part of the meeting.) Because this topic was within the "forbidden sex zone", it could not be considered as a project. However, we agreed that after a molecular modeling calculation (*63*) to be carried out by N. C. Cohen, chemistry would synthesize the estradiol derivative most closely resembling tamoxifen, or OH-tamoxifen (**38**). It appeared that among the three possible candidates we had considered (Figure 5), it was indeed 11β-dimethylaminoethoxyphenyl estradiol, RU 39411 (**39**), a compound we synthesized about 6 months after RU 486 that turned out later to be a partial antiestrogen similar to tamoxifen.

An Antiglucocorticoid Project

Of the various projects we discussed in the prospective meetings, in late 1979 an executive board selected the antiglucocorticoid project. Our formal proposal detailed a number of potential therapeutic applications of antiglucocorticoids: for the effects of ageing in general, and more specifically for hypertension, atherosclerosis, catabolism, diabetes, obesity, depressed immunity, and insomnia. Most of these applications had been suggested previously by Cyr Gaignault and Christian Marchandeau. Daniel Philibert, the project leader, immediately set up the primary screening. With Roger Deraedt, he devised an internally coherent system involving three levels: the cytosolic

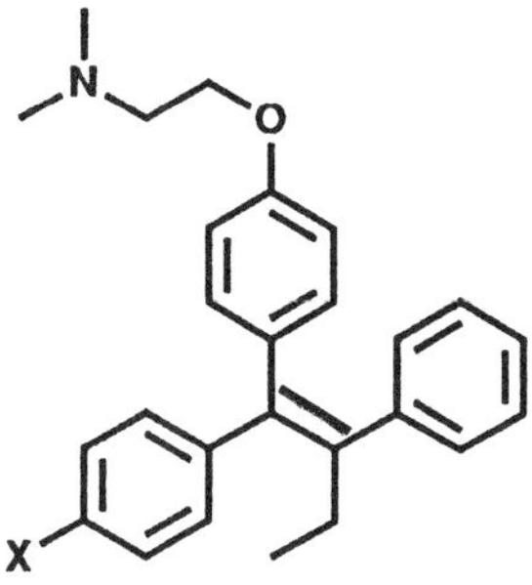

38: X = H, Tamoxifen

X = OH, OH-Tamoxifen

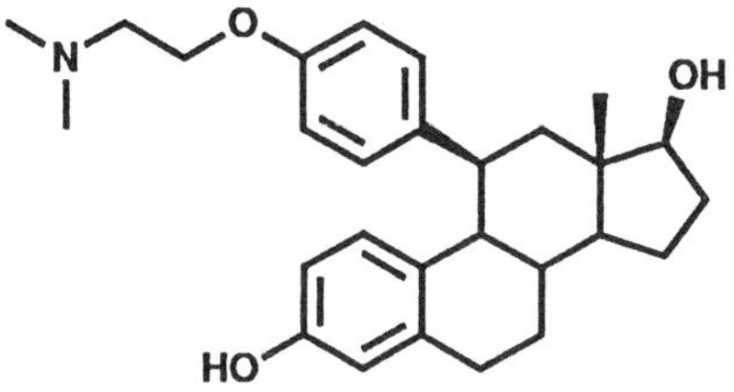

39: RU 39411

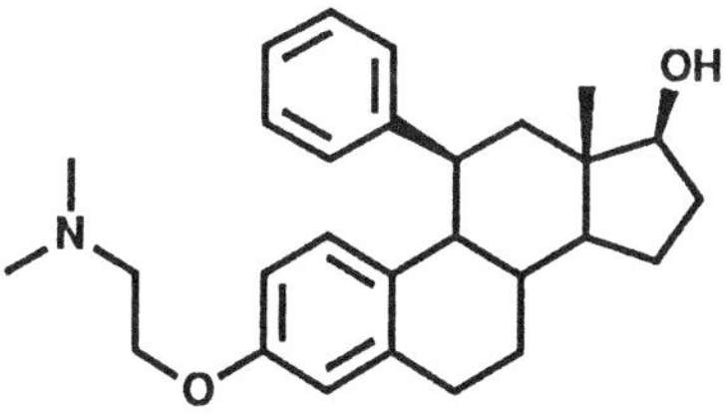

40: RU 39153

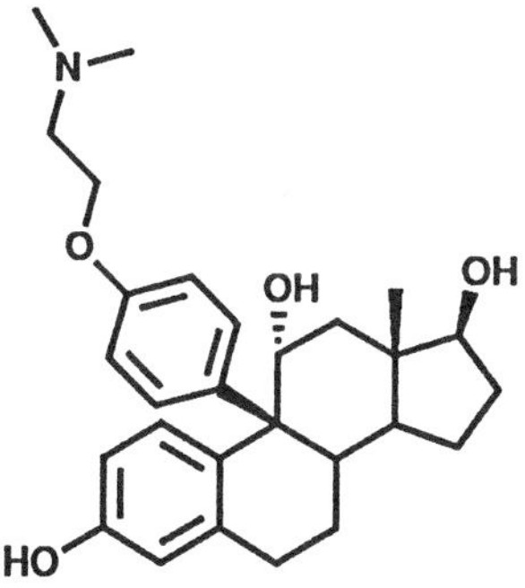

41: not made

Figure 5. Candidate compounds considered for mimicking the antiestrogen Tamoxifen.

receptor (thymus), a cellular response (uridine incorporation into thymocytes), and an in vivo response (inhibition of thymolysis induced by glucocorticoids). In addition to the primary screening, a rat liver tryptophan pyrrolase (TP) assay (65) was set up. This assay has the advantage of being an acute test, which permits investigation of the duration of action and the use of small amounts of test compound (Table 8). This test became the first one, besides the receptor screening, to be operational. The thymocyte assay, which was based on the observation that glucocorticoids reduce the incorporation of tritiated uridine in the RNA of thymocytes (*66*), was ready in July 1980.

The first compounds to be tested in the TP assay were those already available, such as RU 25055 (**30m**), RU 25402 (**35**), RU 25593 (**30l**), and RU 28289 (**36**), as well as progesterone and a few other progestins claimed to possess antiglucocorticoid activity (*56*). Progesterone (**42**) and potent progestins like R 5020 (**43**) (*68*) and RU 25253 (**33**)(*53*) were devoid of any antiglucocorticoid activity up to 100 mg/kg (Table 9), whereas some of the 11β-substituted steroids displayed a significant if low level of antiglucocorticoid effectiveness at the same dose.

Chemistry began work on the new project in February 1980 with the objective of investigating variations of the 11β-substituent in the 17α-propynyl series. I believed that the quickest results would be obtained with substituted aromatic groups in the 11β-position by modulating the electron-donating capacity of the substituent. The couple *p*-fluorophenyl and *p*-methoxyphenyl had already been synthesized by Alain Bélanger in the 17α-

Table 8. Models Used in the Rat for the Evaluation of Antiglucocorticoid Activities

Receptor binding: Binding to the thymus glucocorticoid receptor			
Antiglucorticoid activities versus dexamethasone (dex.)			
In vitro:	Uridine incorporation in thymocytes (vs. 5.10^{-8} M dex.)		
	ACTH secretion by pituitary cells (vs. 10^{-8} M dex.)		
In vivo:	In adrenolectomized animals		
	Single dose:	Test compounds (TC) were given IP or PO, vs. 10 µg/kg dex. IP	
		Hepatic tryptophan pyrrolase (TP)	
		Hepatic tyrosine aminotransferase (TAT)	
		Hepatic glycogen	
		TC by oral route vs. 50 µg/kg dex.	
		Diuresis and potassium excretion (dex. PO)	
		Corticotropic power of the plasma (dex. SC)	
	Repeated doses:	In intact animals, TC vs. 50 µg/kg dex. PO (once daily for 4 days)	
		Thymus weight	
		Cotton-pellet-induced granuloma	

NOTE: The routine screening models are underlined. IP, intraperitoneally; PO, per os; SC subcutaneously.

42: Progesterone

43: R 5020

Table 9. Antiglucocorticoid Activity of Various Compounds on Hepatic Tryptophan Pyrrolase (TP) in Adrenalectomized Rats

Compound	*Dose (mg/kg)*	*Inhibition of Dexamethasone Effects*
RU 25055	50	20
	100	49
RU 25402	100	18
RU 25593	100	62
RU 28289	100	0
RU 25253	100	0
R 5020	100	0
Progesterone	100	0
Cortexolone	100	0
	300	0

NOTE: The test compound and 10 μg/kg of dexamethasone were administered simultaneously intraperitoneally to adrenalectomized male rats. Six hours later the animals were killed and the liver was removed and processed as described previously (67). TP activity was assayed (65) and expressed as micromoles of kynurenin formed per gram of liver per hour. The TP activities in control and dexamethasone-treated rats were respectively 1.2 ± 0.2 and 12.5 ± 2.5 μmol/kg/h.

ethynyl series; these compounds displayed quite high affinity for the glucocorticoid receptors. Consequently, our goal was to prepare the 17α-propynyl analogues of these compounds, as well as some others like methylthiophenyl, dimethylaminophenyl, trifluoromethylphenyl, and so on, in both the para and meta positions. The ortho substitution was not considered because we had previously shown that it reduced receptor binding (Table 3). In addition to substituted phenyls, a number of nonaromatic substituents were planned, to be sure we had not missed anything. In addition to Germain Costerousse, there were now two new technicians in the laboratory, as well as Gérard Cahiez, who was spending a few months working in industry. Cahiez specialized in organometallics, with a particular interest in manganese. For the moment, he was reacting organomanganese reagents with acid chlorides derived from various bile acids (*69*) and my wish was that once finished with these preliminary experiments he would switch to the reaction of copper or manganese reagents with cephalosporins, a topic that was still quite original at the time. But when Cahiez saw us starting on our new project, making potentially active compounds in two steps from a large stock of epoxide we had prepared (Scheme 3), he was taken by the surrounding excitement and asked if he could join the project. Although I was somewhat sad over the aborted cephalosporin experiments, I agreed, and suggested nitrogen-containing substituents to him. He had good experience with the stability of organocopper complexes and for a while the laboratory was the main source of nuisance with regard to the large amounts of sulfur additives he manipulated.

After considering substituents with a ring nitrogen, which presented some difficulties that Costerousse would subsequently solve (*70*), he turned to the *p*-dimethylaminophenyl group, which fitted nicely into our program, while the corresponding Grignard reagent was well-known. At first he took the wrong epoxide from the refrigerator, the one with the protected cyanohydrin on the D-ring. In order to progress quickly, I proposed that he repeat the reaction with the "good" epoxide **44**, the one with the 17α-propynyl, while I transformed his intermediate compound to the 17α-ethynyl analogue. In this way we would have an additional couple of compounds to verify the expected superiority of the propynyl over the ethynyl derivative.

Breakthrough

Within a short time the compounds (code numbers RU 38473 and RU 38486) were submitted for testing. Although RU 38486 was only the sixth

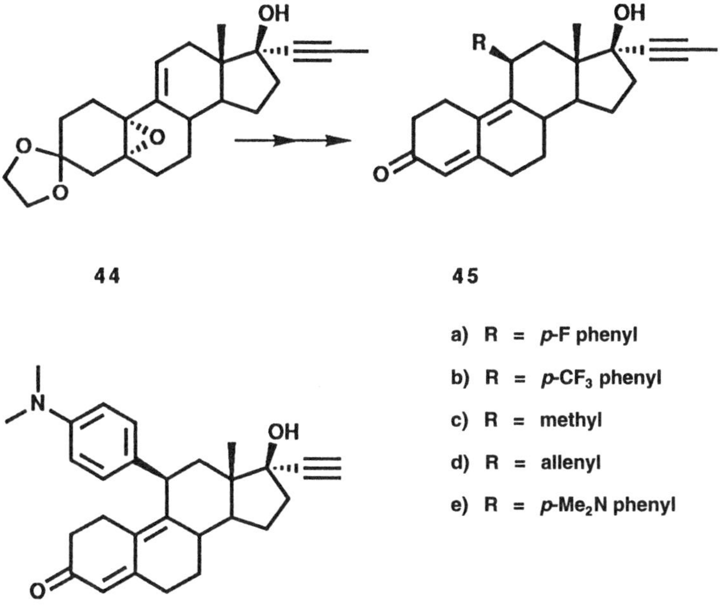

44

45

a) R = *p*-F phenyl

b) R = *p*-CF_3 phenyl

c) R = methyl

d) R = allenyl

e) R = *p*-Me_2N phenyl

46: RU 38473

Scheme 3

compound to be synthesized in the project (Table 10 summarizes the results of these compounds), its activity came as a shock to Daniel Philibert. When he saw the first result on the TP model, he wondered if the dexamethasone had been omitted from the experiment. Indeed, at 100 mg/kg, the first dosage tried, antagonism against dexamethasone (0.01 mg/kg) was total. As the dosage of RU 38486 was progressively lowered, excitement rose. It was the first time that an effect of a glucocorticoid could be antagonized as clearly. Results were even more spectacular in the thymocyte model, which became operational in July 1980: RU 38486 would antagonize dexamethasone on a mole to mole ratio, reversing by 50% the inhibition of uridine incorporation (*67*). This effect could also be visualized under the microscope by counting the number of pycnotic cells (cells in which the genetic material had coalesced)(*71*) appearing after dexamethasone treatment. Our antagonist clearly protected the cells against degeneration, giving rise to some fantasms about an anti-ageing potential.

Table 10. Antiglucocorticoid Activity on Hepatic Tryptophan Pyrrolase in Adrenalectomized Rats

Compound	*Code*	*Dose (mg/kg)*	*Inhibition of Dexamethasone Effect (%)*
45a	RU 38140	10	24
		50	73
		100	89
45b	RU 38199	50	9
		100	89
45c	RU 38275	100	0
45d	RU 38399	100	11
45e	RU 38486	1	18
		2.5	65
		10	94
		100	100
46	RU 38473	10	100

During the months that followed, Philibert would multiply the experimental models, showing that in all of them our compound acted as an antagonist: in reversing the anticorticotropic potential of the plasma of adrenalectomized rats treated with dexamethasone, in reversing the diuretic effect of dexamethasone, in counteracting thymus involution or granuloma inhibition induced by dexamethasone, and so on. He was unable to demonstrate any agonistic activity. Now that we saw what an antiglucocorticoid looked like, previous claims of antagonistic activity for compounds such as cortexolone (**47**) (72), MPA (**48**) (73), and others (74) appeared senseless (Table 11). One other immediate conclusion we drew from these experiments was that the glucocorticoid receptors must be the same in different organs of a same species, a fact not yet clearly established at this time. Etienne Baulieu urged testing the compound in humans to confirm its antiglucocorticoid activity.

The First Antiprogestin

Although I believed that testing should wait until we had a more dissociated compound, one that would not bind as strongly as RU 38486 to the progesterone receptor, it was decided to proceed with toxicology studies.

47: Cortexolone

48: MPA

Table 11. In Vivo Antiglucocorticoid Activities in Different Rat Models

Compound	Dose (mg/kg)	Percentage Inhibition of Dexamethasone Effect on: Plasma Corticotropic Potency[a]	Urinary Potassium[b]	Thymus Weight[c]
RU 38486	3	23	—	35
	5	—	58	65
	10	100	97	92
RU 38473	3	—	—	0
	10	—	73	—
Cortexolone	100	0	0	0
$\Delta^{1-9(11)}$ cortexolone	100	—	—	0
R 5020	100	0	0	0
MPA	100	—	0	0

[a] Male rats adrenalectomized for 15 days received the test compound orally and 1 h later 50 μg/kg of dexamethasone subcutaneously. After 4 h, blood samples were taken from the ophthalmic plexus and the corticotropic potency of the plasma was assessed according to the method of Sayers (75). It consists in incubating an aliquot of plasma with rat fascicular adrenal cells for 2 h at 37 °C and in assaying by radioimmunoassay the corticosterone secreted in the medium. The quantity of corticosterone is generally proportional to the concentration of plasma adrenocorticotropin: the plasma corticotropic potency of control and dexamethasone-treated animals was respectively equal to 565 ng and 70 ng of corticosterone per 30 μL of plasma.

[b] For details see ref. 67. Urinary potassium excretion of control and dexamethasone-treated animals was respectively equal to 0.60 ± 0.04 and 1.7 ± 0.1 mEq/kg of body weight.

[c] For details see ref. 76. Thymus weight of control and dexamethasone-treated animals was respectively equal to 0.34 ± 0.03 and 0.13 ± 0.01 g/100 g body weight.

"Binding to the progesterone receptor is not a problem, all we want to know is the antiglucocorticoid activity of the product," was the statement of our consultant. Not one word was expressed about the possibility of antiprogestational activity. Though concerned about the high affinity of RU 38486 for the progesterone receptor, I do not remember clearly expressing the idea that it could be related to antagonism. Nevertheless, I urged Philibert to test the compound in the usual endometrial proliferation model in rabbits (77). When Philibert reported that the compound was devoid of any progestomimetic activity in this model, the evidence struck me like lightning. It was the same story all over again—RU 38486 had to be an antiprogestin. Although formal demonstration was still lacking, I asked our patent department to include this activity in the patent (78), which was already at an advanced stage of preparation. At the same time, the predictable uses of an antiprogestin were included, among them fertility control and the treatment of hormone-dependent cancers.

From the results already available, we were confident of our conclusion: when given orally, RU 38486 would completely occupy uterine progesterone receptors, with a very slow dissociation and without exhibiting any agonistic activity of its own. This is the exact definition of an antagonist (67). About 2 months later, results of the formal experiment with RU 38486 against progesterone in the Clauberg assay fully confirmed this prediction (Table 12). Immediately thereafter, in February 1981, it was shown that the compound, unlike progesterone, was unable to maintain gestation in pregnant rats, further evidence of its inactivity as a progestational agent and an additional point in favor of the antiprogestin hypothesis. These results were presented to management and our consultant, Etienne Baulieu, on March 18, 1981. We hoped that this second breakthrough would have an even greater impact than the first one, and we were not disappointed: antiprogestational activity received priority over the antiglucocorticoid development, although the consensus was far from general.

Within 6 months Philibert had provided the full pharmacological profile, which was ready at the same time as the 1-month toxicology results in rats and *Cynomolgus* monkeys. Apart from a reduced tolerance at the highest dose in monkeys, which was fully assignable to the antiglucocorticoid activity, the compound appeared clean and ready for a preliminary clinical trial. In the meantime, investigations by B. Vannier of the toxicology department, aimed at demonstrating the antiglucocorticoid activity of the compound on glucocorticoid-induced cleft palate in pregnant mice, brought the first evidence of the abortive activity of RU 486 (June 1981). This result confirmed what we expected for an antiprogestin, and Philibert would confirm it formally somewhat later by showing that the compound

Table 12. Antiprogesterone Activity of RU 38486 on Endometrial Transformation in Rabbits

RU 486 (per os) (mg/kg)	*Progesterone (SC) (mg/kg)*	*MPU*	*Histological Appearance*
0	0	0	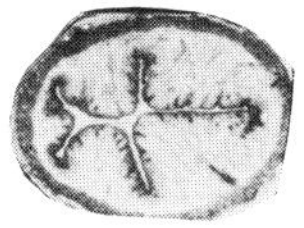
0	0.2	2.8	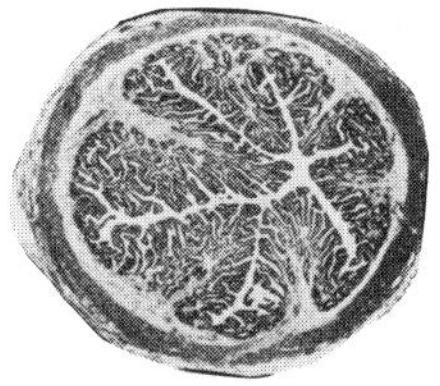
1	0.2	1.9	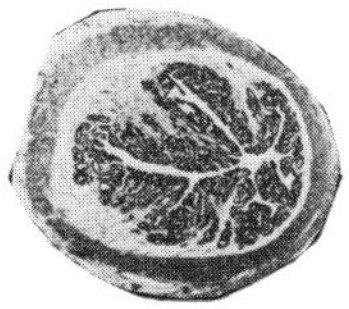
5	0.2	1.3	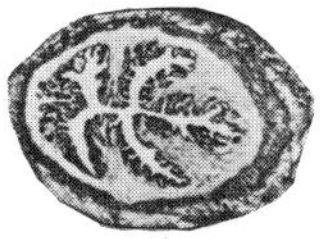
20	0.2	0	
0.3–100	0	0	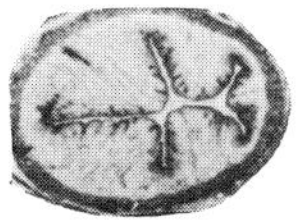

NOTE: Groups of three immature rabbits received a daily subcutaneous injection of 5 μg/kg estradiol from day 1 to day 5. Progesterone (subcutaneously) and RU 486 (per os) were administered as shown in the table from day 7 to day 10. The rabbits were killed on day 11. Uteri were removed, fixed in Bouin solution, cut into transverse sections, and their endometrial transformation was graded according to McPhail's scale (77), the maximum of which is 4 MPU (McPhail units).

Table 13. Antiprogesterone Activity of Mifepristone in Various In Vitro and In Vivo Models

In vitro:

RU 486 (10^{-8} M) totally antagonizes the effect of progesterone (10^{-8} M) on luteinizing hormone secretion by LHRH-stimulated pituitary cells

In vivo:

When given orally, RU 486 totally antagonizes the effect of:

1. Exogeneous progesterone (SC) in rats
   a. on deciduoma formation (3 mg/kg RU vs. 10 mg/kg progesterone)
   b. on giant mitochondria formation in uterine gland cells (10 mg/kg RU vs. 20 mg/kg progesterone)
   c. on maintenance of pregnancy (5 mg/kg RU vs. 75 mg/kg progesterone)
2. Endogeneous progesterone
   a. totally abortive in rats and mice at 10 mg/kg
   b. induces menstruation in monkeys within 48 h when given at a dosage of 16 mg/kg during the midluteal phase

NOTE: For details, see refs. 79 and 80.

antagonized the effect of progesterone on the maintenance of gestation in rats (*79*) and by a range of in vitro and in vivo models (*79, 80*), summarized in Table 13.

Walter Herrmann, in Geneva, was contacted by Baulieu and agreed to perform clinical trials with the compound, based on its antiprogestational activity. We were not given any details about the protocol. We guessed that the compound would be used as a "once-a-month" pill, an inducer of menstruation, which would present many advantages over existing chemical contraception, both for the women who used it and in terms of reducing developmental constraints. When Baulieu gave us the first clinical results in March 1982, however, we were quite surprised: Herrmann had used the compound to interrupt pregnancy, and it proved to be successful in nine women out of nine. This excellent score was subsequently reduced by the results in two additional patients who were included in the first public report (*3*). Although this experiment confirmed the antiprogestational activity, we were not completely satisfied because it did not correspond exactly to our dreams.

The public announcement of the result was made on April 19, 1982, by Baulieu at the French Academy of Sciences (*3*) and at a following press conference, where he introduced the abbreviated term, RU 486, for the full code number, RU 38486. Subsequent developments in the history of mifepristone and the re-emergence of Henshaw's concept of contragestion (5) can easily be traced in the numerous newspaper accounts that have appeared on the subject since, as well as a book by Etienne Baulieu, *Génération pilule* (*The Pill Generation*) (*81*). For the record of drug design, it is worth mentioning that among the hundreds of analogues we made in the following years, none proved very much superior to Mifepristone in terms of active dose. However, progress has been achieved in the specificity of the compounds (*76, 82–85*).

The scientific implications one is tempted to bring away from this story may be the following:

> Never take anything for granted.
>
> The most useful hypotheses are those that turn out to be wrong.
>
> Even if the result is in agreement with the working hypothesis, do not have too much faith in the hypothesis.

Acknowledgments

We thank our collaborators who took an active part in the preparation of this manuscript by checking their laboratory archives; F. Sweet, for linguistic advice; and S. Fritsch for typing.

Literature Cited

1. Philibert, D.; Deraedt, R.; Teutsch, G. *Abstracts of Papers*, 8th International Congress of Pharmacology, Tokyo, 1981; Abstract No. 1463.
2. Philibert, D.; Deraedt, R.; Teutsch, G.; Tournemine, C.; Sakiz, E. *Abstracts of Papers*, 64th Annual Meeting of the Endocrine Society, San Francisco, 1982; Abstract No. 668.
3. Herrmann, W.; Wyss, R.; Riondel, A.; Philibert, D.; Teutsch, G.; Sakiz, E.; Baulieu, E. E. *C.R. Acad. Sci. Paris* **1982,** *294,* 933.
4. Rousseau G. G.; Kirchhoff J.; Formstecher P.; Lustenberger P. *Nature (London)* **1979,** *279,* 158.
5. Applezweig, N. *Steroid Drugs;* McGraw-Hill: New York, 1962; pp 103, 194–195.
6. Pincus, G. *The Control of Fertility;* Academic: New York, 1965; p 128.

7. Aitken, R. J.; Harper, M. J. K. *Contraception* **1977,** *16,* 227.
8. Clark, S. W.; Sweet, F.; Warren, J. C. *Biol. Reprod.* **1974,** *11,* 519.
9. Sakiz, E.; Azadian-Boulanger, G.; Ojasoo, T.; Laraque, F. *Contraception* **1976,** *14,* 275.
10. Grunwell, J. F.; Benson, H. D.; O'Neal Johnston, J.; Petrow, V. *Steroids* **1976,** *27,* 759.
11. Verma, U.; Laumas, K. R. *J. Steroid Biochem.* **1981,** *14,* 733.
12. Bucourt, R. In *Topics in Stereochemistry;* Wiley-Interscience: New York, 1974; Vol. 8, p 159.
13. Branceni, D.; Rousseau, G.; Jequier, R. *Steroids* **1965,** *6,* 451.
14. Heel, R. C.; Brogden, R. N.; Speight, T. M.; Avery, G. S. *Drugs* **1978,** *16,* 302.
15. Teutsch, G.; Weber, L.; Page, G.; Shapiro, E. L.; Herzog, H. L.; Neri, R.; Collins, E. J. *J. Med. Chem.* **1973,** *16,* 1370.
16. Green, M. J.; Bisarya, S. C.; Herzog, H. L.; Rausser, R.; Shapiro, E. L.; Shue, H. J.; Sutton, B.; Tiberi, R. L.; Monahan, M.; Collins, E. J. *J. Steroid Biochem.* **1975,** *6,* 599.
17. Bucourt, R.; Vignau, M.; Torelli, V.; Richard-Foy, H.; Geynet, C.; Secco-Millet, C.; Redewith, G.; Baulieu, E. E. *J. Biol. Chem.* **1978,** *253,* 8221.
18. Nédélec, L. *Bull. Soc. Chim. Fr.* **1970,** 2548.
19. Gasc, J.-C.; Nédélec, L. *Tetrahedron Lett.* **1971,** 2005.
20. Chambers, R. D.; Clark, M. *Tetrahedron Lett.* **1970,** 2741.
21. Gasc, J.-C. *French Demande* **1974,** *220,* 1287.
22. Teutsch G.; Richard, C. *J. Chem. Res. (S)* **1981,** 87.
23. Costerousse, G.; Buendia, J.; Toromanoff, E.; Martel, J. *Bull. Soc. Chim. Fr.* **1978,** 355.
24. Robinson, C. H.; Finckenor, L.; Olivetto, E.; Gould, D. *J. Am. Chem. Soc.* **1959,** *81,* 2191.
25. Laurent, H.; Gerhards, E.; Wiechert, R. *Arzneimittelforschung* **1977,** *27,* 2187.
26. Kapp, J. F.; Koch, H.; Töpert, M.; Kessler, H. J.; Gerhards, E. *Arzneimittelforschung* **1977,** *27,* 2191.
27. Teutsch, G.; Costerousse, G.; Deraedt, R.; Benzoni, J.; Fortin, M.; Philibert, D. *Steroids* **1981,** *38,* 651.
28. Philibert, D.; Moguilewsky, M. *Abstracts of Papers,* 65th Annual Meeting of the Endocrine Society, San Antonio, TX, 1983; Abstract No. 1018.
29. Anderson, R. J. *J. Am. Chem. Soc.* **1970,** *92,* 4978.
30. Herr, R. W.; Johnson, C. R. *J. Am. Chem. Soc.* **1970,** *92,* 4979.

31. Rickborn, B.; Staroscik, J. *J. Am. Chem. Soc.* **1971,** *93,* 3046.
32. Johnson, C. R.; Wieland, D. M. *J. Am. Chem. Soc.* **1971,** *93,* 3047.
33. Teutsch, G.; Bélanger, A. *Tetrahedron Lett.* **1979,** 2051.
34. Elks, J. *J. Chem. Soc.* **1960,** 3333.
35. Kirk, D. N.; Petrow, V. *J. Chem. Soc.* **1961,** 2091.
36. Baran, J. S.; Lennon, H. D.; Mares, S. E.; Nutting, E. F. *Experientia* **1970,** *26,* 762.
37. Gilbert, H. G.; Phillipps, G. H.; English, A. F.; Stephenson, L.; Woollett, E. A.; Newall, C. E.; Child, K. *J. Steroids* **1974,** *23,* 585.
38. Azadian-Boulanger, G.; Bertin, D. *Chim. Ther.* **1973,** *8,* 451.
39. Raynaud, J.-P.; Bouton, M. M.; Gallet-Bourquin, D.; Philibert, D.; Tournemine, C.; Azadian-Boulanger, G. *Mol. Pharmacol.* **1973,** *9,* 520.
40. Azadian-Boulanger, G.; Bouton, M. M.; Bucourt, R.; Nédélec, L.; Pierdet, A.; Torelli, V.; Raynaud, J.-P. *Eur. J. Med. Chem.* **1978,** *13,* 313.
41. Ferland, L.; Labrie, F.; Kelly, P. A.; Raynaud, J.-P. *Biol. Reprod.* **1978,** *18,* 99.
42. Bertin, D.; Pierdet, A. *C.R. Acad. Sci. Paris* **1967,** *264 C,* 1002.
43. Jensen, E. V.; Mohla, S.; Gorell, T. A.; DeSombre, E. R. *Vitam. Horm.* **1974,** *32,* 89.
44. Baulieu, E. E.; Atger, M.; Best-Belpomme, M.; Corvol, P.; Courvalin, J. C.; Mester, J.; Milgrom, E.; Robel, P.; Rochefort, H.; De Catalogne, D. *Vitam. Horm.* **1975,** *33,* 649.
45. Gorski, J.; Gannon, F. *Annu. Rev. Physiol.* **1976,** *38,* 425.
46. Raynaud, J.-P.; Ojasoo, T.; Bouton, M. M.; Philibert, D. In *Drug Design;* E. J. Ariëns, Ed.; Academic: New York, 1979; Vol. 8, pp 169–214.
47. Castaner, J.; Thorpe, P. *Drugs Future* **1977,** *11,* 131.
48. Bélanger, A.; Philibert, D.; Teutsch, G. *Steroids* **1981,** *37,* 361.
49. Velluz, L.; Muller, G. *Bull. Soc. Chim. Fr.* **1950,** 166.
50. Djerassi, C.; Rosenkranz, G.; Iriarte, J.; Berlin, J.; Romo, J. *J. Am. Chem. Soc.* **1951,** *73,* 1523.
51. Raynaud, J.-P.; Bonne, C.; Bouton, M. M.; Moguilewsky, M.; Philibert, D.; Azadian-Boulanger, G. *J. Steroid Biochem.* **1975,** *6,* 615.
52. Neri, R.; Florance, K.; Koziol, P.; Van Cleave, S. *Endocrinology* **1972,** *91,* 427.
53. Teutsch, G.; Bélanger, A.; Philibert, D.; Tournemine, C. *Steroids* **1982,** *39,* 607.
54. Teutsch, J. G.; Costerousse, G.; Deraedt, German Patent 1977, DE 2709078.

55. Dausse, J.-P.; Duval, D.; Meyer, P.; Gaignault, J.-C.; Marchandeau, C.; Raynaud, J.-P. *Mol. Pharmacol.* **1977,** *13,* 948.
56. Giesen, E. M.; Bollack, C.; Beck, G. *Mol. Cell. Endocrinol.* **1981,** *22,* 153.
57. Giesen, E. M.; Beck, G. *Horm. Metab. Res.* **1982,** *14,* 252.
58. Rousseau, G. G.; Baxter, J. D.; Higgins, S. J.; Tomkins, G. M. *J. Mol. Biol.* **1973,** *79,* 539.
59. Bouton, M. M.; Raynaud, J.-P. *J. Steroid Biochem.* **1978,** *9,* 9.
60. Raynaud, J.-P.; Bouton, M. M.; Ojasoo, T. *Biochem. Soc. Trans.* **1979,** *7,* 547.
61. Raynaud, J.-P.; Bouton, M. M.; Ojasoo, T. *Trends Pharmacol. Sci.* **1980,** *1,* 324.
62. Lefebvre, P.; Formstecher, P.; Rousseau, G. G.; Lustenberger, P.; Dautrevaux, M. *J. Steroid Biochem.* **1989,** *33,* 557.
63. Cohen, N. C.; Colin, P.; Lemoine, G. *Tetrahedron* **1981,** *37,* 1711.
64. Jordan, V. C.; Koch, R. *Endocrinology* **1989,** *124,* 1717.
65. Knox; W. E.; Auerbach, V. H. *J. Biol. Chem.* **1955,** *214,* 307
66. Makman, M. H.; Nakagawa, S.; Dvorkin, B.; White, A. *J. Biol. Chem.* **1970,** *245,* 2556.
67. Philibert, D. In *Adrenal Steroid Antagonism;* Agarwal, M. K., Ed.; Walter de Gruyter: Berlin and New York, 1984; pp 77–101.
68. Raynaud, J.-P. In *Progress in Cancer Research and Therapy;* McGuire, W. L.; Raynaud, J.-P.; Baulieu, E. E., Eds.; Raven: New York, 1977; Vol. 4, pp 9–21.
69. Cahiez, G. *Tetrahedron Lett.* **1981,** 1239.
70. Teutsch, G.; Costerousse, G. *J. Chem. Res. (S)* **1983,** 294.
71. Wyllie, A. H. *Nature (London)* **1980,** *284,* 555.
72. Cutler, G. B., Jr.; Barnes, K. M.; Sauer, M. A.; Loriaux, D. L. *Endocrinology* **1979,** *104,* 1839.
73. Guthrie, G. P., Jr.; John, W. J. *Endocrinology* **1980,** *107,* 1393.
74. Chrousos, G. P.; Barnes, K. M.; Sauer, M.A.; Loriaux, D. L.; Cutler, G. B., Jr. *Endocrinology* **1980,** *107,* 472.
75. Sayers, G. *Ann. N.Y. Acad. Sci.* **1977,** *237,* 220.
76. Philibert, D.; Hardy, M.; Gaillard-Moguilewsky, M.; Nique, F.; Tournemine, C.; Nédélec, L. J. *Steroid Biochem.* **1989,** *34,* 413.
77. McPhail, M. K. *J. Physiol. (London)* **1934,** *83,* 145.
78. Teutsch, G.; Costerousse, G.; Philibert, D.; Deraedt, R. Eur. Patent 005115 A2, 1982.
79. Philibert, D.; Moguilewsky, M.; Mary, I.; Lecaque, D.; Tournemine, C.; Secchi, J.; Deraedt, R. In *The Antiprogestin Steroid RU 486 and Human*

Fertility Control; Baulieu, E. E.; Segal, J., Eds.; Plenum: New York, 1985; pp 49–68.
80. Secchi, J.; Lecaque, D. *Cell Tissue Res.* **1984,** *238,* 247.
81. Baulieu, E. E. *Génération pilule;* O. Jacob: Paris, 1990. The English version has been published under the title *The Abortion Pill* (Simon & Schuster: New York, 1991).
82. Teutsch, G. In *Adrenal Steroid Antagonism;* Agarwal, M. K., Ed.; Walter de Gruyter: Berlin and New York, 1984; pp 43–75.
83. Teutsch, G. In *The Antiprogestin Steroid RU 486 and Human Fertility Control;* Baulieu, E. E.; Segal, J., Eds.; Plenum: New York, 1985; pp 27–47.
84. Teutsch, G.; Ojasoo, T.; Raynaud, J.-P. *J. Steroid Biochem.* **1988,** *31,* 549.
85. Philibert, D.; Costerousse, G.; Gaillard-Moguilewsky, M.; Nédélec, L.; Nique, F.; Tournemine, C.; Teutsch, G. In *Antihormones in Health and Disease;* Agarwal, M. K., Ed.; S. Karger: Basel, 1991; p 1.

Ranitidine

John Bradshaw
Glaxo Group Research Ltd.

Ranitidine (**1**), first prepared in the summer of 1976 in the research laboratories of Allen & Hanburys Ltd., in Ware, England, is a member of the class of compounds known as histamine H_2-receptor antagonists. Histamine H_2-receptor antagonists are used extensively in the treatment of acid peptic disease, as they increase the pH of the stomach by lowering the output of acid, allowing ulcers to heal naturally. The first commercial compound of this class, cimetidine (**2**), was launched in the United Kingdom as Tagamet in November 1976; its development was discussed in Volume I in this series (*1*). Within 5 years of its launch as Zantac, in 1981, ranitidine had displaced cimetidine as the world's best-selling prescription drug, a position it still maintains at the time of writing, and it was in large part owing to ranitidine that Glaxo became the second largest pharmaceutical company in the world, based on sales. Although much has been published on

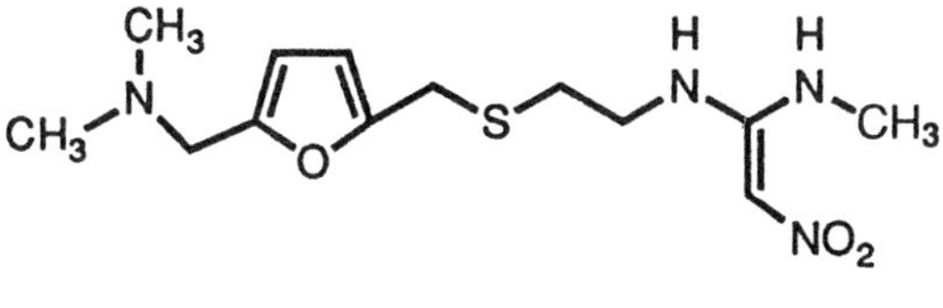

1

2523–0/93/0045$10.25/0

the pharmacology of such an important commercial drug, little has appeared on the synthesis of ranitidine or the paradigm that gave rise to it. This chapter is an attempt to redress that omission.

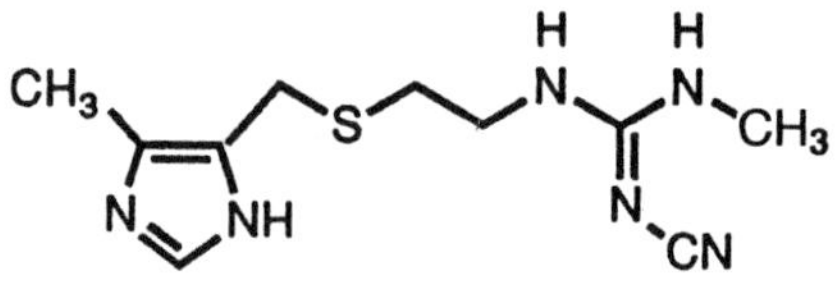

2

Historical Setting

Allen & Hanburys Ltd., became part of the Glaxo organization in 1958. Probably the single event that propelled Allen & Hanburys onto the world pharmaceutical stage was the decision in 1961 to employ David Jack as Research Director (*2*). Along with Roy Brittain, Alec Ritchie, and others, he would build up drug research within the company. More important, he would define a paradigm for drug research that would change Allen & Hanburys from a "reformulator of other people's drugs" into an innovative drug discovery machine (*3*). Thus, within the Glaxo Group of companies, there were two separate research facilities developing along quite distinct lines, the Allen & Hanburys research group at Ware, under David Jack's direction, and the research facility at Glaxo, Greenford, which continued its work on steroids and antibiotics. The two facilities were finally unified as Glaxo Group Research in 1979, headed by David Jack.

At Allen & Hanburys, Jack set up multidisciplinary groups of chemists and biologists with defined therapeutic objectives. Each objective was to be approached through a defined biological mechanism, usually entailing the modification of action of one of the body's natural mediators. This approach had an advantage, from a chemist's point of view, in that there would always be a lead molecule. One could modify the structure of the natural compound and use the congeneric molecules to "ask the receptor questions". At this time there were no receptor structures known. Singer and Nicholson had yet to publish the fluid mosaic model of membranes (*4*), and it would be a further 15 years before Huber's group would publish a reasonably resolved structure of an integral membrane protein (*5*). Perhaps more surprising, it was only in 1957 that Sutherland's work on adenylate cyclase (*6*) had identified 3,5-cyclic adenosine monophosphate (**3**), an important secondary transmitter for many of the natural mediators of interest to Jack's group of scientists. Even as late as 1968 it was thought that catecholamines such as epinephrine (**4**) acted *directly* on adenylate cyclase (7). There was, therefore, little accurate knowledge about the true nature of receptors for the natural mediators. Hence the ability to design small molecules to ask specific questions was of paramount importance.

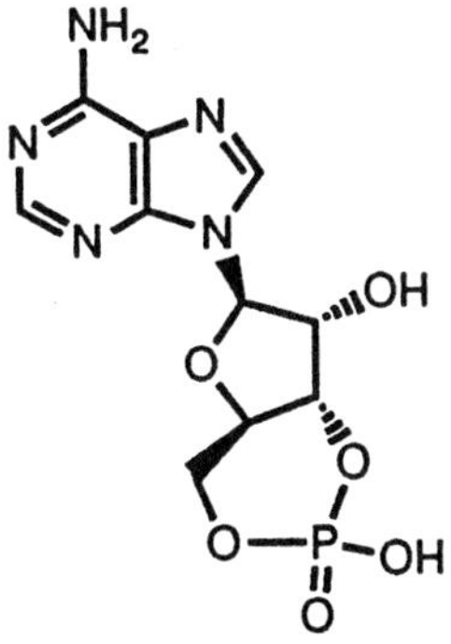

3 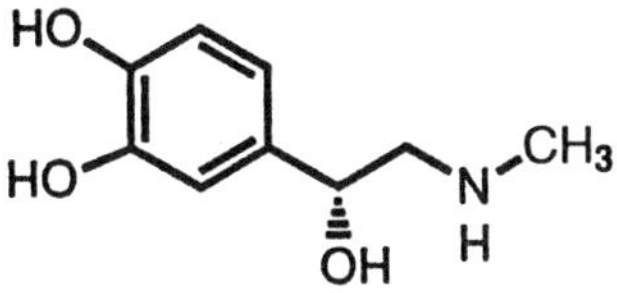 **4**

The major single concept that Jack insisted on was selectivity of action of drugs. The body gets selectivity of action by local release of specific mediators, which are then rapidly deactivated. Only when many organs need to be supplied with the same transmitter, such as epinephrine in the fright-flight-fight reflex, do circulating blood levels become significant. In systemic drug therapy, because the blood circulation is used to distribute the drug, it is difficult to target specific organs. Selectivity of action needs to be built into the compound along with stability, as specific mechanisms exist in the body to remove the natural mediators. It may also be necessary to be able to reverse or antagonize the effect of the natural compounds.

In order to test whether these needs are satisfied and whether the objective is achievable, researchers must be able to answer the following questions:

> Can we make compounds?
> Can we test for selectivity of action in vitro?
> Can we test for the disease state in a valid animal model in vivo?

At the time of Jack's appointment, at least three possible targets had been established as true natural mediators: acetylcholine (**5**), norepinephrine (**6**), and dopamine (**7**). The roles of histamine (**8**), 5-hydroxytryptamine (**9**), and γ-aminobutyric acid (**10**) were still a matter of debate.

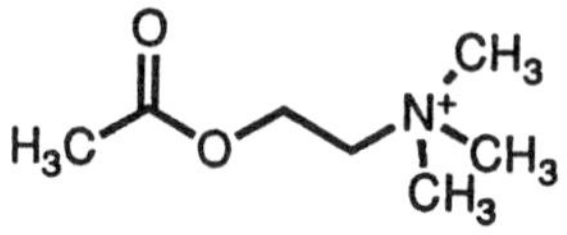

5 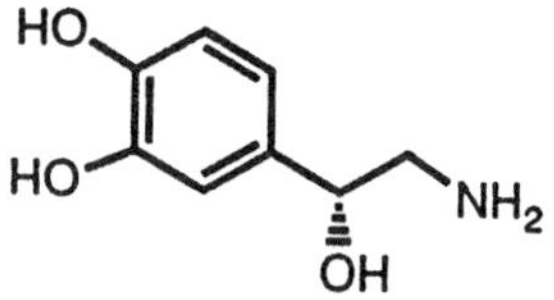 **6**

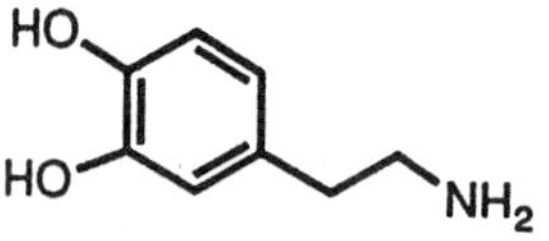

7 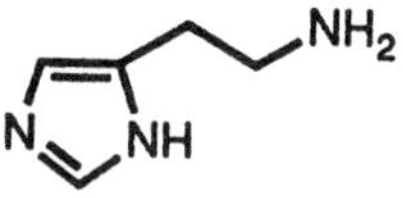 **8**

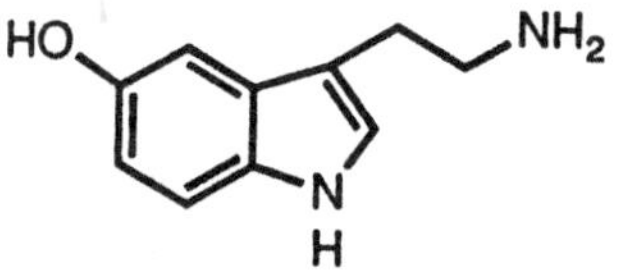

9 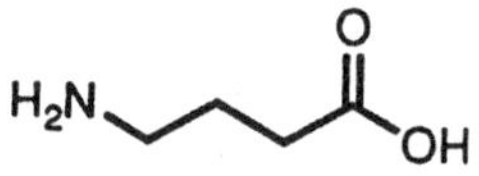 **10**

The initial choice for the application of the paradigm was norepinephrine (**6**). This would give rise, in the late 1960s, to two series of compounds, selective β_2-agonists, typified by albuterol (**11**, AH3365, Ventolin) and salmefamol (**12**, AH3923) for the treatment of asthma (*8*), and combined α- and β-antagonists such as labetalol (**13**, AH5158, Trandate) for use as antihypertensives (*9*). Ten to fifteen years later, the same process of "asking the receptor questions" would give rise to salmeterol (**14**, GR33343, Serevent) (*10*), a long-lasting β_2-agonist.

The concept of selectivity could also be applied to steroids used to treat asthma and allergic rhinitis. Beclomethasone dipropionate (**15**) did not depress the adrenal cortex and therefore did not produce the side effects associated with corticosteroids, particularly Cushing's syndrome, in children.

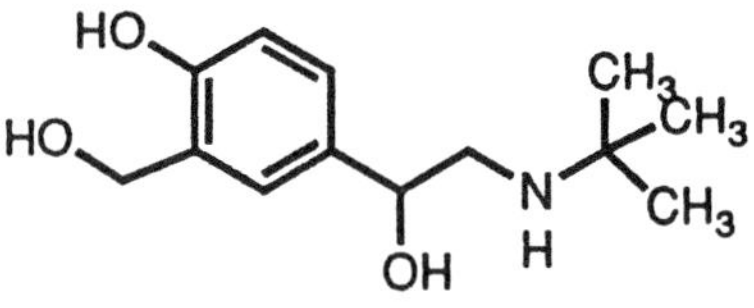

11

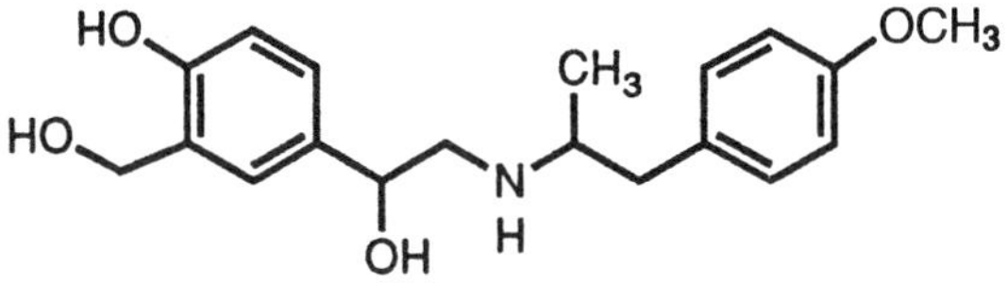

12

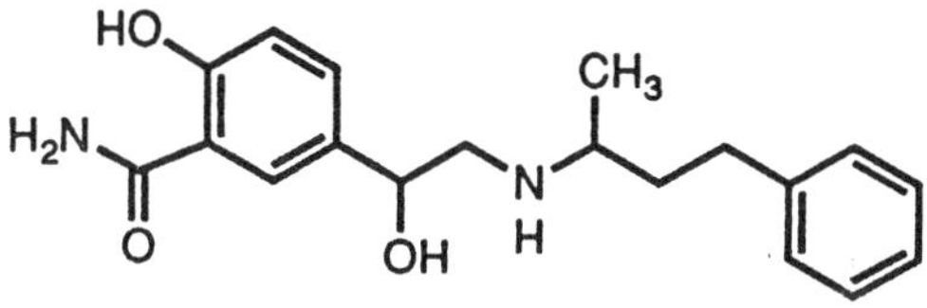

13

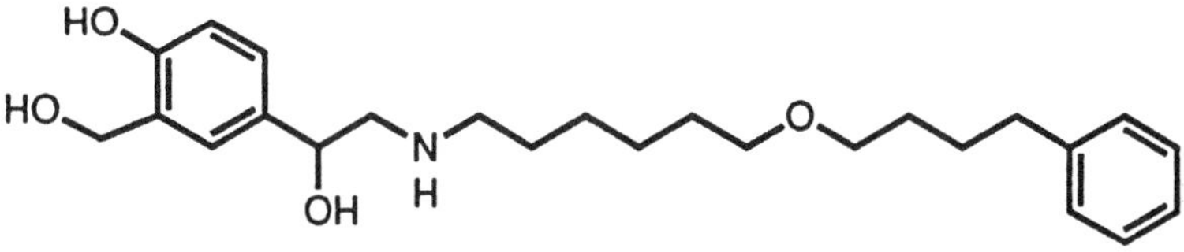

14

Two products based on this compound, Becotide and Beconase, were launched in 1972 and 1974, respectively.

By the late 1960s the physiological role of both histamine (**8**) and 5-hydroxytryptamine (**9**) as natural mediators had been confirmed, and they became valid targets for this approach. Selective analogues of 5-hydroxytryptamine would eventually be developed, such as AH21467 (**16**, 5-CT) and GR38032 (**17**, ondansetron, Zofran) as an antiemetic (*11*) and GR43175 (**18**, sumatriptan, Imigran) for migraine (*12*). It was in this emerging culture of successful drug discovery that the search for an improved treatment for peptic ulcers began.

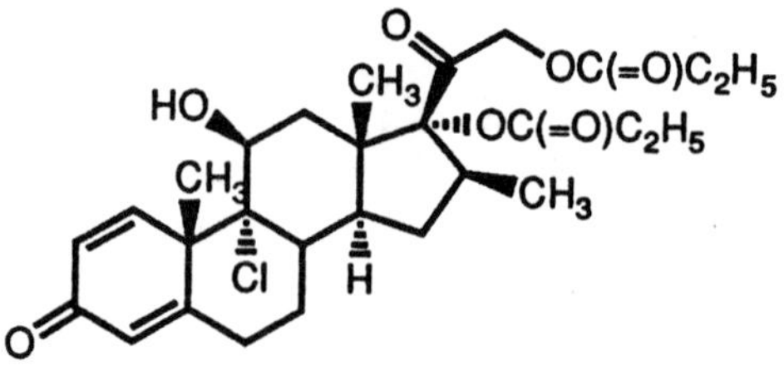

15

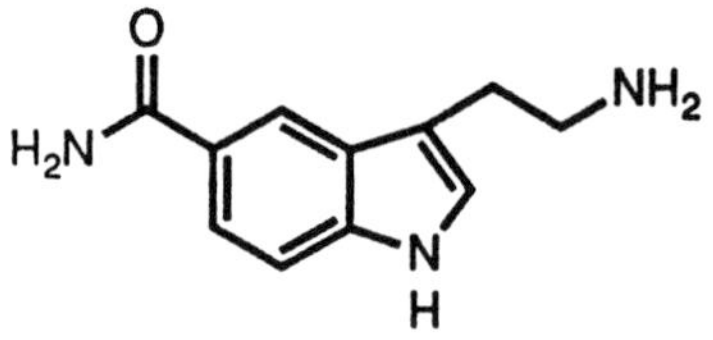

16

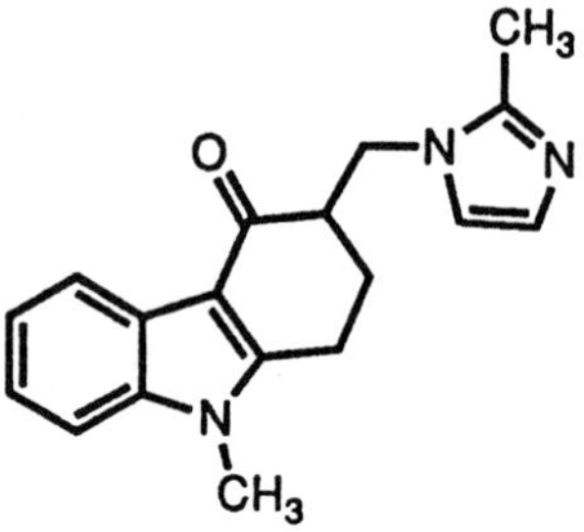

17

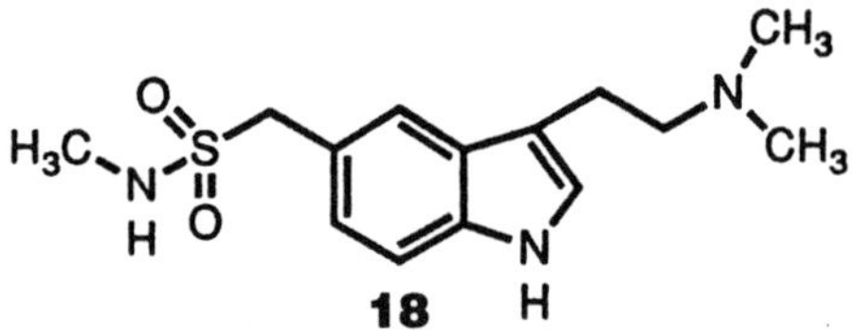

18

Application of the Drug Discovery Paradigm to the Treatment of Peptic Ulcers

In the spirit of these volumes (*13*), I will try to reconstruct the thinking prevalent in the Peptic Ulcer Project Group in the early 1970s, and how the approach to drug discovery outlined in the previous section was applied to the treatment of peptic ulcers.

Clearly there was a defined disease, and the disease was poorly treated, requiring surgery or a major change in life-style for the patient to obtain symptomatic relief. Indeed, the edition of Goodman and Gilman's *The Pharmacological Basis of Therapeutics* current at that time (1970) (*14*) listed no suitable drug remedies other than antacids. It was also known that reducing gastric acid, either surgically or by using drugs such as anticholinergics, would allow ulcers to heal.

We also had a defined method of drug delivery. To reduce the need for hospitalization or attendance by health-care professionals, oral delivery was the route of choice.

In terms of a mechanism to treat the disease, there were two major alternatives: (1) a systemic approach to reduce acid secretion, which should work for both gastric and duodenal ulcers, or (2) increasing the protective mucous layer in the stomach for gastric ulcer treatment at least. This treatment could be topical, as evolution kindly provides external access to the stomach.

The second alternative did not accord well with David Jack's philosophy. The mechanisms for mucous secretion were ill-understood at a receptor level, although there were some indications that prostaglandins such as **19** might be important and that liquorice extracts and carbenoxelone (**20**, Biogastrone) might work by this mechanism (*15*). There were no good in vitro biological tests available at the time, so this approach was put on the back burner.

Reducing acid secretion had many more advantages as an approach. Animal models were available, and known chemicals that occurred in the body, whether or not they really were natural mediators, were known to affect the level of acid secretion, both in humans and in experimental animals. Of these, five were candidates for consideration: gastrin (**21**), epinephrine (**4**), acetylcholine (**5**), histamine (**8**), and prostaglandin E_2 (**19**).

The various pros and cons of these approaches were discussed over the years, and factors that were considered during a typical discussion session are listed in Table 1. Note that there was input from all the drug discovery disciplines, in both chemistry and biology. The table also indicates the spectrum of beliefs as to the correct way to proceed. The dialectical approach encouraged by Jack ensured that everyone's opinion was considered.

A histamine H_2-antagonist or a selective anticholinergic clearly would fulfill the criteria for application of the paradigm, but the other approaches were not rejected out of hand. If compounds were available and there was capacity in the biological screens, the compounds would be tested. The distinction between basing the project on the treatment of peptic ulcer disease and handling it as an H_2-antagonist project was real. The ability to

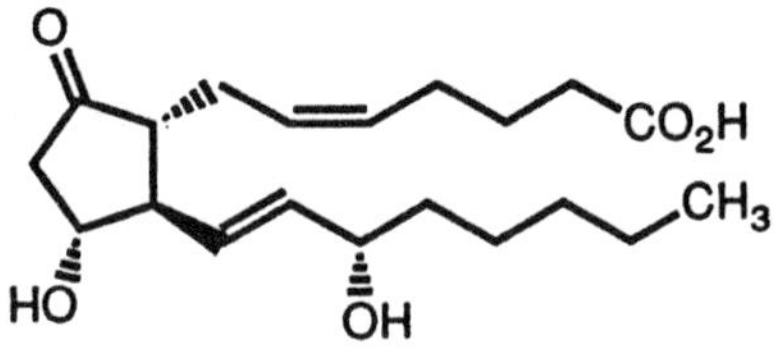

19

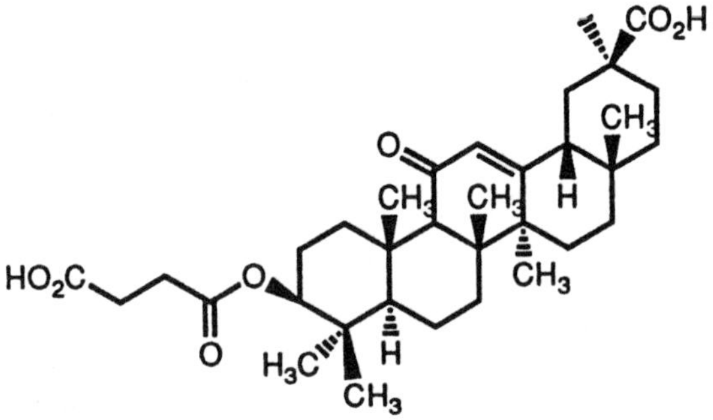

20

5-oxo-Pro-Gly-Pro-Trp-Leu-Glu-Glu-Glu-Glu-Glu
|
H_2N-Phe-Asp-Met-Trp

21

$(CH_3)_3COC(=O)CH_2CH_2NH$
|
H_2N-Phe-Asp-Met-Trp

22

change quickly between approaches should the biology so dictate was another characteristic of the Jack era at Allen & Hanburys. In the end the histamine approach won out, for by 1972 Black and his colleagues had established histamine as the ultimate extracellular physiological mediator of gastric acid secretion by the parietal cells in the gastric pits (*16*).

Table 1. Approaches Considered in Developing a Drug for Treating Peptic Ulcer Disease

Approach	*Pros*	*Cons*
Gastrin antagonists	None known—would be novel In vivo tests were available that used pentagastrin (**22**) as agonist	Peptides—unlikely to be orally active No in vitro tests available No chemical experience in making peptides.
β_2-agonists (epinephrine)	Considerable experience in the area In vitro and in vivo tests were available Compounds for testing were available from another project group	Many known—novelty may be a problem Problems with skeletal tremor with β_2-agonists Another project group was already working in the area
Acetylcholine antagonists	In vitro and in vivo tests were available Chemistry was straightforward Starting structures were available	Many known—novelty may be a problem Could we find selectivity? If we found one could we sell one against the prejudice of lack of selectivity in humans?
Histamine antagonists	No commercial drugs In vitro and in vivo tests were available Selecivity had been established for agonists Classic antihistamines did not affect acid secretion Considerable experience in heterocyclic chemistry	Area increasingly well patented by SK & F
Prostaglandin E_2 agonists	No commercial drugs In vitro and in vivo tests were available	Could we breed out the abortifacient action? Another project group was already working in the area Chemistry was difficult despite the Newton–Roberts synthesis (*61*)

Application of the Drug Discovery Paradigm to H_2-Antagonists

The paper by Black et al. was pivotal in deciding the direction of the research program, so it is perhaps worth emphasizing some of the main points in the background of histamine-receptor typing. By the mid-1940s it had become clear that not all the actions of histamine were antagonized by the large number of antihistamines that had become available following introduction of the initial compound piperoxan (**23**, 933F), described by Bovet and Staub (*17*). This observation led Folkow et al. (*18*) to postulate that there was more than one histamine receptor. Alles et al. (*19*) and Grossman et al. (*20*) had shown that alkyl-substituted histamines did not activate all histamine receptors equally, and Ash and Schild (*21*) subsequently defined the receptors that were sensitive to mepyramine as H_1-receptors. The receptors they classified as H_1 existed in the guinea pig isolated ileum, whereas the rat isolated uterus and rat stomach contained receptors that were activated by histamine but were not antagonized by mepyramine.

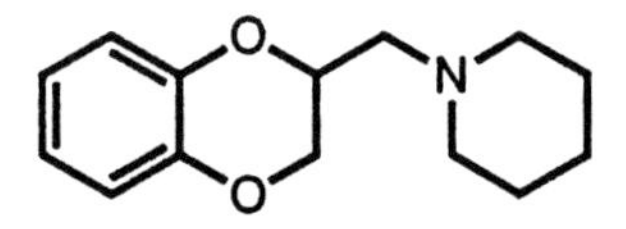

23

The notable paper by Black et al. (*16*) announced the full characterization of these "other" receptors, which were termed H_2-receptors. Black and his colleagues showed that, in addition to the uterus and stomach preparations described by Ash and Schild (*21*), H_2 receptors were present in the guinea pig right atrium, and that this tissue was the tissue of choice to investigate the action of compounds on H_2-receptors. Following the approach that had been used for adrenergic (epinephrine) receptors, they synthesized all possible methyl-substituted histamines in an attempt to find selectivity. Alles et al. (*19*) in 1943 had shown that 4-methyl histamine (**24**) had little or none of the peripheral actions of histamine, whereas Grossman et al. 9 years later showed it was nevertheless a potent stimulator of acid secretion (*20*). Black's group quantified this result, showing that 4-methyl histamine had 40% of the activity of histamine in vivo for stimulating acid secretion in the rat and in vitro for stimulating spontaneous atrial frequency in guinea pig isolated right atrium. However, in confirmation of Alles's

$$\text{24: 4-methylhistamine (}CH_3\text{ at position 4 of imidazole, side chain }CH_2CH_2NH_2\text{)} \qquad \text{25: 2-methylhistamine (}CH_3\text{ at position 2)}$$

$$\text{General structure: imidazole (positions 2, 3, 4) with side chain } \beta, \alpha \text{ and NHR}$$

Table 2. In Vitro H_1 and H_2-Receptor Activities of Methyl-Substituted Histamines

Position of Methyl Group							*Equipotent Concentration (Histamine = 1)*	
1	*2*	*3*	*4*	*α*	*β*	*R*	*Atria*	*Ileum*
						CH_3–	1.3	1.4
					CH_3–		167	167
				CH_3–			125	125
			CH_3–				Inactive	
		CH_3–					22.7	6.1
	CH_3–						Inactive	
CH_3–							2.5	500

result, Black et al. found that 4-methyl histamine had only 0.2% of the activity of histamine for stimulation of contractions of guinea pig isolated ileum. In contrast, 2-methyl histamine (**25**) showed the reverse selectivity, having only 2% of the activity of histamine in stimulating acid secretion but 17% of the activity of histamine in stimulating contraction of guinea pig isolated ileum. 2-Methyl histamine was described as inactive in the acid secretion model studied by Grossman et al. (*20*), and was described by Lee and Jones (*22*) as having 30% of the activity of histamine on guinea pig ileum. Table 2 shows the complete set of data for H_2-receptor tissues and for H_1-receptor tissues for the monomethyl histamines (*16*).

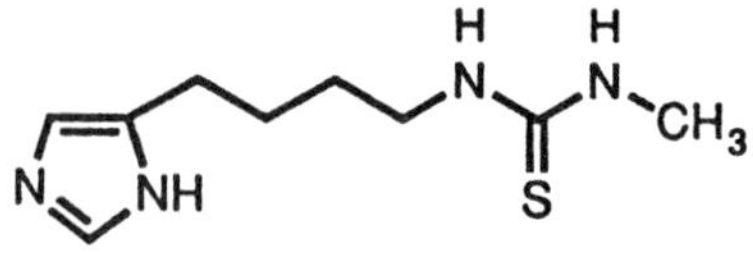

26

Despite the careful nature of this work with agonists, it was the synthesis and testing of the compound burimamide (**26**), shown to be a selective antagonist of these newly defined H_2-receptors, that confirmed their classification (*23*). More important, burimamide inhibited acid secretion, in humans as well as in animals, induced by the infusion of histamine (**8**) or pentagastrin (**22**), or by the ingestion of food. Burimamide itself would not have made a good medicine, for whereas it was potent when given intravenously, it was only poorly absorbed when given orally.

Black had established that H_2-receptor antagonism in a chemical compound is most conveniently determined in vitro by using the chronotropic response of guinea pig right atrium or the relaxation of rat uterus caused by histamine. The nature of the antagonism is "both determined and quantified by the dose-dependent rightward shift of the histamine concentration effect curve" (*23*). Tests in a battery of other tissues confirmed that the compound exhibits only H_2-receptor antagonism (*16*). The inhibition of acid secretion in vivo is measured using the perfused rat stomach preparation in anaesthetized rats (*24*) in which the effect of intravenously administered drug on histamine- or pentagastrin-induced gastric acid secretion is determined. Oral activity and the duration of action are measured in conscious dogs with Heidenhein pouches, a model in which it is possible to study the effect of compounds on acid secretion induced by histamine (**8**), bethanechol, or pentagastrin (**22**) (*25*). Conscious dogs with gastric fistulas are used to study responses to test meals (*26*).

The value of these tests in predicting activity in humans was confirmed by the report by Black and his colleagues that burimamide was active in humans against acid secretion induced by pentagastrin, histamine, or food (*16*). These tests fulfilled two of the criteria outlined earlier for setting up a project: we could test for selectivity of action in vitro, and we could test for the disease state in a valid animal model in vivo. The only question remaining was whether we could make the compounds. In the early 1970s there were two major approaches. We could start again from histamine (**8**), as the scientists at Smith Kline and French Laboratories (SK & F) had, or we could make use of the information starting to appear in the literature, that compounds such as burimamide (**26**) were selective antagonists of histamine. In the end we avoided the decision and approached the problem from both directions.

Histamine as a Starting Point

There were reports as early as 1943 (*19*) and 1952 (*20*) that 5-methyl histamine and 5,α-dimethyl histamine showed selectivity as agonists for gastric acid secretion. If selective agonists could be made, it was not unreasonable that one could also make selective antagonists by suitable manipulation of the imidazole nucleus. We therefore set out to make a series of substituted imidazoles that would maximize the variation in physicochemical properties of the imidazole ring.

In the early 1970s Topliss (*27*) had proposed rational design sequences for substituted benzenoid and aliphatic systems. Martin and Dunn had shown their utility in the post facto rationalization of selective β-stimulant activity (*28*). What was required, in this case, was the corresponding set of substituents for substitution in an imidazole system. In general, the corresponding physicochemical parameters used by Topliss were not available for heterocyclic systems. However, it was reasonable to assume that, for carefully chosen substituents, the extrathermodynamic principle should hold, and there should be a simple relationship between the parameters for heterocyclic systems and the known values in the benzene system (*29*).

With this basic premise in mind, we drew up a scheme (Figure 1) for the design of 4-substituted analogues of histamine of the general formula **27**. The choice of substituents was such as to exclude those that would interact tautomerically with the imidazole ring, such as OH or NH_2. We also neglected, at that time, the effect of the substituent on the tautomerism of the imidazole itself, as reported by Ganellin (*30*), which played such an important part in the development of metiamide (**28**) (*1*).

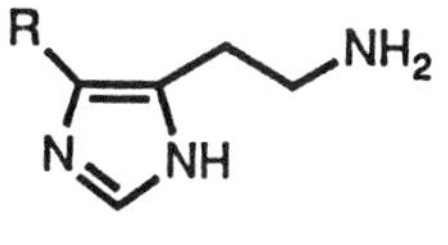

R, N, NH, NH_2

27

H_3C, N, NH, S, H N, H N, CH_3, S

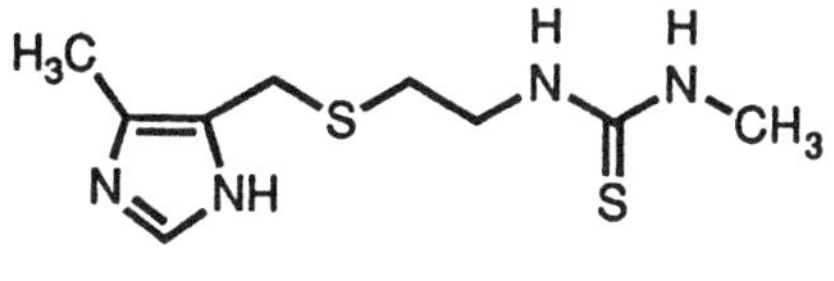

28

The scheme shown in Figure 1 was designed to be used like Topliss's original. Thus, if the chloro compound were less potent than the parent, one would prepare the methoxy derivative; equal activity would call for the

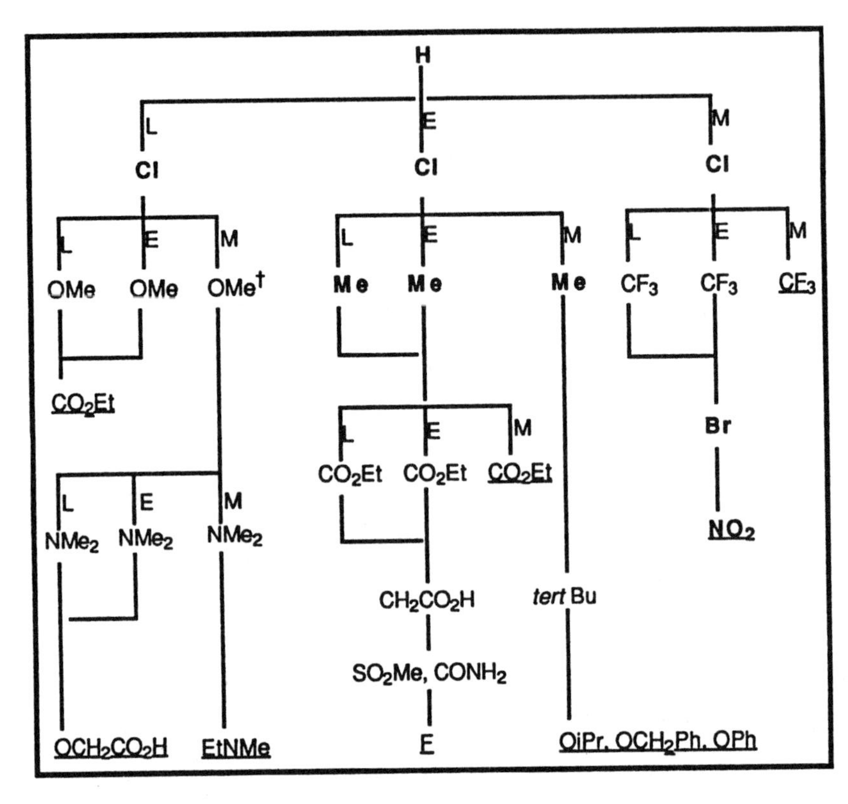

*Figure 1. Topliss tree for 4-substituted histamines (**27**). Boldface indicates results published (Table 3) (30).*

methyl compound; and greater activity would suggest the trifluoromethyl derivative. The activities of some of these compounds were already available in the literature. Ganellin (*30*) had reported the H_1- and H_2-receptor activities for 4-methyl-, 4-chloro-, 4-bromo-, and 4-nitrohistamines (Table 3). These compounds had different levels of H_2-receptor agonist activity, as would be required if such a design scheme were to be useful. However, if this scheme were correct, it would imply that SK & F had explored the wrong region of parameter space. Ganellin (*30*) chose to explain the variation in activity by correlating it with the changes in tautomer populations. In 1986, however, Weinstein et al. (*31*) would show that the parameter population was affected at least as much by the interaction of the ammonium cation on the side chain with an anion in the putative receptor.

Table 3. H_2-Receptor Agonist Activity of Compounds of General Formula 27 on Atria

R	*H_2 Activity*
H–	1.00
CH_3–	0.43
Cl–	0.11
Br–	0.09
O_2N–	0.006

NOTE: Values are relative to histamine (=1.00).

As we had parameter estimates for the physicochemical properties of substituents on the imidazole, we could do a primitive experimental design. We chose four substituted imidazoles that would cause large variation in the physicochemical properties of the imidazole ring and thus of the corresponding histamine. The compounds chosen were **29–32**. At the time (1976), structures **29–32** were all novel. In the event, we were unsuccessful in synthesizing any of the four key compounds, and further work in the area was abandoned as the alternative approach of using the known H_2-receptor antagonist, burimamide (**26**), as a starting point began to yield exciting results.

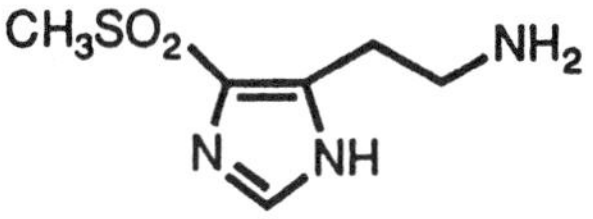

CH_3SO_2 NH NH_2 N

29

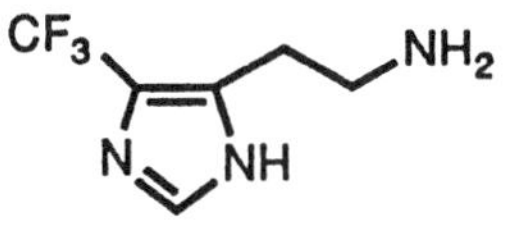

CF_3 NH NH_2 N

30

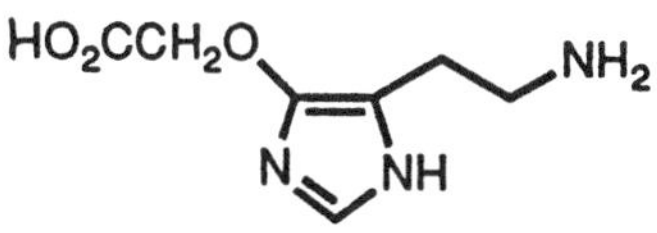

HO_2CCH_2O NH NH_2 N

31

C_6H_5O NH NH_2 N

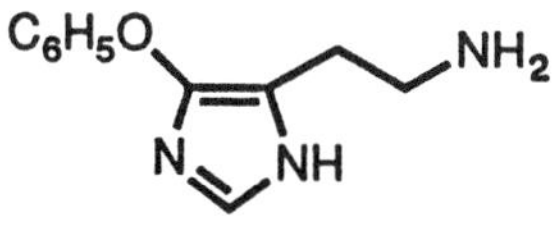

32

Burimamide as a Starting Point

Many accounts are available of the elegant work undertaken by scientists at SK & F that led from burimamide (**26**) to the highly successful drug cimetidine (**2**, Tagamet) (e.g., ref. 1). This section reviews the steps we took at Allen & Hanburys to get from burimamide to ranitidine (**1**). The process is an excellent illustration of the postulate of May (*32*), namely, that novel ideas need the four elements of time, place, person, and culture to flourish. Ranitidine resulted from the efforts of a small group of people of different backgrounds working together at the same time in a corporate culture geared to the discovery of new drugs.

A useful nomenclature grew up, among the project group, to describe the structures of H_2-receptor blockers. Like Gaul, they were divided into three parts: a left-hand side (LHS), a chain, and a right-hand side (RHS) or end group. The LHS in the preferred SK & F compounds was a (substituted) imidazole, with, one was led to believe at the time, a basic pK_a of about 6.8 (*33*). At this time, the preferred RHS was an *N*-methyl thiourea group, as in burimamide (**26**). The LHS and RHS were separated by a four-carbon chain. Later publications by SK & F would relate how the thiourea was derived rationally from a guanyl group (*34*). We took a more pragmatic view. We had had a group working on gastric acid secretion since the late 1960s. Among other approaches (*35*), the group was trying to exploit the known activity of thioamides as supposed inhibitors of gastrin. Indeed, the edition of Goodman and Gilman available in that era (*14*) suggested that thioamides held the main hope for the treatment of peptic ulcers. We were screening a large number of compounds and had developed experience in tetrazole chemistry. This encouraged David Bays and Roger Hayes to try to incorporate the features of burimamide into these tetrazoles. This effort gave rise to, among other compounds, AH15475 (**33**), which had a 2-linked 5-amino-tetrazole replacing the imidazole of burimamide. AH15475 was equipotent with burimamide in both the perfused rat model and as an H_2-antagonist on isolated guinea pig atria. It had no effect on isolated guinea pig ileum but, in contrast to burimamide, it was orally active in the Heidenhein pouch dog model at concentrations corresponding to its in vitro activity. We thus had a novel starting point for development of our own H_2-antagonists.

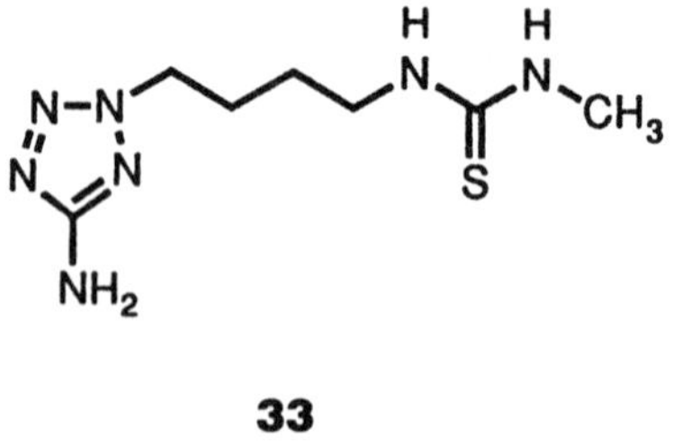

33

A large number of 5-substituted analogues of AH15475 (**33**) were prepared (Table 4). Other than AH15475, however, only AH17047, the 5-trifluoromethyl derivative, showed any marked inhibition of acid secretion. This inhibition, however, was occurring by an unknown mechanism and was not of sufficient magnitude to pursue further.

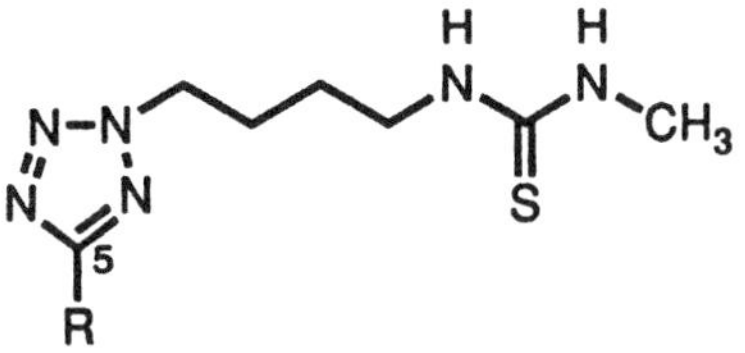

Table 4. Examples of 5-Substituted-2-Tetrazolyl-Butyl-Thioureas

Registry Number	*5-Substituent*
AH15475	$-NH_2$
AH15555	$-NHC(=S)NHCH_3$
AH16169	$-C_6H_5$
AH16221	$-H$
AH16452	$-NHC(=O)CH_3$
AH16469	$-CH_3$
AH16645	$-N(CH_3)_2$
AH16673	$-NHCH_2C_6H_5$
AH16674	$-OCH_2CH_3$
AH16784	$-SCH_3$
AH17047	$-CF_3$
AH17085	$-CH_2CH_3$
AH17088	$-CH(CH_3)_2$
AH17200	$-OCH_3$
AH17445	$-S(=O)(=O)CH_3$
AH17885	$-CF_2CF_2CF_3$
AH17890	$-CH_2F$
AH18008	$-N(CH_3)CH_2C_6H_5$
AH18009	$-NHCH_3$
AH19061	$-CH_2N(CH_3)_2$

The alteration of the RHS or end group was carried out by replacing the thiourea by a range of formic acid derivatives (Table 5). Most notably, replacing the thiourea with a cyanoguanidine, as SK & F had done in making cimetidine, did not cause any increase in activity (AH17100). Neither did the "nitrovinyl" group present in AH17857, which we synthesized, as SK & F had, using the chemistry described by Gomper and Schaefer (36).

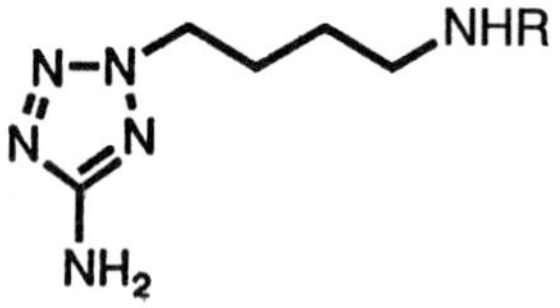

Table 5. Variation of the End Group in AH15475 (33)

Registry Number	*R Substituent*	*Melting Point (°C)*
AH15475	$-C(=S)NHCH_3$	87–89
AH16250	$-C(=NH)NH_2$	166–168 as HNO_3 salt
AH17100	$-C(=NCN)NHCH_3$	oil
AH17159	$-C(=S)NH$-*cyclo*$-C_6H_{11}$	103–104
AH17160	$-C(=S)NH(CH_2)_3CH_3$	65–68
AH17161	$-C(=S)NHC(CH_3)_3$	135–137
AH17232	$-C(=S)NHCH_2CH_3$	80–83
AH17542	$-C(=NCH_3)SCH_3$	138–139 as HI salt
AH17853	$-C(=CHNO_2)SCH_3$	123–126
AH17857	$-C(=CHNO_2)NHCH_3$	171–175 decomp.
AH18082	$-C(=NCH_3)NHCH_3$	142–143 as HI salt
AH18265	$-SO_2CH_3$	Oil
AH18479	$-C(=O)NHCH_3$	124–126
AH18541	$-C(=C(CN)_2)NHCH_3$	127–130

Nevertheless, AH17857 would prove to be a key compound in the preparation of ranitidine. AH17857 was a highly crystalline solid with a melting point 90 °C higher than the corresponding thiourea, AH15475. The cyanoguanidine, AH17100, was an oil. This increase in crystallinity would be remembered 9 months later when we had a prototype compound in Chemical Development causing some difficulties. None of the other end groups markedly increased activity. Replacement of the four-carbon chain by shorter (C-3) or longer (C-5) chains did not improve activity. We were unable to find any increased activity by replacing nitrogens in the ring by carbon atoms to give progressively triazoles and pyrazoles.

Hence, in early 1976, we seemed unable to increase the potency of AH15475. SK & F, meanwhile, had produced metiamide (**28**) and cimetidine (**2**) as orally active H_2-antagonists. Cimetidine was being developed in preference to metiamide, as the latter had resulted in a few cases of reversible agranulocytosis in humans (37). This effect was believed (34) to be due to the thiourea group. At Allen & Hanburys the number of chemists preparing compounds was reduced to three, and consideration was given to abandoning H_2-antagonists as a primary target, to be replaced by the investigation of selective anticholinergics. Several arguments favored this change in strategy. The work on pirenzepine (**34**, LS519) had been published, so we knew it was possible to prepare compounds that selectively antagonized bethanechol-induced acid secretion (38). The chemistry team was now led by John Clitherow, who had extensive experience in working with anticholinergic compounds, a fact that, perversely, would ensure that we not only would be able to continue with H_2-antagonists but also would produce compounds with activity equivalent to the activity of metiamide (**28**) and cimetidine (**2**).

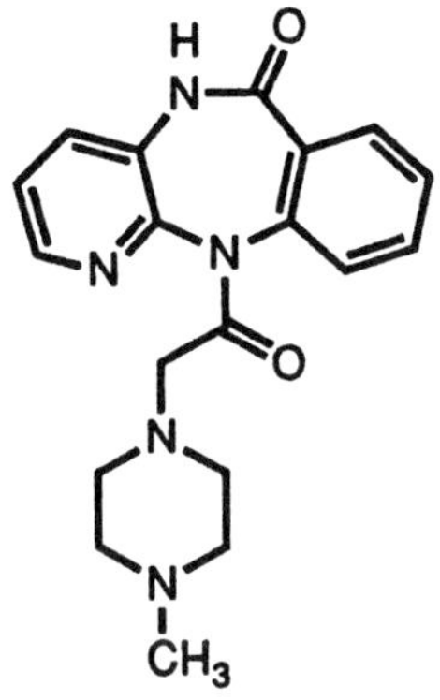

34

As far as the H_2-antagonists were concerned, we had established that AH15475 (**33**) was equipotent with burimamide (**26**), and we concluded that (1) the LHS ring need not be overtly basic (amino tetrazoles have a pK_a of about 1), and (2) the side chain need not be linked via the C atom of the heterocycle.

A decision was then made to prepare the furan (**35**, AH18166). The thiourea was chosen as the end group for convenience, as we had no evidence at the time that changing the end group from thiourea would increase the potency. The 2-isomer of furan was chosen, as 2-furfuryl mercaptan was commercially available. The butyl chain was also abandoned in favor of the methylthioethyl chain for synthetic ease.

AH18166 (**35**) was not totally devoid of activity on atria at 10 mg/mL, the standard test dose, but the pharmacologists observed that it was quite insoluble and gave variable results. We knew that furans underwent the Mannich reaction in the 2(5)-position. An example is provided by the dimethylamino compound **36**. If the Mannich reaction were carried out on

AH18166, it should give a product with a pK_a of approximately 8.5. This compound should at least be in solution under the test conditions. The Mannich product AH18665 (**37**) was produced via the route shown in Scheme 1.

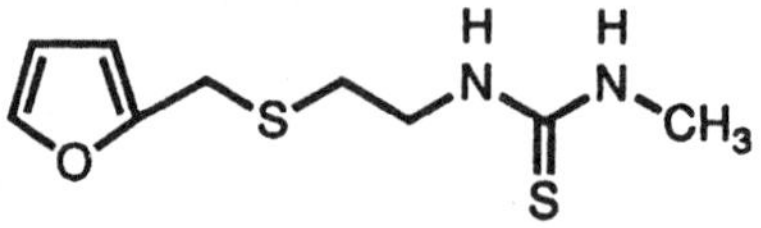

35 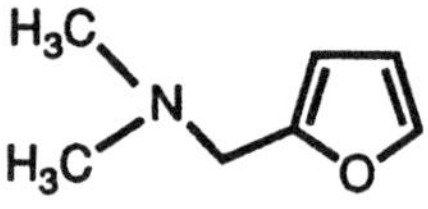 **36**

This was the breakthrough we had been waiting for. AH18665 (**37**) was as active as metiamide in both atrial tests and the perfused rat stomach model. Quite clearly this product, as it contained a thiourea, was likely to produce the same side effect of agranulocytosis (*37*) as metiamide (**28**) did. Hence a decision was made to prepare the cyanoguanidine derivative (**38**). SK & F was close to marketing cimetidine without any reports of adverse effects, and experience with AH15475 (**33**) and a comparison of the activities of metiamide (**28**) and cimetidine (**2**) indicated that we would not lose activity by making the change. In the event, AH18801 (**38**) proved to be as active as cimetidine with a similar profile of action both in vivo and in vitro.

Time was now of the essence. In the drug industry, "good enough soon enough" is more critical to the choice of a compound for development than whether the compound has optimal structure. A decision was made to prepare large quantities of AH18801 for further pharmacological, biochemical, and toxicological studies. One problem that faced the development chemists was the lack of crystallinity and the low melting point observed for AH18801.

As mentioned earlier, the nitrovinyl group markedly increased the crystallinity without adversely affecting the (low) potency of AH15475 (**33**) in a series in which the corresponding cyanoguanidine, AH17100, was an oil (Table 5). We decided to investigate if the new furan series behaved equivalently. As quite often occurs in drug research, 1 plus 1 does not equal 2. This transformation was an example of this paradox. Far from being more crystalline, AH19065 (**1**) was isolated, in early August 1976, as a dark red oil. The initial sample proved resistant to all attempts to crystallize it. However, it was an order of magnitude more active than anything else we had seen and rapidly replaced AH18801 (**38**) as the development candidate. In the perfused rat model, AH19065 (**1**) was active at 0.18 mg/kg (*39*), compared with an active dosage of 1.39 mg/kg for AH18801 (**38**).

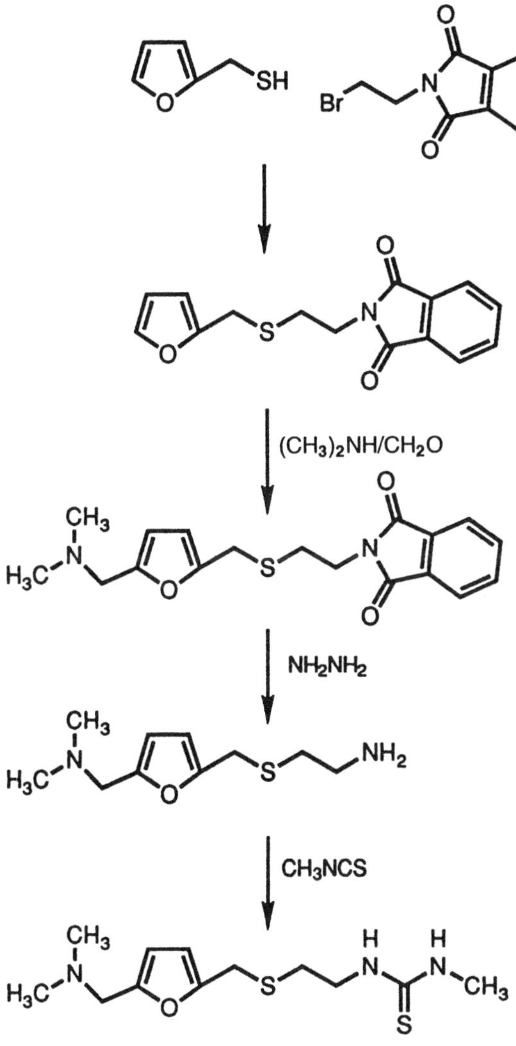

*Scheme 1. Synthesis of AH18665 (**37**)*

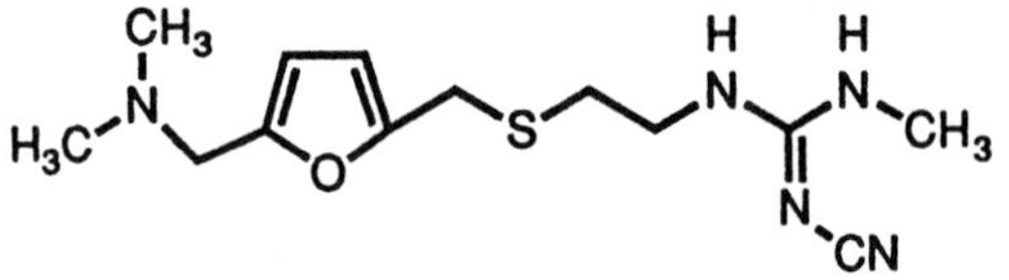

38

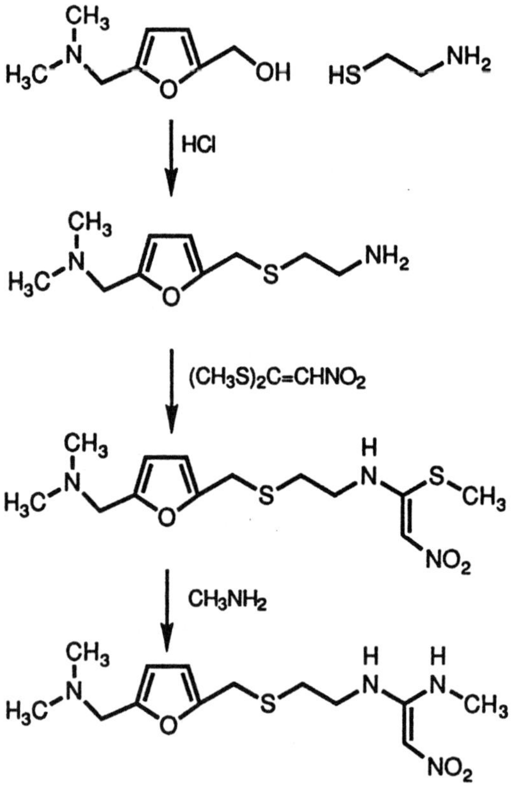

HCl

$(CH_3S)_2C{=}CHNO_2$

CH_3NH_2

1

Scheme 2. Synthesis of AH19065

Painstaking work in Development Chemistry provided crystals, both of the free base and the hydrochloride salt. The hydrochloride was crystallized, first as polymorph I, then as polymorph II. Within 5 years the second polymorph would be marketed as Zantac. A synthesis of AH19065 is shown in Scheme 2.

The transition from the cyanoguanidine to the nitrovinyl derivative proved fortuitous in other ways. When cimetidine was marketed later in 1976, as Tagamet, reports began to appear of side effects in some patients taking high doses of the drug. Cimetidine inhibits cytochrome P450, an important drug-metabolizing enzyme (*40*). This interaction has the effect of inhibiting the metabolism of drugs such as propranolol (**39**), warfarin (**40**), and diazepam (**41**), thus producing effects equivalent to an overdose of these medicines (*41, 42*). Because these drugs are often co-administered to ulcer patients, it is a distinct disadvantage.

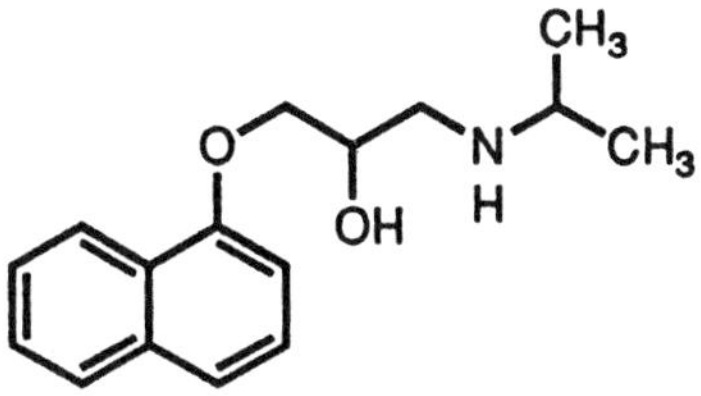

39

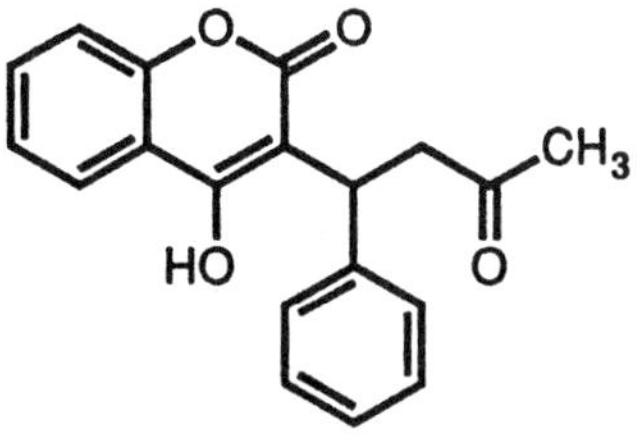

40

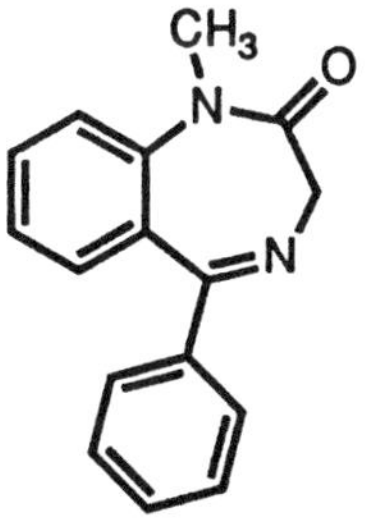

41

Angus Bell and his colleagues (*43*) in our laboratories showed that ranitidine does not interact with these drug-metabolizing enzymes, whereas AH18801 (**38**) is as active as cimetidine (**2**) in these tests. Hence this interaction presumably occurs via the nitrogen of the cyano group (Figure 2). Thus, AH18801 would have offered no advantage over cimetidine in this respect.

Another side effect of cimetidine that began to be reported was an antiandrogen effect, leading, in extreme cases, to gynecomastia in men. Cimetidine in high doses reduces prostate and seminal vesicle weight in developing rats (*44*). The scientific basis for this observation has been

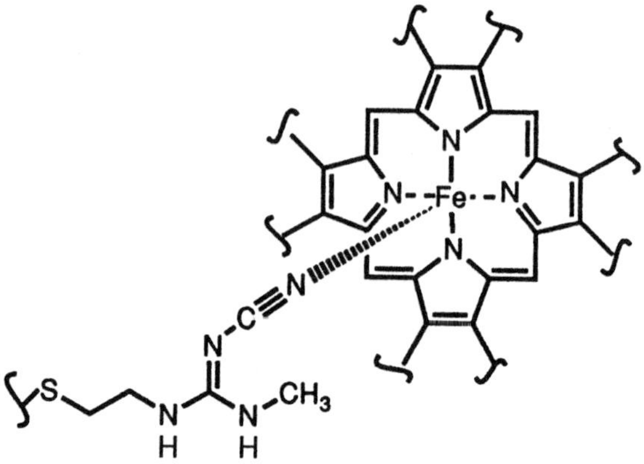

Figure 2. Interaction of cyanoguanidines with the heme unit of cytochrome P450.

established by studies on the binding of cimetidine and ranitidine to androgen receptors (45). Cimetidine displaces ^{3}H-dihydrotestosterone (**42**) from androgen receptors, but ranitidine is without effect. Similar results were obtained both in vivo and in vitro, and the authors concluded that the action of cimetidine at androgen receptors is not linked to its H_2-antagonist activity. This difference has been confirmed in rats in vivo, where ranitidine, unlike cimetidine, does not reduce prostate weight (26, 46).

Confusional states in elderly patients being treated with cimetidine were also reported. The lower lipophilicity (43) and lower dosing schedule ensures

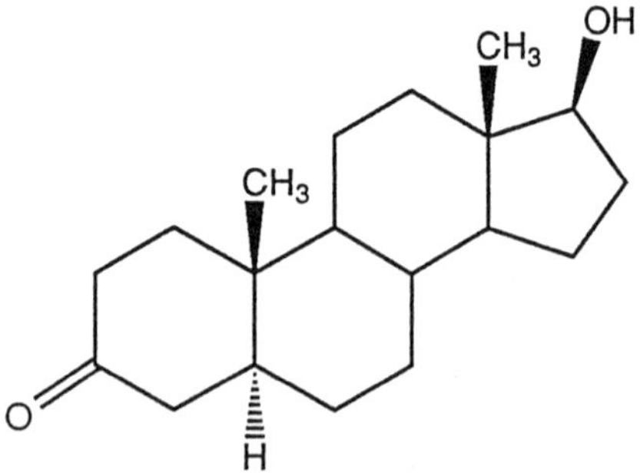

42

much lower cerebrospinal fluid levels of ranitidine, thereby markedly reducing the likelihood of any CNS effects.

Structure–Activity Studies

The structure–activity relationships of ranitidine have been published elsewhere (47). The salient points do, however, bear repeating, if only to reinforce the adage that 1 plus 1 does not always make 2.

As was remarked earlier, substitution of the nitrovinyl group for the cyanoguanidine in AH18801 (**38**) improved the physical properties of AH18801. We had no evidence that changing the RHS of H_2-antagonists from thiourea to either cyanoguanidine or nitrovinyl would have any effect on potency. In SK & F's series there is a slight decrease in potency in the perfused rat stomach preparation, in the order thiourea > cyanoguanidine > nitrovinyl (Table 6). In our series the order is reversed, with a large difference between the nitrovinyl and cyanoguanidine.

Table 6. Effect of Changing the RHS Group on Two Series of H_2-Antagonists

	Left-hand Side	
Right-hand Side	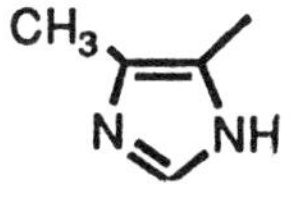 CH_3 / N NH (5-methylimidazol-4-yl)	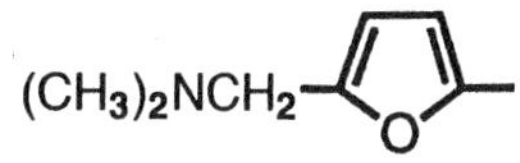 $(CH_3)_2NCH_2$–(furan-2-yl, O)
$-NHC(=S)NHCH_3$	0.52	2.32
$-NHC(=NCN)NHCH_3$	1.12	1.39
$-NHC(=CHNO_2)NHCH_3$	1.75	0.18

NOTE: Data are activity (in milligrams per kilogram) in the perfused rat stomach preparation.

Further evidence for this difference in structure-activity relationships between cimetidine analogues and ranitidine analogues was the reported need for a basic group in the LHS (37) with a pK_a of about 6.8. Although we knew from our work with tetrazoles that a basic group was not essential, we were unaware of highly potent compounds without such a feature. Changes to the basic amine structure in ranitidine (**1**) showed that the pK_a of the amine could adopt a value as low as about 2 in AH20507, through to approximately 8.5 in AH20261, without a decrease in potency (Table 7).

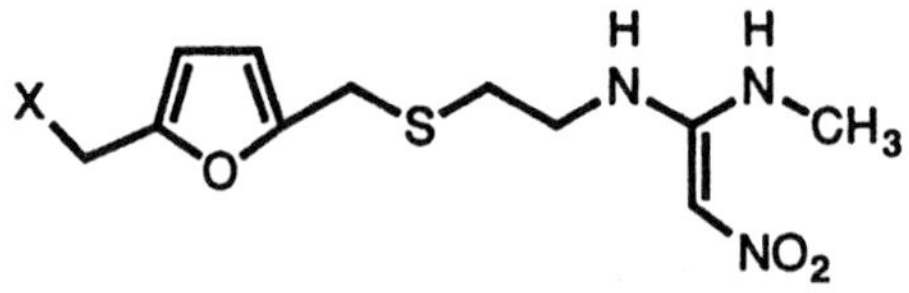

Table 7. Effect of Changing LHS Amine in AH19065 (1)

Registry Number	*X*	*Activity*
AH19691	CH_3NH-	0.23
AH20646	CH_3CH_2NH-	0.52
AH20672	$CH_2{=}CHCH_2NH-$	0.45
AH21085	CF_3CH_2NH-	0.30
AH19065	$CH_3N(CH_3)-$	0.18
AH20261	$CH_3CH_2N(CH_3)-$	0.48
AH20507	$HON(CH_3)-$	0.15

NOTE: Activity is reported in milligrams per kilogram in the perfused rat stomach preparation.

All of this evidence clearly indicated that the structure-activity relationships observed for cimetidine analogues were different from those for ranitidine analogues, and that the RHS contributed differing amounts to the activity, depending on the LHS. This example is an excellent example of the need for molecules to be designed as a whole (*48*).

The choice of the 2,5-isomer of the furan was driven by the chemistry. 2-Furfuryl mercaptan was commercially available, and the Mannich reaction produced the 5-isomer. It was necessary, for our own peace of mind, to investigate the other isomers. The sometimes difficult chemistry leading to the other isomers has been described (*47*). In the event, as the thioureas all the compounds in Table 8 were substantially less active in the perfused rat stomach model ($ED_{50} > 10$ mg/kg) than the compound with the 2,5-substitution pattern, AH18665 (**37**).

In some of SK & F's series, the effect of methyl substitution on the ring adjacent to the side chain was shown to give a compound with enhanced activity (*49*). Thiaburimamide (**43**) has a K_b value on atria of 3.2 μm, compared with a K_b of 0.8 μm for metiamide (**28**). The lack of a similar increase in K_b value between burimamide (**26**) and methylburimamide (**44**) was rationalized in terms of tautomer populations (*1*). Activities for all these compounds are given in Table 9.

We decided that we too should look at the effect of ring substitution in ranitidine. We already knew that we would not have the equivalent

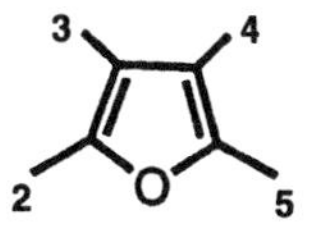

Table 8. Substitution Pattern of Ranitidine Isomers

Registry Number	Position of $(CH_3)_2NHCH_2$–		Position of $-CH_2S(CH_2)_2NHC(=S)NHCH_3$			
	2	3	2	3	4	5
AH18665	*					*
AH19808	*				*	
AH19467	*			*		
AH19466		*	*			
AH19622		*			*	
AH19827		*				*

SOURCE: Data are from ref. 47.

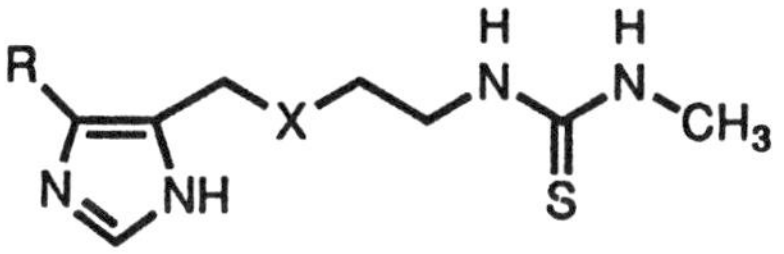

Table 9. Contrasting Effect of Methyl Substitution in Various H_2-Antagonists (Smith Kline and French Compound)

Compound	*R*	*X*	*Activity*
26	H–	$-CH_2-$	7.8
44	CH_3–	$-CH_2-$	8.9
43	H–	–S–	3.2
28	CH_3–	–S–	0.9

NOTE: Activity is K_b on atria in micromoles.

tautomeric and pK_a effects, for we had a different structure. Nevertheless, some in our group wished to relate the two structures in the manner shown in Figure 3. This relationship indicates that the equivalent substitution for the alkyl group would be in the 3-position rather than the 4-position of the furan. In the event, the 3-methyl derivative (**45**, AH20264) was inactive, whereas activity was maintained in the 4-methyl compound (**46**, AH20430)

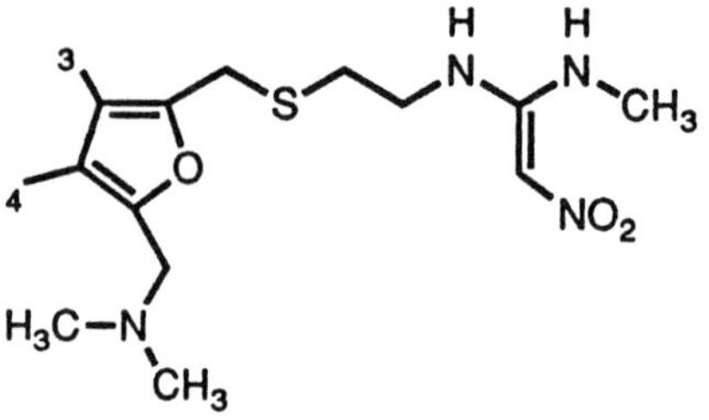

*Figure 3. Possible relationship between the orientation of ranitidine (**1**) and cimetidine (**2**) at the receptor, implying the need for 3-substitution in the furan ring of ranitidine.*

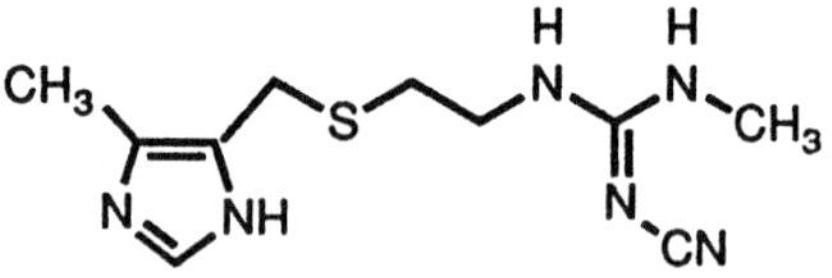

(*47*). This finding is another salutary lesson in the complexities of drug-receptor interactions much ignored by the growing breed of "Nintendo chemists" (*50, 51*).

Duncan Judd (*52*) synthesized seven such 4-substituted analogues of ranitidine, which formed the basis of a quantitative structure–activity relationship (QSAR) study. The compounds were chosen to maximize the spread of physicochemical properties consistent with synthetic accessibility (Table 10). The atria results are reported as DR_{10}, the dose of compound (in micrograms per milliliter) necessary to shift the histamine dose-response curve 10-fold to the right. The ED_{50} is as described earlier for the perfused rat stomach preparation. Physicochemical data were taken from the corresponding benzenoid system (*29*), a practice I would no longer recommend (*48, 53*). The relative lipophilicity of the substituent is measured by π, the inductive electronic effect by $\mathfrak{I}$, and the mesomeric effect by $\mathfrak{R}$. The overall bulk of the system is measured by the molar refractivity, MR.

There is no simple relationship between biological activity and any of the parameters in Table 10 taken singly, and the use of two or more parameters is invalid (*54*). However, the directional steric parameters of

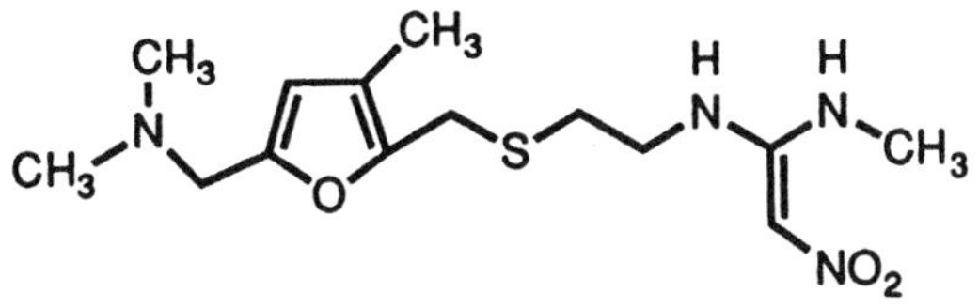

45

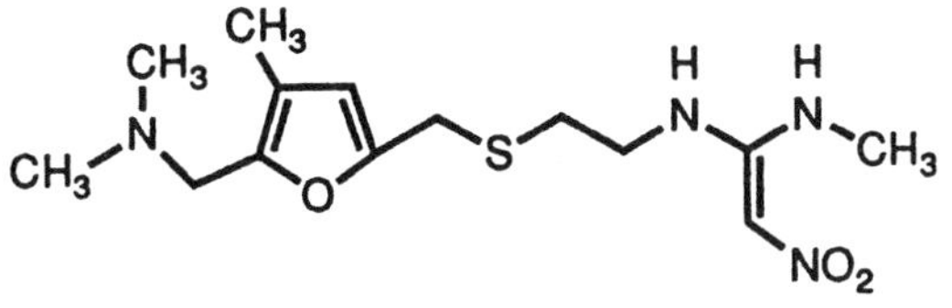

46

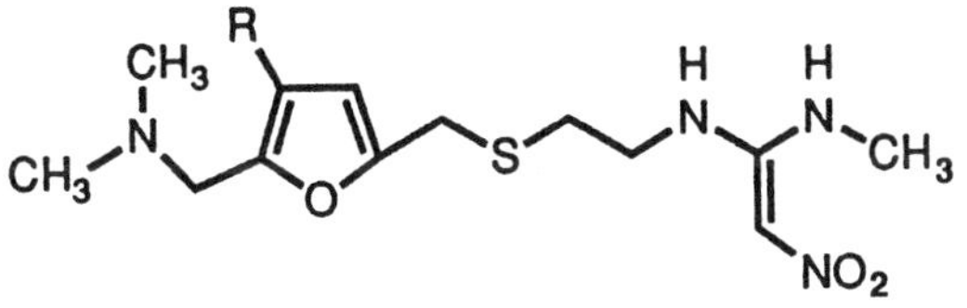

Table 10. 4-Substituted Analogues of AH19065

Registry Number	*R*	*ED_{50} (mg/kg)*	*DR_{10} (µg/mL)*	π	$\mathfrak{F}$	$\mathfrak{R}$	*MR*
AH19065	H–	0.18	0.14	0.00	0.00	0.00	1.03
AH20403	CH_3–	0.25	0.24	0.56	−0.04	−0.13	5.65
AH21061	$CH_3CH_2OC(=O)$–	2.70	>10	0.51	0.33	0.15	17.47
AH21514	$(CH_3)_2CH$–	0.33	0.20	1.53	−0.05	−0.10	14.96
AH21561	$HOCH_2$–	1.72	1.00	−1.03	0.00	0.00	7.19
AH21830	Br–	0.29	0.46	0.86	0.44	−0.17	8.80
AH21936	CH_3OCH_2–	0.21	0.13	−0.78	0.01	0.02	12.06

Verloop (55) did provide a possible correlation. Using the minimum van der Waals radius of the substituent B_1, a correlation was found, with the ester AH21061 and hydroxymethyl derivative AH21561 as outliers, using both in vitro and in vivo data. However, following the practice established by Verloop for nonsymmetrical tops, the radius value for the radius "opposite" to the B_1 value may need to be used for the rogue points.

This procedure gave acceptable equations

$$pED_{50} = 0.646(\pm 0.488) - 0.366(\pm 0.188)B_1$$
$$n = 7,\ r = 0.913,\ s = 0.198$$

and

$$pDR_{10} = 1.211(\pm 0.457) - 0.624(\pm 0.192)B_1$$
$$n = 7,\ r = 0.966,\ s = 0.203$$

where *n* is the number of compounds, *r* is the correlation coefficient, and *s* is the standard error. Values in parentheses are the 95% confidence limits.

These equations were not inconsistent with the idea that the group R (Table 10) was forcing the Mannich base into an unfavorable conformation for activity. One option, which I rejected, was that the substituents were orienting themselves so that their minimum radius was toward the receptor. Although this possibility could be true for the compounds in which B_1 was used, it was unlikely that the ester in AH21061 was both out of conjugation *and* presenting its maximum radius to the receptor. The hydroxymethyl derivative AH21561 would have to be accommodated by assuming hydrogen bonding to the receptor. No doubt some of these effects will be rationalized in detail as models of the H_2-receptor are developed.

In the context of commercial drug discovery, it was unlikely that the activity of AH19065 (**1**) would be markedly increased by increased substitution of the furan ring. However, equally it should be possible to alter the gross transport properties of AH19065 (**1**) without markedly affecting the potency at histamine H_2-receptors. In the event this alteration proved unnecessary, for a new class of H_2-receptor antagonists came into being via a second QSAR study. These compounds, typified by lamtidine (**47**, AH22216), had an increased duration of action and have been described in detail elsewhere (56).

The compounds with the guanidino RHS, exemplified by AH18801 (**38**), were much easier to handle chemically than the ethene diamine derivatives such as ranitidine. We knew that our series was more sensitive to the nature of the RHS group than was the imidazole series of SK & F. Hence it was not unreasonable that we should pursue the objective of making a more potent "guanidine". Two series were available, one in which the nature of the R group in structure **48** had been changed and one in which the X group was altered. Although by tautomerism these groups are equivalent, the formal

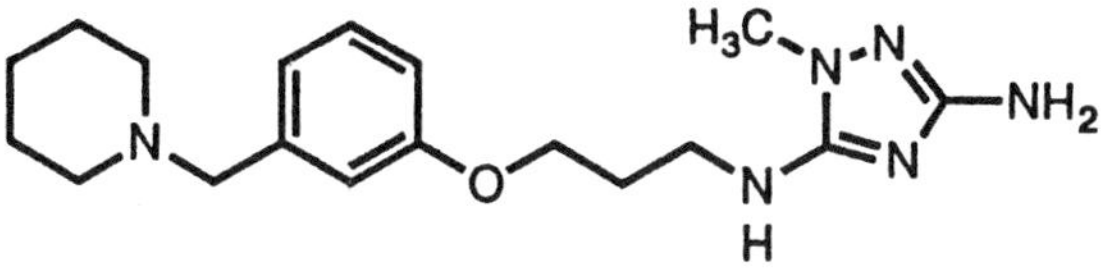

47

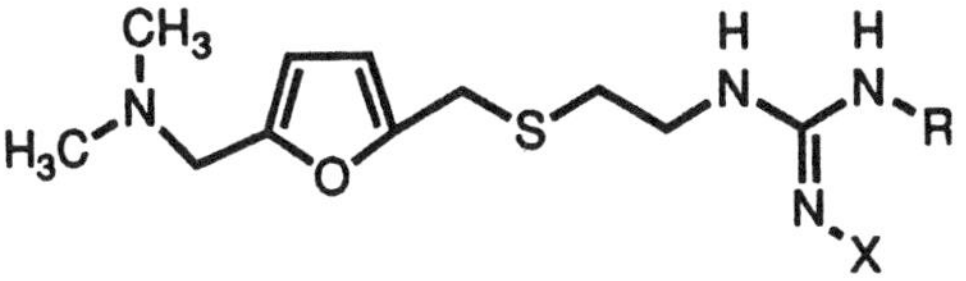

48

Table 11. Effect of Substitution on Guanidines Related to Structure 48, Where R = CH_3–

Registry Number	*X*	*ED_{50} (mg/kg)*	σ	$\sigma^{-}-\sigma$
AH18801	–CN	1.62	0.69	0.19
AH19818	–H	5.01	0.00	0.00
AH19942	–C(=O)CH_3	15.14	0.84	0.38
AH19960	–CH_3	7.94	−0.12	0.00
AH19998	–C_6H_5	22.91	−0.01	0.12
AH19999	–SO_2CH_3	3.80	0.69	0.29
AH20044	–$SOO_2C_6H_5$	3.39	0.70	0.26
AH20072	–SO_2CF_3	50.1	0.93	0.43
AH20435	–$CONH_2$	6.17	0.43	0.18

NOTE: Activity was measured in the perfused rat stomach preparation.

arrangement of the bonds was known (57). Naturally, one would not expect the effects of X and R to be independent. In the first series the group R was a methyl group. Data for these compounds are shown in Table 11.

Assuming that there was no change in mechanism between the electron-withdrawing X substituents and the electron-donating ones, I fitted the Yukawa–Tsuno-type equation below (58):

$$pED_{50} = 0.02 + 2.64(\sigma - 2.90\ (\sigma^- - \sigma))$$

$$n = 9,\ r = 0.903,\ s = 0.241.$$

In the original work, carried out before AH20435 was synthesized, an almost identical equation had been obtained that predicted the activity of AH20435 to within 7%.

Biphasic Hammett plots of this type are known for biological systems (*59, 60*). This particular relationship would indicate that activity is related to the ability of the guanidino function to accommodate a developing negative charge. This is consistent with the receptor acting in a nucleophilic manner—for example, via a serine hydroxyl function. Amidines and guanidines are known to interact with chymotrypsin in this fashion. It would also appear that the negative charge should not be localized on the substituent X.

As discussed earlier, there is no reason why the R substituent should not behave equivalently. Eight compounds were available at the time from the cyanoguanidine series (**48**, X = –CN) (Table 12).

Table 12. Effect of Substitution on Guanidines Related to Structure 48, Where X = –CN

Registry Number	*R*	*ED_{50} (mg/kg)*	σ^*
AH18801	$-CH_3$	1.62	0.00
AH19711	$-CH(CH_3)_2$	12.88	−0.19
AH19712	$-CH_2CH_2OCH_3$	11.22	0.23
AH19863	$-C_6H_5$	Inactive	0.60
AH19953	–*cyclo*-C_6H_{11}	45.00	−0.20
AH19954	$-CH_2CH{=}CH_2$	3.24	0.13
AH20132	$-C_7H_{11}$	8.32	−0.16
AH20342	–H	1.91	0.50

NOTE: Activity was measured in the perfused rat stomach preparation.

Although no quantitative relationship emerged from the data in Table 12, there was a potential for a biphasic relationship. The intermediate-valued structure where X was a carboxamide, which confirmed the biphasic nature of the relationship was required, has been referred to earlier. Also missing were compounds with first-row heteroatoms attached directly to the guanidine unit, and compounds with electron-withdrawing groups on both of the nitrogen atoms. To fulfill both these requirements, I began to look at the effect of hydrazine derivatives on the isothiourea (**49**). Initially hydrazine itself was used; however, the required compound (**50**) cyclized to give the aminotriazole (**51**, AH20285). This was a potent novel lead and was the progenitor of a range of new H_2-antagonists (56) (Scheme 3).

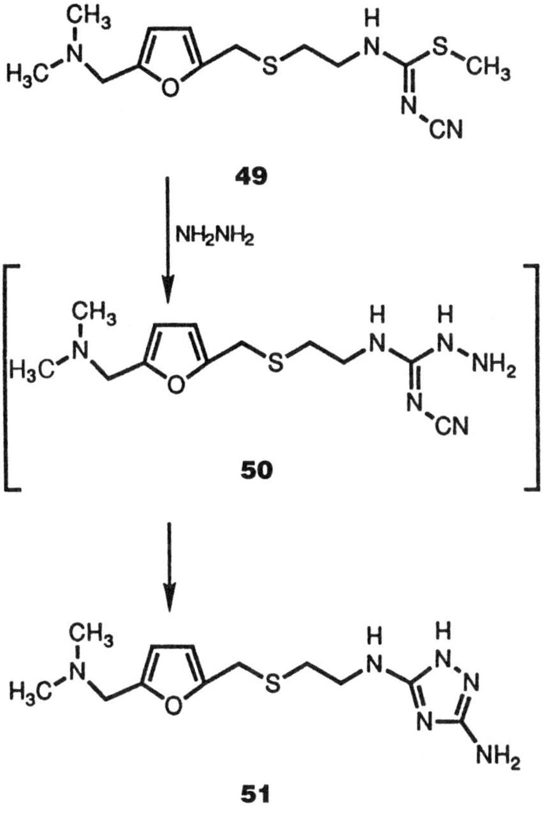

49

NH_2NH_2

50

51

*Scheme 3. Synthesis of AH20285 (**51**)*

Summary

The synthesis of AH20285 (**51**) occurred near the end of my formal involvement with the peptic ulcer project, and further work will be discussed by others. The 1970s were exciting times to be a medicinal chemist at Allen & Hanburys. It is rare to develop a compound that makes it to market, and even rarer to develop one as commercially successful as Zantac. It is also true that good drugs do not sell themselves. During this period, the development, production, and marketing arms of the Glaxo Group expanded worldwide, testing, developing, synthesizing, and marketing Zantac. These coordinated activities helped provide the firm financial platform on which today's Glaxo is built.

Acknowledgments

As I said at the beginning of the chapter, the whole ethos of the Jack approach to drug discovery relied on teamwork and group effort. Certain individuals must be mentioned in the research phase. John Clitherow and Lena Elliston-Ball were the other two chemists who were there at the beginning. Most of the early pharmacology was done by Mike Daly, Janet Humphray, and Roger Stables. The project group was managed by Barry Price and Roy Brittain, and the whole was overseen by David Jack.

I thank Barry Price, David Bays, and Roy Brittain for reading the manuscript and providing helpful suggestions as to content, and the current directors of Glaxo Group Research Ltd., for permission to publish this account.

Literature Cited

1. Ganellin, C. R. In *Chronicles of Drug Discovery*; Bindra, J. S.; Lednicer, D., Eds.; Academic: New York, 1982; Vol. 1, p 1.
2. Tweedale, G. In *At The Sign of the Plough*; John Murray: London, 1990; p 206 et seq.
3. Tweedale, G. In *At The Sign of the Plough*; John Murray: London, 1990; p 208.
4. Singer, S. J.; Nicholson, G. L. *Science (Washington, DC)* **1972,** *175,* 720–731.
5. Deisenhofer, J.; Epp, O.; Miki, K.; Huber, R.; Michel, H. *Nature (London)* **1985,** *318,* 618–624.
6. Sutherland, E. W.; Rall, T. W. *J. Am. Chem. Soc.* **1957,** *79,* 3608.
7. Robison, G. A.; Butcher, R. W.; Sutherland, E. W. *Ann. N.Y. Acad. Sci.* **1967,** *139,* 703–723.
8. Hartley, D.; Jack, D.; Lunts, L. H. C.; Ritchie, A. C. *Nature (London)* **1968,** *219,* 861–862.
9. Brittain, R. T.; Levy, G. P. *Br. J. Clin. Pharmacol.* **1976,** *3,* 681–694.

10. Bradshaw, J.; Brittain, R. T.; Coleman, R. A.; Jack, D.; Kennedy, I; Lunts, L. H. C.; Skidmore, I. F. *Br. J. Pharmacol.* **1987,** *92,* 590P.
11. Brittain, R. T.; Butler, A.; Coates, I. H.; Fortune, D. H.; Hagan, R.; Hill, J. M.; Humber, D.C.; Humphrey, P. P. A.; Ireland, S. J.; Jack, D; Jordan, C. C.; Oxford, A.; Straughan, D. W.; Tyers, M. B. *Br. J. Pharmacol.* **1987,** *90,* 87P.
12. Brittain, R. T.; Butina, D; Coates, I. H.; Feniuk, W.; Humphrey, P. P. A.; Jack, D.; Oxford, A. W.; Perren, M. J. *Br. J. Pharmacol.* **1987,** *90,* 102P.
13. Bindra, J. S.; Lednicer, D. In *Chronicles of Drug Discovery*; Bindra, J. S.; Lednicer, D., Eds.; Academic: New York, 1982; Vol. 1, p vii.
14. Harvey, S. C. In *The Pharmacological Basis of Therapeutics,* 4th ed.; Goodman, L. S.; Gilman, A., Eds.; Macmillan: New York, 1970; p 1002.
15. Robson, J. M.; Sullivan, F. M. In *Carbenoxelone Sodium*; Butterworths: London, 1970.
16. Black, J. W.; Duncan, W. A. M.; Durant, C. J.; Ganellin, C. R.; Parsons, E. M. *Nature (London)* **1972,** *236,* 385–390.
17. Bovet, D.; Staub, A. M. *Comp. Rend. Seances Soc. Biol. Paris* **1937,** *124,* 547.
18. Folkow, B.; Haeger, K.; Kahlson, G. *Acta Physiol. Scand.* **1948,** *15,* 264.
19. Alles, G. A.; Wisegarnier, B. B.; Shull, M. A. *J. Pharmacol. Exp. Ther.* **1943,** *77,* 54.
20. Grossman, M. I.; Robertson, C.; Rosiere, C. E. *J. Pharmacol. Exp. Ther.* **1952,** *104,* 277.
21. Ash, A. S. F.; Schild, H. O. *Br. J. Pharmacol.* **1966,** *27,* 427–439.
22. Lee, H. M.; Jones, R. G. *J. Pharmacol. Exp. Ther.* **1949,** *95,* 71.
23. Arunlakshana, O.; Schild, H. O. *Br. J. Pharmacol. Chem.* **1959,** *14,* 48.
24. Parsons, M. E. Ph.D. Thesis, University of London, 1969.
25. Daly, M. J.; Humphray, J. M.; Stables, R. *Gut* **1980,** *21,* 408–412.
26. Brittain, R. T.; Daly, M. J.; Sutherland, M. *J. Pharm. Pharmacol.* **1980,** *32,* 76P.
27. Topliss, J. G. *J. Med. Chem.* **1972,** *15,* 1006.
28. Martin, Y. C.; Dunn, W. J. *J. Med. Chem.* **1973,** *16,* 578.
29. Leo, A.; Hansch, C.; Elkins, D. *Chem. Rev.* **1971,** *71,* 525.
30. Ganellin, C. R. In *Molecular and Quantum Pharmacology*; Bergman, E. D.; Pullmann, B., Eds.; Reidel: Boston, 1974; p 44.
31. Weinstein, H.; Mazurek, A. P.; Osman, R; Tipiol, S. *Mol. Pharmacol.* **1986,** *29,* 28–33.
32. May, R. M. *Proc. R. Soc. A* **1987,** *413,* 27–44.
33. Black, J. W.; Durant, G. J.; Emmett, J. C.; Ganellin, C. R. *Nature (London)* **1974,** *248,* 65.

34. Ganellin, C. R. *J. Med. Chem.* **1981,** *24,* 913.
35. Ashby, P.; Curwain, B. P.; Daly, M. J.; McIsaac, R. L. *Br. J. Pharmacol.* **1976,** *57,* 440P.
36. Gomper, R.; Schaefer, H. *Chem. Ber.* **1967,** *100,* 591.
37. Forrest, J. A. H.; Shearman, D. J. C.; Spence, R.; Celestin, L. R. *Lancet* **1975,** *i,* 392.
38. Toldy, L.; Toth, I.; Borsy, J. *Acta Sci. Hung. Tomus* **1967,** *53,* 279.
39. Bradshaw, J.; Brittain, R. T.; Clitherow, J. W.; Daly, M. J.; Jack, D.; Price, B. J.; Stables, R. *Br. J. Pharmacol.* **1979,** *66,* 464P.
40. Rendic, S.; Sunjic, V.; Toso, R.; Kajfez, F. *Xenobiotica* **1979,** *9,* 555–564.
41. Serlin, M. J.; Sibeon, R. G.; Mossman, S.; Breckenridge, A. M.; Williams, J. R. B.; Atwood, J. L.; Willoughby, J. M. T. *Lancet* **1979,** *ii,* 317.
42. Klotz, U.; Reimann, I. *N. Engl. J. Med.* **1980,** *302,* 1012.
43. Bell, J. A.; Gower, A. J.; Martin, L. E.; Mills, E. N. C.; Smith, W. P. *Biochem. Soc. Trans.* **1981,** *9,* 113.
44. Leslie, G. B.; Walker, T. F. In *Cimetidine,* Proceedings of the Second International Symposium on Histamine H_2-Receptor Antagonists; Burland, W. L.; Simkins, M. A., Eds.; Excerpta Medica: Oxford, 1977; pp 24-37.
45. Pearce, P.; Funder, J. W. *Clin. Exp. Pharmacol. Physiol.* **1980,** *7,* 442.
46. Brittain, R. T.; Daly, M. J. *Scand. J. Gastroenterol.* **1981,** *69,* 1.
47. Bradshaw, J.; Butcher, M. E.; Clitherow, J. W.; Dowle, M. D.; Hayes, R.; Judd, D. B.; MacKinnon, J. M.; Price, B. J. In *Chemical Regulation of Biological Mechanisms*; Creighton, A. M.; Turner, S., Eds.; Royal Society of Chemistry: London, 1982; p 45.
48. Bradshaw, J.; Latour, K.; Maliski, E. G. *Drug Des. Discovery* **1992,** *9,* 1–9.
49. Ganellin, C. R.; Durrant, G. J.; Emmett, J. C. *Fed. Proc.* **1976,** *35,* 1924.
50. Bartlett, P. Lecture presented at the Cambridge Crystallographic Data Centre, Cambridge, England, April 19, 1991.
51. Taylor, P. In *Comprehensive Medicinal Chemistry*; Hansch, C.; Sammes, P. G.; Taylor, J. B., Eds.; Pergamon: New York, 1989; Vol. 4, p 242.
52. Judd, D. B. Master's Thesis, Council for National Academy Awards: London, 1980.
53. Bradshaw, J.; Taylor, P. *J. Quant. Struct. Act. Rel.* **1989,** *8,* 279–287.
54. Topliss, J. G.; Costello, R. J. *J. Med. Chem.* **1972,** *15,* 1066.
55. Verloop, A.; Hoogenstraaten, W.; Tipker, J. In *Drug Design*; Ariens, E. J., Ed.; Academic: New York, 1976; Vol. VII, p 165.

56. Bays, D. E.; Finch, H. *Nat. Prod. Rep.* **1990,** *7,* 365–458.
57. Charton, M. *J. Org. Chem.* **1965,** *30,* 3346.
58. Yukawa, Y.; Tsuno, Y. *Bull. Chem. Soc. Jpn.* **1959,** *32,* 971.
59. Deitrich, R. A.; Hellman, L.; Wein, J. *J. Biol. Chem.* **1962,** *237,* 560–564.
60. Mares-Guia, M.; Nelson, D. L.; Rogana, E. *J. Am. Chem. Soc.* **1977,** *99,* 2331.
61. Newton R. F.; Roberts, S. M. In *Chemistry, Biochemistry and Pharmacological Actions of Prostanoids*; Roberts, S. M.; Scheinmann, F., Eds.; Pergamon: New York, 1979; p 61.

Loratadine

Allen Barnett and Michael J. Green
Schering-Plough Research Institute

There are no hard-and-fast rules to work from in drug discovery. Each new search requires a creative approach and brings its own peculiar set of problems to be overcome. It also requires scientists of many different disciplines to work together toward a common goal. This chapter describes the background conditions and the contributions of different people at Schering-Plough Corporation that enabled us to find and develop loratadine as a non-sedating antihistamine for use in treating various allergic diseases.

Background

In 1973, when the first report on the pharmacology of terfenadine was presented at the American Society for Pharmacology and Experimental Therapeutics meetings (*1*), the popular view held that a non-sedating antihistamine was unobtainable and perhaps a contradiction in terms. Schering-Plough's most recent prior experience had been with azatadine (Optimine), a potent antihistamine. It had been predicted from preclinical data, mainly on the basis of a single behavioral test in cats, that azatadine would be non-sedating at therapeutic doses, but it was not. Other companies had had similar experiences with their drug candidates. Thus, when it was suggested that terfenadine might in the future pose a threat to Schering-

2523–0/93/0083$06.00/0

Plough's considerable antihistamine business, there was no significant internal response. One reason for the lack of research response was that sedative antihistamines did not produce sedation in common laboratory animal species; another reason was the lack of validated methods to predict the sedative liability of a new antihistamine. The work force at Schering-Plough was deployed on other projects, and there was only a limited, sporadic effort toward the synthesis and evaluation of new potential antihistamines.

Early Clinical Results with Terfenadine

Following the 1973 report on the pharmacology of terfenadine, the next significant publication on terfenadine did not appear until 1977 (2). In that article the drug was reported to inhibit histamine-induced skin wheals in humans, with 60 mg twice daily as an effective dose regimen and no central nervous system (CNS) side effects. The lapse of 4 years between reports suggested that terfenadine was yet another non-sedating failure. However, in 1978, several important clinical publications appeared that confirmed the non-sedating profile of terfenadine (3–6).

Early Stages of the Project

In 1978, after the aforementioned publications appeared, we started to explore what preclinical methods might be used to evaluate the sedative liability of antihistamine candidates, as a major step toward initiating a project. A variety of test procedures were studied, among them the inhibition of electroconvulsive-shock-induced seizures in mice (7), hexobarbital potentiation, and antagonism of acetic acid-induced writhing in mice. The major criteria were that diphenhydramine show activity at doses near the antihistamine dose range and that terfenadine be inactive or weakly active at high multiples of its therapeutic dose. In January 1979, a meeting involving chemists and biologists was held to try to put together a project in this area. The biologists left that meeting with the assignment of developing an appropriate strategy to screen and evaluate antihistamine candidates. The chemists agreed to work on two approaches: one to reduce the lipophilicity (log P) of selected existing antihistamine molecules in order to reduce brain penetration, the other to systematically modify the structure of azatadine, the most potent antihistamine available, to reduce its CNS effects. Both synthetic strategies are described in greater detail later in the chapter.

Development of a Testing Strategy

Several key decisions had to be made with respect to the testing strategy. It was apparent that no single test could be used to confidently predict that an antihistamine would be non-sedating. Moreover, the clinical efficacy of terfenadine and its lack of sedation were not widely accepted findings. Studies on objective measures of CNS performance had not yet been published, and there were rumors about formulation difficulties with terfenadine. Nevertheless, in our own studies we found terfenadine to be an antihistamine that was inactive or weakly active in conventional CNS screening models in mice, and we used it as our standard. As such we used it to validate our strategy, along with data on a series of sedating antihistamines. Thus, lacking a single good predictive test, we selected a battery of CNS tests that included different measures and different animal species. Our drug candidate would have to be no worse than terfenadine in all of the seven tests we chose. This was a rigorous criterion, but we were able to meet it.

When our initial testing strategy was put together, it became obvious that we had introduced an unexpected variable. We were screening for oral antihistamine activity using as a measure the prevention of histamine-induced lethality in guinea pigs, and we were evaluating CNS effects using three procedures in mice—antagonism of pentylenetetrazole-induced seizures, acetic acid-induced writhing, and physostigmine-induced lethality. If we found a potent antihistamine in guinea pigs that was not CNS-active in mice, we would not know if the result reflected a species difference, such as lack of bioavailability in mice. Therefore, we needed either an antihistamine test in mice or a CNS measure in guinea pigs. We elected the former, and an assay in mice, the antagonism of histamine-induced paw edema (*8*), was developed expressly for this purpose. The assay was perfect for its intended purpose as a follow-up test but had at least one anomaly that prevented its use as a primary screen. Chlorpheniramine, the prototype H_1-receptor antihistamine, was very weak in this test, with an oral ED_{50} of 9.6 mg/kg, as opposed to an ED_{50} of 0.15 mg/kg in guinea pigs and a recommended clinical dose of 0.08 mg/kg (*8*). This problem, as well as the unproven utility of the mouse model, led us to perform screening in guinea pigs and to use the mouse model after compounds had proved to be free of CNS activity in mice.

This decision was not reached without extended debate and disagreement. The group doing the mouse test was in a different department (Biochemistry) from the group doing the guinea pig test (Pharmacology). What followed was a series of vigorous exchanges over the merits of screening with the new, unproven mouse test (lower cost, less compound needed, same species as would be used for evaluation of CNS effects) versus the old, reliable guinea pig test (long history of proven utility, potency of standards closer to potency in human, end point respiratory, at least in part). The dispute was finally resolved operationally, but the debate continued in a friendly fashion even after we had found the drug we were looking for.

It is clear that the project succeeded because of the combined efforts of many people in Pharmacology, Biochemistry, and Chemistry. Loratadine was discovered by Frank J. Villani and Charles V. Magatti in Medicinal Chemistry, and the biological evaluations were done by a team headed by Allen Barnett. Studies of antihistamine activity were done by Sal Tozzi and William Kreutner and their respective groups; the additional parts of the CNS evaluation, which included gross behavioral studies in mice, rats, dogs, and monkeys, were performed by the CNS group of Louis Iorio; and the receptor-binding studies were done by the Biochemistry group (Ho-Sam Ahn and Arax Gulbenkian). The group was quite effective in coordinating data obtained with each compound (pre-computerization era) and in strategic planning.

Development of a Chemical Strategy

The chemical origins of loratadine lie in Frank Villani's fascination with the tricyclic structures of drugs like promethazine and amitriptyline. This interest had already led Villani to the discovery of azatadine (*9*) as a cyclized analogue of pheniramine, an earlier Schering antihistamine. Azatadine, a potent but sedating antihistamine, was introduced in 1973. Buoyed by this success, Villani continued synthetic work around the structure, with, however, a focus on biological activities other than antihistaminic (e.g., the amitriptyline-like antidepressants). This effort waxed and waned as the years passed; however, tricyclic structures were never very far from his mind when he made plans for future synthetic work.

In the mid-1970s, as part of a program to develop an antiulcer drug, Villani undertook a systematic study of the azatadine structure in order to try and introduce H_2-receptor antagonist activity into the molecule. The theory at the time was that a dual H_1/H_2-receptor antagonist would be a useful profile for antiulcer therapy. As part of this study, Villani explored the effect of changing the basicity of the two nitrogen atoms in the molecule on the H_2-receptor antagonist activity (or rather lack thereof) of azatadine. Thus, for the piperidine nitrogen, the methyl group was removed and various urea, sulfonamide, and carbamate derivatives, among others, were prepared in his laboratory. These analogues showed a uniform lack of H_2-receptor antagonist activity.

In January 1979 a formal effort to identify a non-sedating antihistamine that would compete with terfenadine was initiated. Two chemical approaches were suggested by Medicinal Chemistry, and work began. The first approach was to synthesize H_1 antagonists with low log P values, on the basis that hydrophilic compounds are less likely to cross the blood-brain barrier and cause CNS effects. To this end, a number of analogues of tripelennamine and diphenhydramine were prepared that were calculated (*10*) to have decreased log P values; some of these are shown in Figure 1. These and other modifications did indeed generate compounds with lower log P values; however, they were devoid of antihistaminic activity. Our theory remained untested.

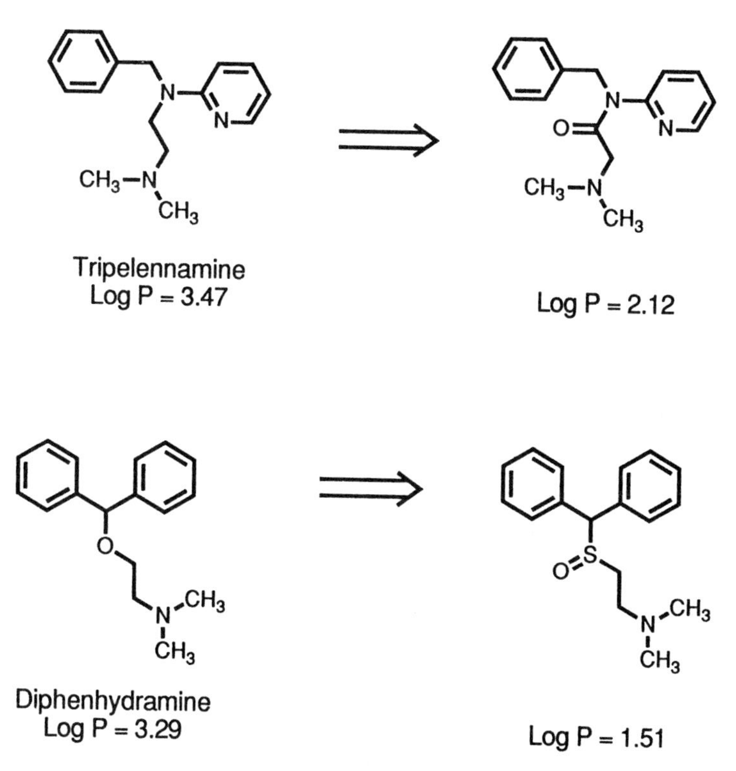

Figure 1. Structures of tripelennamine, diphenhydramine, and low log P analogues.

The second approach, based on traditional medicinal chemistry, was to make analogues of terfenadine, and this work was undertaken by Frank Villani and his group. By combining the unusual structural features of terfenadine with those of Schering-Plough's antihistamines, Villani hoped to marry the desirable attributes of both. For example, a pheniramine analogue with a terfenadine tail, compound **II** (Figure 2), was prepared, but it proved to be a fairly weak antihistamine and was not pursued.

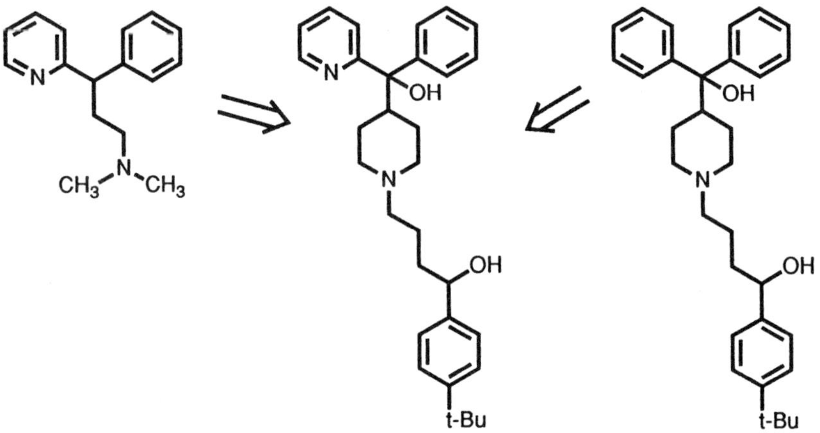

Pheniramine II Terfenadine

Figure 2. Structures of pheniramine, terfenadine, and the hybrid molecule **II**.

While this synthetic effort was proceeding, a variety of antihistamine-like structures from our compound file were being tested. Among these was the carbamate, compound **I** (Figure 3), which had been prepared two years earlier as part of the antiulcer project. This compound was an active antihistamine in the guinea pig lethality assay and was equipotent to terfenadine. Further, it was active in the mouse antihistamine assay and was virtually inactive in the battery of three CNS mouse tests. The other analogues with substituents that lowered the basicity of the nitrogen were all much weaker antihistamines. It therefore seemed that the salient structural feature had been found: the COOEt removed the CNS activity of azatadine while retaining much of its antihistamine potency.

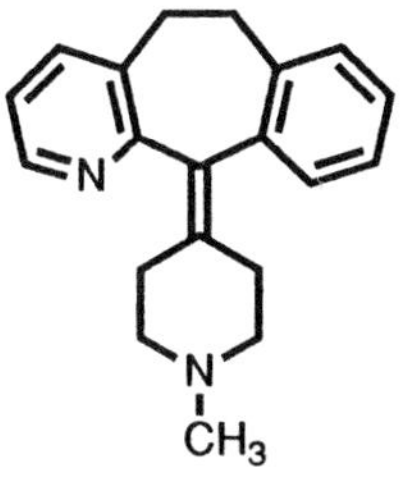

Azatadine

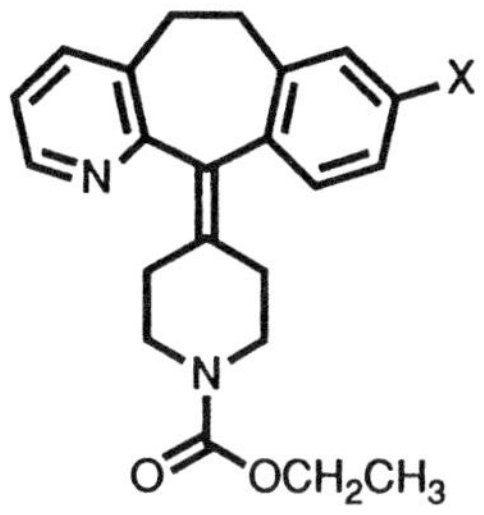

X = H; I
X = Cl; Loratadine

*Figure 3. Structures of azatadine, loratadine, and deschloro loratadine (**I**).*

At this point work on the two directed approaches was stopped and all efforts in Villani's laboratory were directed toward optimizing the properties of carbamate analogues of azatadine. The results of this study have been summarized by Popper and colleagues (*11*); some of them are shown in Table 1. The size of the alkyl group of the carbamate does not have a great effect on potency; however, it appears that the larger groups, such as *tert*-Bu, are less potent than smaller ones, with Et being the optimum size. Of interest, the phenyl carbamate is inactive as an antihistamine at the usual screening dose of 5 mg/kg. The reasons for the inactivity are not readily apparent.

The End in View

At this point, support for the project was waning and we began to feel some pressure to bring it to a conclusion: to quit, or to name a clinical candidate. How close was carbamate **I** to becoming a clinical candidate? It was as potent as terfenadine, so that was not a major issue, although a more potent compound would be desirable. Lack of sedating properties had been proved, but duration of action was a problem. From studies in guinea pigs we predicted a duration in humans of only 6–8 h, and we felt that at least a twice-daily dosing regimen was a necessity for an antihistamine to compete with terfenadine. To extend the duration of action of carbamate **I**, attention was focused on the benzene ring as the most likely site of metabolic inactivation of the molecule. Substitution at the 8-position would block a probable site of metabolism, and we knew from earlier structure–activity studies with azatadine that the 8-chloro derivative had a longer duration of

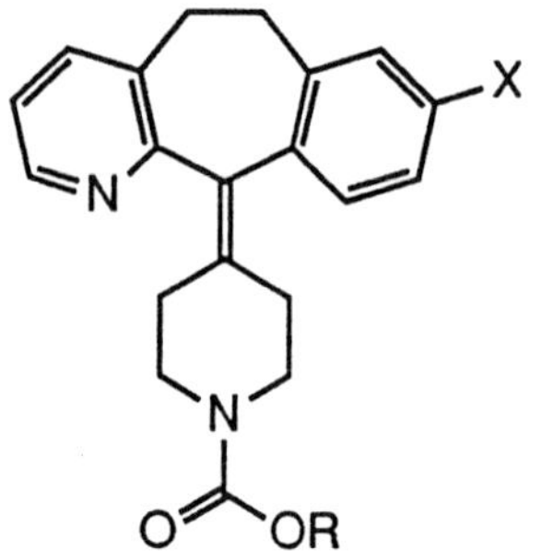

Table 1. Antihistaminic Potencies of Carbamate Analogues of Azatadine

R	*X*	*Inhibition of Histamine-Induced Lethality in Guinea Pigs*[a]	*Inhibition of Histamine-Induced Paw Edema in Mice*[a]
CH_3	H	≥5	—
CH_2CH_3	H	0.79	4.1
$CH_2CH_2CH_2CH_3$	H	1.17	—
$CH(CH_3)_2$	H	1.13	5.3
$C(CH_3)_3$	H	2.50	—
Cyclobutyl	H	≥5	—
Ph	H	>5	—
CH_2Ph	H	1.38	5.2
CH_2CH_3 (loratadine)	Cl	0.19	1.3
Azatadine	—	0.009	0.068
Terfenadine	—	0.81	3.8

[a] Data are ED_{50} values (in milligrams per kilogram) following oral administration.

action than the parent. The drawback was that 8-chloro azatadine was threefold less potent than azatadine. Nonetheless, this modification seemed the most logical course to take, and the 8-chloro derivative of **I** was made the last synthetic target of the project. This compound was synthesized and became known as loratadine. To the delight of all in the project, the prediction that duration of action would be improved was borne out, and as a bonus, loratadine was unexpectedly fourfold more potent than the deschloro compound.

Synthesis of Loratadine

The synthesis of loratadine from the *N*-methyl precursor was straightforward, and preparation of the latter compound was reported in 1972 by Villani and co-workers (*9*) (Figure 4). However, the formation of the azaketone **III** was not very satisfactory, especially in the last two steps, where isomeric mixtures are formed. Consequently, the bulk of the early research material was prepared by a process similar to the azatadine process (*12*). This route, however, was also not very satisfactory in that it required a catalytic reduction step that partially removed the chlorine. An elegant solution to these problems was subsequently found by D. Schumacher and her colleagues (*13*) in the Chemical Development group and is shown in Figure 5. Thus the intermediate **IV** was readily prepared from 2-cyano-3-methyl pyridine by conversion of the nitrile to the *t*-butyl amide and alkylation of the derived bisanion with *m*-chlorobenzyl chloride. Conversion of the amide back to the nitrile was then accomplished with $POCl_3$. The overall yield by this route was better than 83%, with no reduction step involved.

The isomer problem in the formation of azaketone **III** was solved by first attaching the *N*-methyl piperidine to the nitrile derived from intermediate **IV** before closing the seven-membered ring. This step, which requires a superacid (HF/BF_3), produces only the desired 8-chloro isomer. Presumably the increased steric bulk around the carbonyl group destabilizes the transition state leading to the 10-chloro compound. Overall this process is very efficient (57% yield) and has enabled large quantities of loratadine to be made.

Early Clinical Evaluation of Loratadine

In January 1981, Schering-Plough Corporation agreed to proceed with the development of loratadine to an early clinical decision point. Several follow-up meetings led to the decision to clinically test a high dose of loratadine for its sedative liability in normal volunteers. A single dose of 160 mg, or four to eight times the projected daily antihistamine dose, was chosen. The study was designed solely to answer the question of whether loratadine would clearly be sedating at multiples of its projected antihistamine dose. This strategy was to make sure that we were not fooled by a false-negative result. If 160 mg had been sedating, that would have been the end of the compound as a drug candidate. Because 10 mg was later established as the recommended

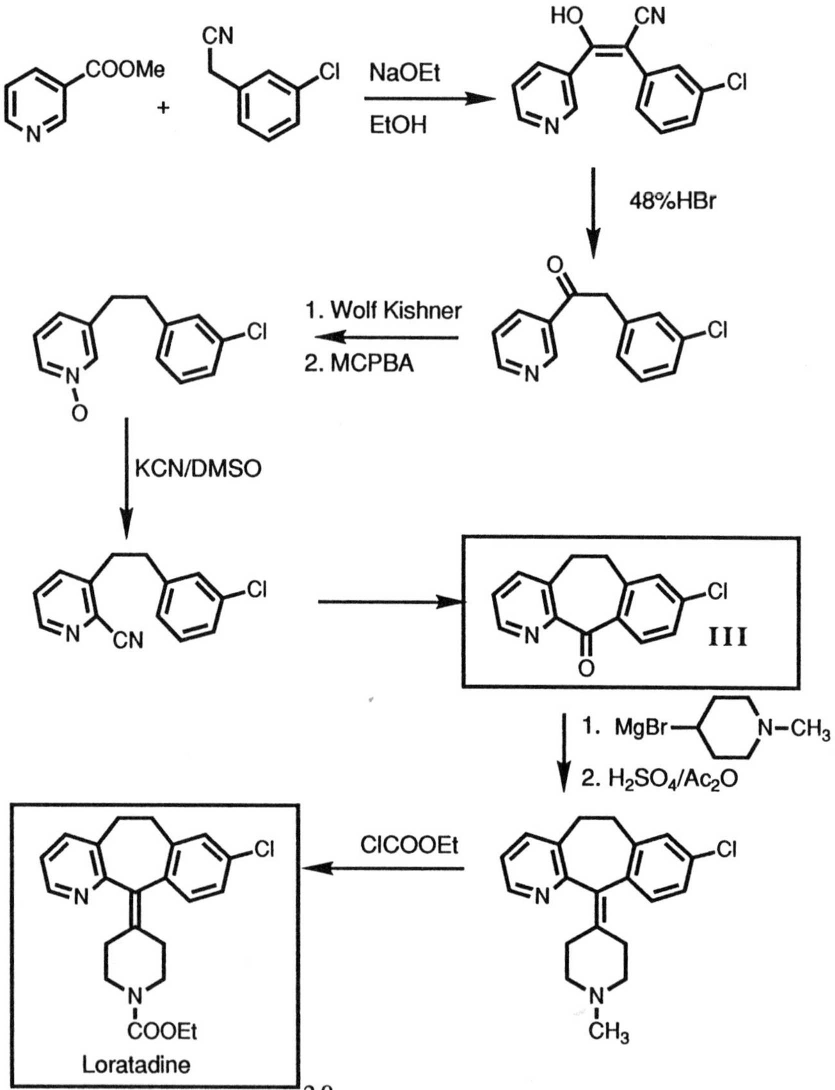

COOMe
CN
Cl
+
NaOEt
EtOH
HO
CN
Cl
48%HBr
O
Cl
1. Wolf Kishner
2. MCPBA
Cl
O
KCN/DMSO
Cl
CN
Cl
III
O
1. MgBr—N—CH_3
2. H_2SO_4/Ac_2O
ClCOOEt
Cl
COOEt
Loratadine
20
Cl
CH_3

Figure 4. Early synthesis of loratadine.

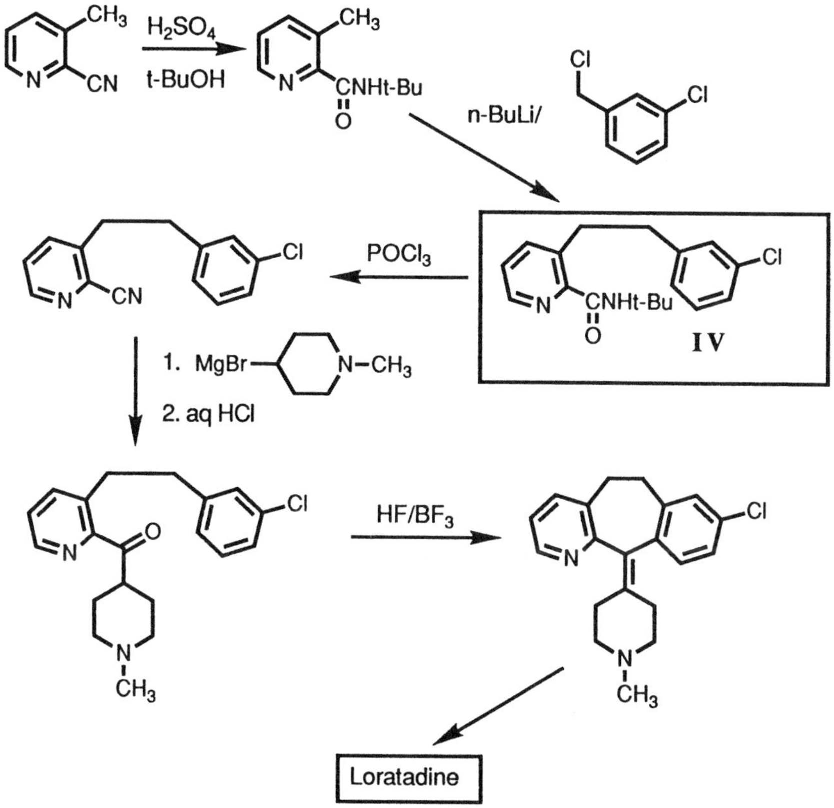

Figure 5. Current synthesis of loratadine.

daily dose, in retrospect we had tested our drug at 16 times its therapeutic dose.

The lack of sedation in this study led to two other questions: (1) Is loratadine absorbed orally? (2) Is it an active antihistamine? A flurry of activity ensued in which significant plasma and urine drug levels were found after oral dosing, and a histamine wheal-and-flare study demonstrated its antihistamine efficacy (*14*). Loratadine was now officially on its way through the new drug application process, and skepticism within the company evaporated.

A Retrospective Look at Loratadine and Its Discovery

In reviewing the development of a non-sedating antihistamine, we analyzed which tests were most predictive of the major properties of loratadine. With respect to predicting antihistamine potency and duration, antagonism of histamine-induced lethality in guinea pigs was of great value. We have retrospectively looked at a variety of antihistamines in this test versus 48/80-induced lethality in rats and histamine-induced paw edema in mice, and the guinea pig test seemed to be most predictive (Table 2) (*8, 15, 16*). The ED_{50} for loratadine in the guinea pig lethality test was 0.2 mg/kg, which translates into a predicted human dose of 10 mg, based on 50 kg as the average body weight. This is exactly the dose of loratadine that is recommended around the world. It is highly unusual to have this sort of carryover from animal dosage to human dosage on a weight basis. The duration of action of loratadine in this test was 18–24 h at twice its minimal effective dose and was clearly longer than the duration of action of two drugs that are used clinically in a twice-daily regimen, terfenadine and azatadine. The prediction of once-daily dosing was also borne out clinically and did not necessitate an artificial increase from 10 mg to achieve greater duration.

Table 2. Potency of Selected Antihistamines Following Oral Administration

Drug	*Recommended Human Dose (mg/kg)*	*Inhibition of Histamine-Induced Lethality in Guinea Pigs*[a]	*Inhibition of 48/80-Induced Lethality In Rats*[b]	*Inhibition of Histamine-Induced Paw Edema in Mice*[a]
Acrivastine	0.16	0.44	—	—
Astemizole	0.2	0.62	0.11	0.3
Cetirizine	0.2	0.16	—	1.0
Loratadine	0.2	0.19	—	1.3
Mequitazine	0.1–0.2	1.4	1.02	4.0
Terfenadine	1.2	0.81	2.69	3.8
Azatadine	0.02	0.009	0.26	0.07
Chlorpheniramine	0.08	0.15	24.7	9.6
Diphenhydramine	1.0	8.0	32.6	55.0
Ketotifen	0.02	0.002	0.59	0.4
Promethazine	0.5	0.43	0.89	5.9

[a] Data are ED_{50} values (in milligrams per kilogram), taken from ref. 8.
[b] Data are ED_{50} values (in milligrams per kilogram), taken from ref. 16.
SOURCE: Reproduced with permission from ref. 15. Copyright 1990, Birkhaüser Verlag.

In addition to potency and duration, the issue of side effects was critical. Several different measures, routes of administration, and animal species were used to assess sedative liability. Some of these data, along with data on several sedating and non-sedating antihistamines, are shown in Table 3. We are convinced that the use of different test conditions and different measures is the best strategy for uncovering potential side effects. However, two measures were particularly useful, and we would now employ them at a relatively early stage of decision-making. One involves inhibition of in vivo ^{1}H-mepyramine binding to brain in mice after oral administration of a test drug (*17*). This study measures a combination of brain penetration and affinity for brain histamine receptors and shows that neither loratadine nor terfenadine interfered with binding at doses more than 1,000 times their respective ED_{50} values in the mouse histamine-induced paw edema test. Other non-sedating antihistamines, such as astemizole and cetirizine, did show displacement in this test, albeit at high doses (Figure 6).

A second test that was useful and discriminating involved changes in operant responding (fixed ratio) in trained monkeys (*18, 19*). This procedure has the advantage of involving a primate species, an intravenous route of administration, and an objective measure of behavioral response. In contrast to the in vivo binding procedure, this test is more elaborate to set up but could possibly be done at an outside laboratory.

Another aspect of the side-effect profile of loratadine was the absence of autonomic effects, particularly anticholinergic effects such as dry mouth and blurred vision. Here we relied more on gross observations in multiple species, which were systematically scored on a blind basis. In these procedures antihistamines with significant anticholinergic effects (e.g., diphenhydramine) show effects, particularly increases in pupil size.

We have speculated on the reason for loratadine's lack of sedation. Using receptor-binding and autoradiographic techniques, we have shown that loratadine penetrates into the CNS very poorly. Yet it is highly lipophilic based on measurement of its octanol:water partition coefficient. We learned in our loratadine experience that most people equate high lipophilicity with good brain penetration, and there is a large "dropoff" after that with respect to the importance of other factors that may determine brain penetrability. Terfenadine is also highly lipophilic and penetrates poorly into the CNS (*20*). Other factors must be considered for both drugs before we might identify why neither penetrates into the brain. Because both drugs are highly protein-bound, a study of protein binding might be a good place to start.

Table 3. Effects of Loratadine on Four CNS Measures in Mice

Treatment	*Inhibition of ECS-Induced Convulsions*[a]	*Inhibition of Acetic Acid-Induced Writhing*[a]	*Inhibition of Physostigmine Lethality*[a]	*Inhibition of in vivo Mepyramine Binding in Mice*[b]
Loratadine	>320	>320	>320	>2,600
Terfenadine	>320	>320	>320	>7,600
Astemizole	>320	100	>320	100
Diphenhydramine	47	14.5	102	11
Chlorpheniramine	>160	46	>160	19
Promethazine	>160	11.8	55	—
Azatadine	>80	8.9	6.1	—

[a] Data are ED_{50} values (in milligrams per kilogram) for oral administration.
[b] Data are Minimum Effective Dose values (in milligrams per kilogram) for oral administration.

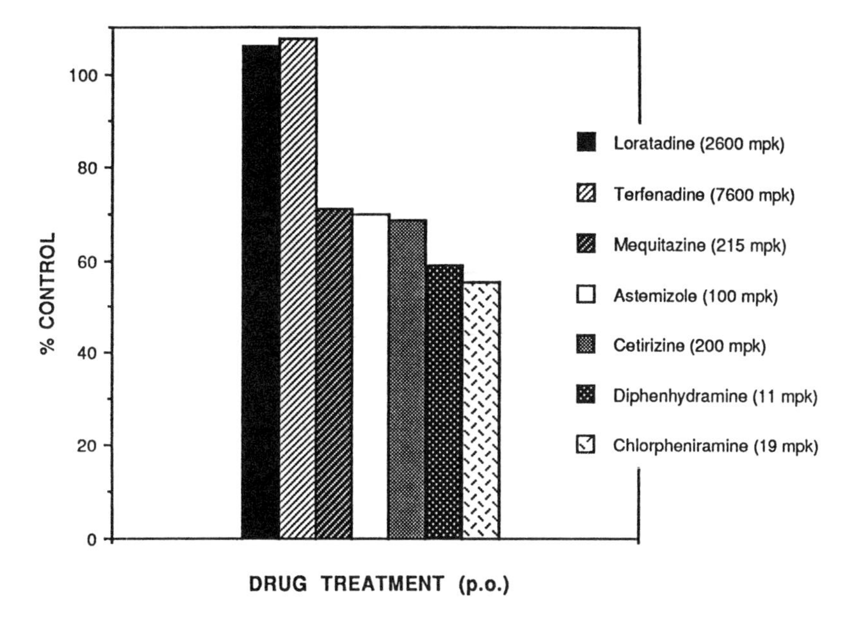

Figure 6. Percent inhibition of in vivo binding of ^{3}H-mepyramine to mouse whole brain. Each antihistamine was tested at several doses, including their respective antihistamine ED_{50} values (mouse paw edema test) and logarithmic multiples. The data are for the lowest dose of each drug that produced statistically significant inhibition of binding, or, in the case of loratadine and terfenadine, the highest dose tested. The drugs used and the doses (in parentheses) are indicated on the right and correspond to loratadine (far left in the graph) to chlorpheniramine (far right). Each determination was made 1 h post dosing. (Reproduced with permission from ref. 15. Copyright 1990 Birkhaüser.)

Summary

The discovery and subsequent commercial success of loratadine was a very satisfying outcome of the project. Because few of the projects we undertake have such successful conclusions, we have tried to identify what factors contributed to the success and how they might be applied to other projects.

First, the extensive experience with antihistamines at Schering-Plough, accumulated since the early 1950s, was a major factor. This experience supplied knowledge and the confidence that we knew what we were doing in the antihistamine area. We also had a storehouse of antihistamines and compounds structurally related to them in our compound files ready to be screened or to be used for the synthesis of new analogues. All this put us in a good position to succeed when we started the project.

Second, the project focus was clear and narrowly defined and the availability of terfenadine as a standard compound for comparative studies was clearly helpful. Third, the strategy of following several different chemical approaches simultaneously, without emphasizing any one approach until a good lead had been identified, was a good one. The approaches that we took, namely, a theoretical design approach (low log P), a medicinal chemistry, analogue approach (terfenadine-azatadine hybrids), and the intelligent selection of file compounds for testing, appears to be a good mix for any new project. This strategy was successful in the development of loratadine, and we will use it for developing new drugs in the future.

Literature Cited

1. Kinsolving, C. R.; Munro, N. L.; Carr, A. A. *Pharmacologist* **1973,** *15,* 221.
2. Hüther, N. J.; Renftle, G.; Barraud, N.; Burke, J. T.; Koch-Weser, J. *Eur. J. Clin. Pharmacol.* **1977,** *12,* 195–199.
3. Clarke, C. H., Nicholson, A. N. *Br. J. Clin. Pharmacol.* **1978,** *6,* 31–35.
4. Kulshrestha, V. K., Gupta, P. P.; Turner, P.; Wadsworth, J. *Br. J. Clin. Pharmacol.* **1978,** *6,* 25–29.
5. Moser, L.; Hüther, K. J.; Koch-Weser, J.; Lundt, P. V. *Eur. J. Clin. Pharmacol.* **1978,** *14,* 417–423.
6. Reinberg, A.; Levi, F.; Guillet, P.; Burke, J. T.; Nicolai, A. *Eur. J. Clin. Pharmacol.* **1978,** *14,* 245–252.
7. Halperin, J. M.; Iorio, L. C. *Pharmacol. Biochem. Behav.* **1980,** *13,* 299–301.
8. Barnett, A.; Iorio, L. C.; Kreutner, W.; Tozzi, S.; Ahn, H. S.; Gulbenkian, A. *Agents Actions* **1984,** *14,* 590–597.
9. Villani, F. J.; Daniels, P. J. L.; Ellis, C. A.; Mann, T. A.; Wang, K.-C.; Wefer, E. A. *J. Med. Chem.* **1972,** *15,* 750–754.
10. Rekker, R. F. *The Hydrophobic Fragmental Constant*; Elsevier: Amsterdam, 1977.
11. Villani F. J.; Magatti, C. V.; Vashi, D. B.; Wong J.; Popper, L. T. *Arzneimittelforschung* **1986,** *36(II),* 1311–1314.

12. Villani, F. J.; Wefer, E. A.; Mann, T. A.; Mayer J.; Peer L.; Levy, A. S. *J. Heterocycl. Chem.* **1972,** *9,* 1203–1207.
13. Schumacher, D. P.; Murphy, B. L.; Clark, J. E.; Tahbaz, P.; Mann, T. A. *J. Org. Chem.* **1989,** *54,* 2242–2244.
14. Batenhorst, R. L.; Batenhorst, A. S.; Foster, T. S.; Groves, D. A.; Kung, M. *Drug Intell. Clin. Pharm.* **1984,** *18,* 505.
15. Barnett, A.; Kreutner, W. In *New Perspectives in Histamine Research;* Timmerman, H.; Vander Goot, H., Eds.; *Agents Actions* **1990,** Suppl. 33, 181–196.
16. Niemegeers, C. J.; Awouters, F. H.; Janssen, P. A. *Drug Dev. Res.* **1982,** *2,* 559–566.
17. McQuade, R. D.; Richlan, K.; Duffy, R. A.; Chipkin, R. E.; Barnett, A. *Drug Dev. Res.* **1990,** *20,* 301–306.
18. Bergman, J.; Spealman, R. D. *J. Pharmacol. Exp. Ther.* **1988,** *245,* 471–478.
19. Bergman, J.; Spealman, R. D. *FASEB J.* **1989,** *3,* A441.
20. McTavish, D.; Goa, K. L.; Ferrill, M. *Drugs* **1990,** *39,* 552–574.

Misoprostol

Paul W. Collins
Searle Research and Development

Misoprostol is a synthetic 15-deoxy-16-hydroxy-16-methyl analogue of naturally occurring prostaglandin E_1 (PGE_1) (Figure 1). Discovered at G. D. Searle & Co. in 1973, misoprostol has undergone extensive clinical evaluation for the treatment of peptic ulcer disease and related conditions. Although misoprostol is effective in healing both gastric and duodenal ulcers (*1*), its primary indication is for the prevention and healing of nonsteroidal antiinflammatory drug (NSAID)-induced gastric and duodenal injury (*2*). Misoprostol has been approved in most countries and in December 1988 was approved by the U.S. Food and Drug Administration (FDA) for use in preventing NSAID-induced gastric ulcers. It is the first prostaglandin to become available for the treatment of peptic ulcer disease, and the first drug found effective against NSAID-induced gastropathy.

Historical Perspective

The history of prostaglandins began in the 1930s when Kurzrok and Lieb (*3*) noted that human semen caused uterine contractions and von Euler (*4*) coined the term *prostaglandin* to describe the biologically active substance(s) in seminal extracts. The area remained dormant until after World War II, when Bergstrom and other scientists at the Karolinska Institute in Sweden

2523–0/93/0101$07.00/0

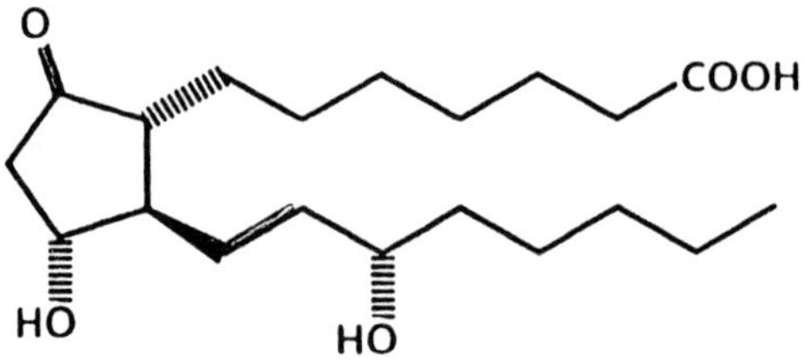

Prostaglandin E_1

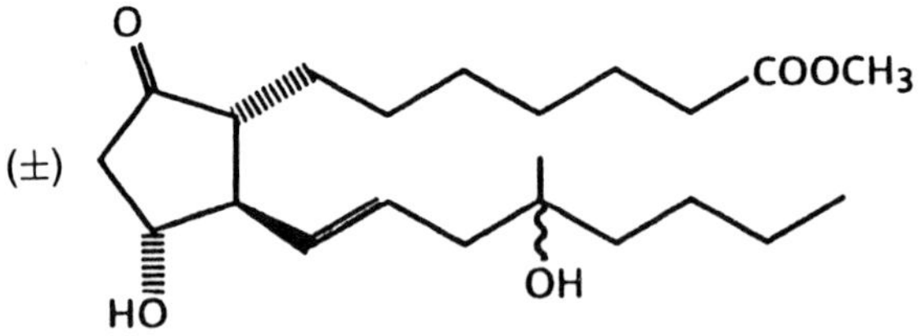

Misoprostol

Figure 1. Structures of PGE_1 and misoprostol.

reinitiated research in the area. Later work by Samuelsson and co-workers (5) at Karolinska led to the elucidation of the chemical structures of prostaglandins and their biosynthetic pathways. By the mid-1960s, intense efforts were being made in both academic and industrial laboratories to synthesize these complex molecules and to study their biological activities. There was great enthusiasm and anticipation throughout the pharmaceutical industry that prostaglandins or their analogues would be therapeutically useful in a myriad of diseases because they were naturally occurring substances, extremely potent, and possessed a broad spectrum of biological effects. In the ensuing years, however, the initial enthusiasm was gradually dampened as more detailed animal studies and clinical trials with the more promising candidates failed to show efficacy or demonstrated unacceptable side effects. In fact, prostaglandins have thus far affected only two therapeutic areas. First, they have found modest utility in the induction of labor and for related gynecological indications, and in the synchronization of estrus in farm animals. Second, a number of synthetic prostaglandin analogues have been investigated clinically for the treatment of peptic ulcer disease (6). With few exceptions, however, these analogues have been dropped because of severe side effects, stability problems, or lack of superior efficacy to the established H_2-receptor blocking drugs. The presence of misoprostol as the first and currently the only antiulcer prostaglandin in the U.S. market is in part due to the redirection of its primary therapeutic indication from standard peptic ulcer disease to NSAID-induced gastropathy. This repositioning decision, made in 1986 by Searle's upper management, enabled misoprostol to serve an unmet clinical need and accelerated its FDA approval and acceptance by the medical community.

The Synthetic Challenge

In the mid-1960s prostaglandins held an extraordinary attraction for synthetic chemists, not only because of their potential therapeutic value, but also because of the challenges presented by their structural and stereochemical complexity. Over the next 10 years considerable synthetic activity led to the development of numerous processes to prepare natural and modified prostaglandins (7). Much of this work required the development of new synthetic methodologies and stereochemical control strategies by organic chemists.

I joined Searle in 1967 as a young chemist fresh from postdoctoral study with Alfred Burger at the University of Virginia. I was assigned to work with Raphael Pappo and given the task of developing a synthesis of natural prostaglandins, substances unknown to me. Raphael Pappo was a world-renowned steroid chemist, a man of tremendous intellectual capacity, with a love for organic chemistry and a dedication to drug discovery. My early experience with Pappo was at times humbling but vastly rewarding and educational. Our working relationship endured until he retired in 1979.

In those early days, the biologists at Searle and elsewhere had to rely on biosynthetic sources of the natural prostaglandins for their research pursuits. Using literature procedures, biochemists at Searle incubated arachidonic acid and other fatty acids with sheep seminal vesicles and then laboriously isolated the prostanoid products from the vile-smelling mélange. This was a very inefficient method for obtaining natural prostaglandins and was not amenable to preparing analogues. Yet the biochemists were convinced that biosynthesis was the only way to prepare large quantities of prostaglandins and constantly criticized our total synthesis approach as untenable and impractical. Thus another incentive was added to the synthetic challenge.

Our initial efforts involved 1,2 Grignard addition of the acetylenic side chain **2** to enol ether derivatives of substituted cyclopentanediones **1** (8) (Figure 2) to generate PGB-like structures. One of our first contributions was the preparation of racemic natural PGB_1 by reaction of **1** (R = H) with **2** and reduction of the acetylene bond with a zinc-lead couple, a mild stereo- and chemoselective reagent (9). PGB_1 and compounds **3** (X = H or OH) were bioassayed in a wide number of tests, but generally showed little or no biological activity. However, compound **3** (X = OH) did show weak antiulcer and gastric antisecretory activity in rats, and we then prepared a series of analogues in which the omega chain was modified. One of these compounds was a 16-hydroxy analogue that showed comparable antisecretory and antiulcer activity as its 15-hydroxy parent. I made a mental note to test this structural change in E-type prostaglandins once we were able to prepare them.

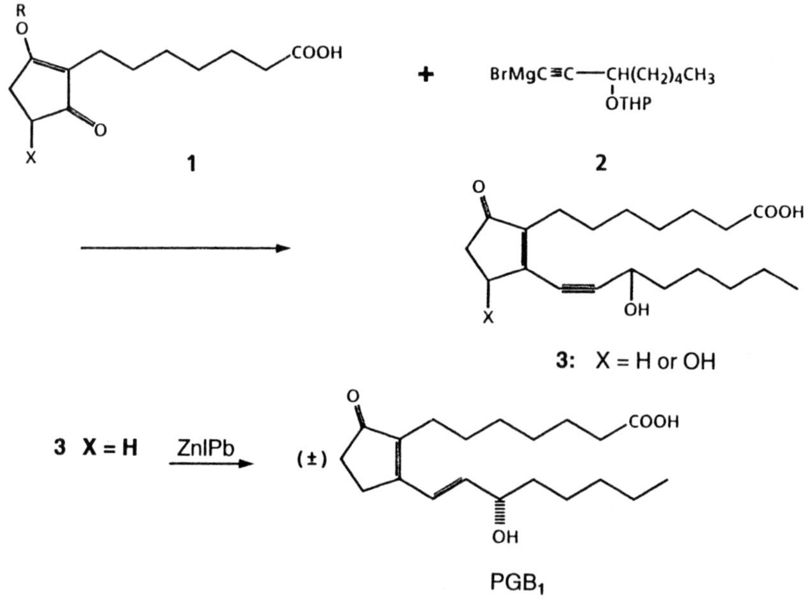

Figure 2. Synthesis of PGB_1.

Realizing that the conformational restrictions imposed by the $\Delta^{8,12}$ double bond of compounds **3** probably was the cause of their weak biological activity, we sought methods for the selective reduction of this double bond in the presence of Δ^{13} unsaturation. After repeated failures, we decided to take a different approach to the problem. At this point Pappo conceived of the conjugate addition approach. This strategy, which involves addition of the prostaglandin omega side chain **5** to a substituted cyclopentenone **4** (Figure 3), was independently researched by several other laboratories (*10–12*) and today represents the preferred manufacturing process for misoprostol and several other synthetic prostaglandins (*6*).

Much of our initial conjugate addition research involved investigation of simple organometallic derivatives of the omega chain, for present-day organocuprate methodology was unknown at the time. We investigated both alkenyl and alkynyl aluminum reagents as well as alkenyl copper compounds (*13*). Interestingly, we found that the aluminum reagents (**5**, M = Al) added to **4** in a conjugate manner only if the hydroxy group was unprotected (**4**, R = H). Further, only cis addition, giving the compound **7** with an 11-epi configuration, was observed, suggesting that the free hydroxy was involved in directing the addition (*14*). However, when simple alkenyl copper reagents (**5**, M = Cu) were added to **4** (R = THP or R_3Si), only the product **6** with the desired prostaglandin stereochemistry was obtained. The stereospecific nature of this organocopper reaction was an important finding and is a major advantage of the conjugate addition approach. In contrast to protected **4**, a mixture of ring geometries is produced when the hydroxy group of **4** is exposed.

The next hurdle was the preparation of the appropriate oxygenated omega side chain derivatives (**10**, Figure 4). Again, modern-day hydrozirconation and hydrostannation procedures were unknown, so we investigated hydroboration with catechol borane and hydroalumination with diisobutyl aluminum hydride of the acetylene **8** followed by iodination to give the (*E*)-vinyl iodides **9**. Neither of these approaches was very satisfactory because of poor yields due to incomplete reaction and loss of protecting groups. In addition, the catechol borane reaction required a laborious workup procedure to separate the vinyl boronic acid from the liberated catechol. Nevertheless, we were able to produce sufficient amounts of **9** to allow the preparation of PGE_1 via the copper species **10**. Initially we employed cuprous iodide to generate the vinyl copper reagent, but Corey and Beames (*15*) soon reported the use of copper pentyne to give simple cuprate reagents, and we then switched to this procedure. Thus, we had met the synthetic challenge of natural prostaglandins, and in 1971 we turned our attention to identifying therapeutically useful analogues.

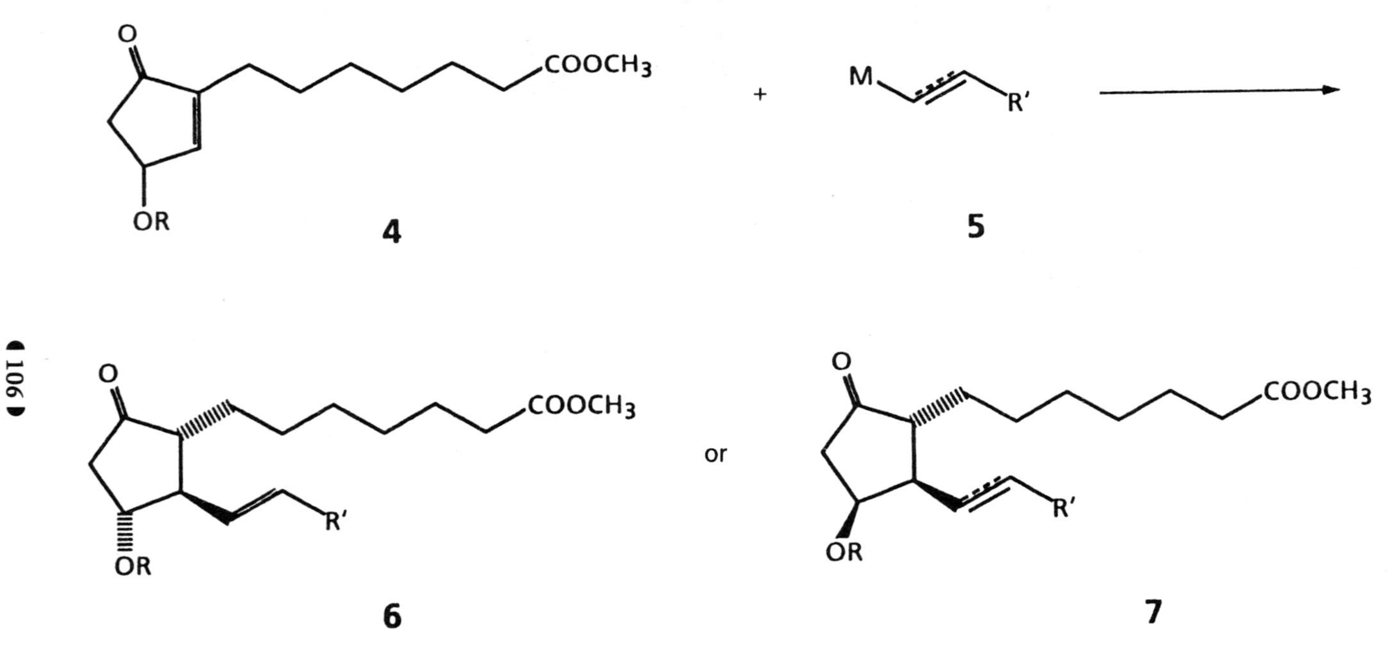

Figure 3. Conjugate addition approach.

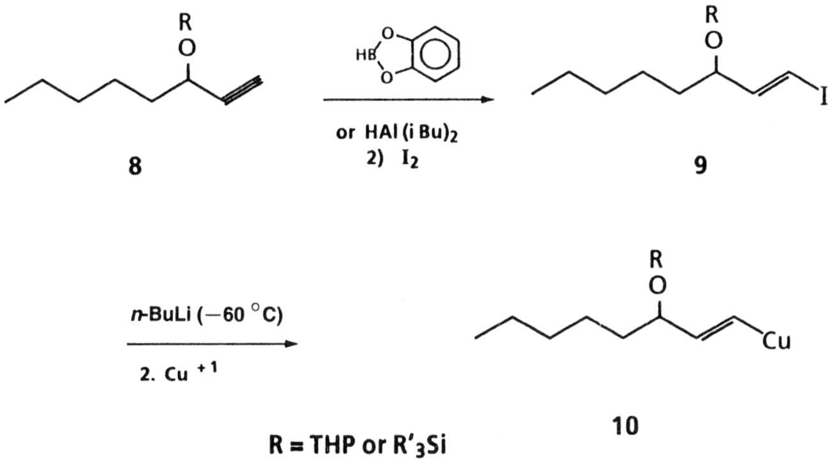

R = THP or R'$_3$Si

Figure 4. Side-chain derivatization.

Discovery of Misoprostol

Disadvantages of Natural Prostaglandins

The discovery that naturally occurring prostaglandins of the E series inhibit gastric acid secretion was made in 1967 by Robert et al. (*16*). Because of the prevailing theory of "no acid, no ulcer," these prostaglandins became logical candidates for ulcer therapy. Three critical drawbacks of the natural prostaglandins soon surfaced, however, that thwarted their therapeutic entry and whose circumvention has been the objective of many synthetic modification programs. These problems were (1) rapid metabolism, manifested as a lack of oral activity and a short duration of action when given parenterally, (2) the incidence of numerous side effects, and (3) chemical instability.

The natural prostaglandins are subject to three major modes of enzymatic degradation (Figure 5) in animals and humans. The most rapid of these processes is oxidation of the 15-hydroxy group to the corresponding ketone and subsequent reduction of the 13,14-double bond. The resulting metabolite **11** is virtually devoid of biological activity. The second process is β-oxidation of the carboxylic acid chain (alpha chain), a reaction common to fatty acids in general. This reaction sequence involves dehydrogenation at

C-2 and C-3, followed by oxidation to give the 3-ketone and finally cleavage to produce the dinor metabolite **12** and acetic acid. A second sequence usually occurs to generate the tetranor metabolite. The third point of attack is the omega-chain terminus. Oxidation occurs either at C-20, to give the corresponding alcohol and subsequently the acid **13**, or at C-19, to produce the 19-hydroxy metabolite. The major urinary metabolite of natural prostaglandins is **14**, which is the product of all three of these oxidative processes plus the reduction of the carbonyl group at C-9.

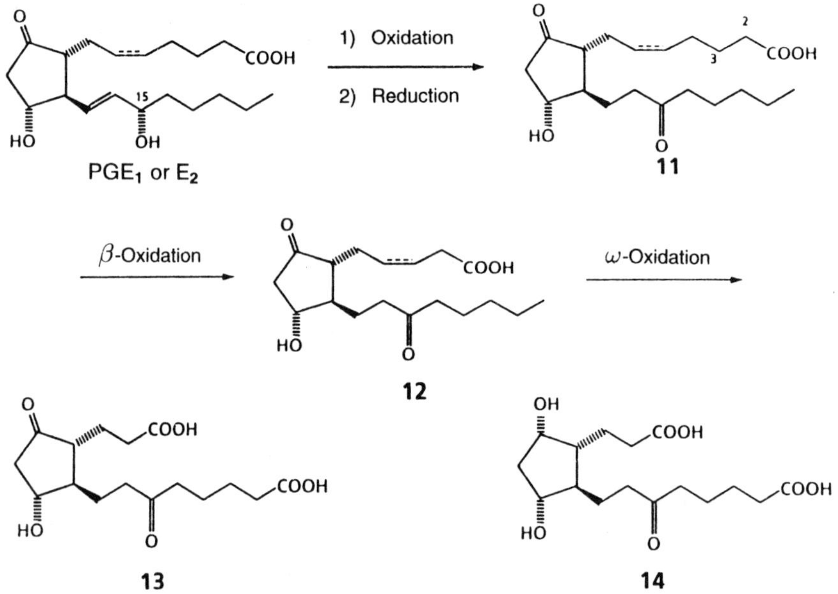

Figure 5. Metabolism of natural prostaglandins.

Because prostaglandins play important cellular roles in both health and disease, occur in virtually every tissue and organ, and are locally acting substances, it is not surprising that they display a wide variety of side effects when administered systemically. In laboratory animals, effects such as rhinorrhea, trembling, retching, emesis, and diarrhea are routinely observed with PGE_1 and occur even at effective gastric antisecretory doses (*17*). In humans, the symptoms of prostaglandin administration are erythema of the face, headache, abdominal cramps, diarrhea, hypotension, hyperthermia, and shivering (*18, 19*). Uterotonic properties, blood pressure and platelet effects, and cortical hyperostosis (*20*) are other worrisome side effects of natural prostaglandins and a source of potential concern with synthetic analogues.

Both natural and synthetic E-type prostaglandins are chemically unstable. The primary reason for their lability is the susceptibility of the β-hydroxy ketone system in the cyclopentane ring to acid- or base-catalyzed elimination of water to give the more stable α,β-unsaturated ketone system of PGA (Figure 6). The A form can also isomerize to the PGB derivative under the same conditions. In general, esters or other carboxylic acid derivatives are more stable than the corresponding free acids, which are sufficiently acidic to catalyze their own dehydration. E-type prostaglandins also suffer from susceptibility to inversion of the alpha chain under alkaline or thermal conditions to give the corresponding 8-epimer. These stability problems are quite serious and have been a major stumbling block in the development of many of the synthetic prostaglandin agents, including misoprostol, for ulcer therapy.

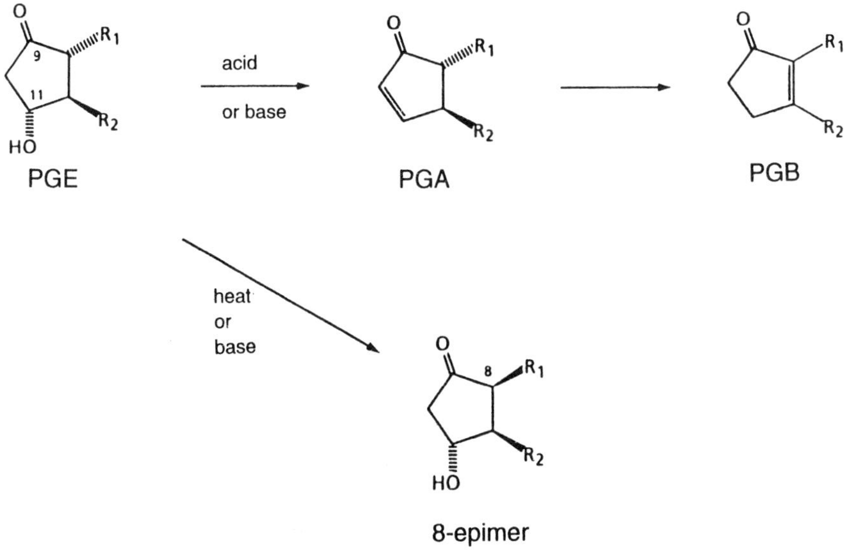

Figure 6. Degradation products of E prostaglandins.

Modification of PGE_1

The specific goal of our synthetic prostaglandin program was to identify an orally active compound with gastric antisecretory activity equal to or greater than that of PGE_1, a longer duration of action, and fewer side effects than the natural compound. Though mindful of the chemical stability problem, we did not fully appreciate its magnitude at the time and chose not to address it through chemical modification.

The probable reason for the lack of oral activity of PGE_1 is the very rapid metabolic oxidation of the C-15 hydroxy group to the corresponding ketone. In fact, other investigators (*21*) achieved oral activity by blocking the oxidation of this hydroxy with placement of either a methyl group at C-15 or two methyl groups at C-16 (steric inhibition). The resulting compounds were potent, orally active, and long-acting inhibitors of gastric acid secretion. Unfortunately, the selectivity of the compounds was not improved.

This finding convinced us that a more profound structural modification had to be attempted in order to separate undesirable side effects from the primary pharmacological activity. Structural thought in the early 1970s held that a hydroxy group at C-15 was essential for biological activity. This belief was based on the fact that either oxidation or removal of the hydroxy group produced pharmacologically inactive compounds. However, remembering that the 16-hydroxy compound in the early $\Delta^{8,12}$ series possessed gastric antisecretory activity equal to that of its 15-hydroxy counterpart, we decided to test the effect of this modification in the E series. Thus, I prepared the 15-deoxy-16-hydroxy analogue of PGE_1 (**15**, Figure 7) in the hope of reducing side effects while preserving antisecretory activity.

Just prior to the synthesis of this compound, Esam Dajani joined Searle as a gastrointestinal pharmacologist. Dajani, an extremely knowledgeable and capable scientist, instituted a Heidenhain pouch dog model to test the gastric antisecretory activities of the prostaglandin analogues. Unlike the existing rat assays at Searle, the dog model was acutely sensitive to prostaglandin effects and provided an excellent method for assessing potency, duration, and side effects. After testing the 16-hydroxy compound **15** with a group of other prostaglandin analogues, Dajani reported that **15** was about as active as PGE_1 itself but was short-acting and very weakly active following oral administration. Nevertheless, these chemical findings confirmed the original observation in the $\Delta^{8,12}$ series and set the stage for rational medicinal chemical manipulations to improve duration and potency. Enzymatic studies performed by our biochemical colleagues indicated that the C-16 hydroxy group was also a substrate for the 15-dehydrogenase enzyme that inactivates natural prostaglandins. Because of the dramatic increase in oral potency and duration of action achieved by placing a methyl group at C-15 in natural prostaglandins, we decided to add a methyl group to C-16 to block the metabolic oxidation of our 16-hydroxy prostaglandin. This structural change produced misoprostol (coded as SC-29333), which was approximately 35 times more active than the 16-hydrogen compound when given

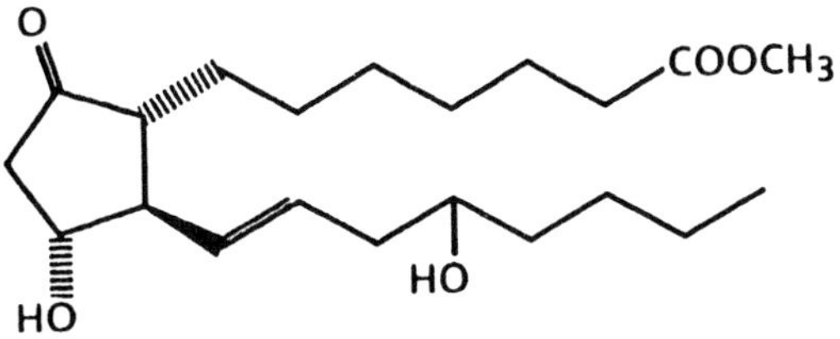

O COOCH3 HO HO

15-Deoxy-16-H, 16-OH

15

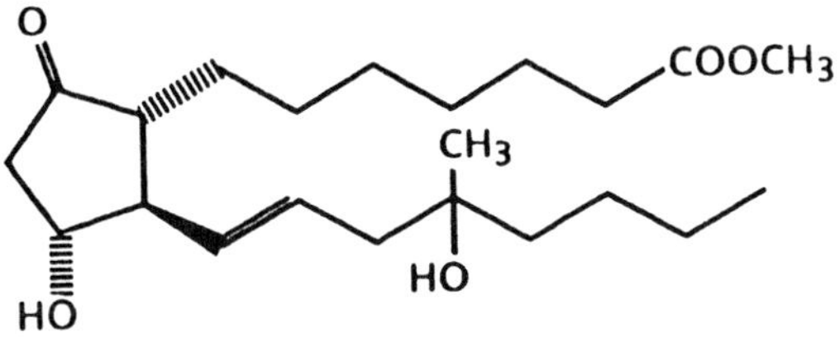

O COOCH3 CH3 HO HO

(16-methyl, 16-OH)

Misoprostol

Figure 7. Discovery of misoprostol.

intravenously, had good oral activity, and had an increased duration of action (*14, 22*). Dajani also observed that misoprostol did not produce the rhinorrhea, trembling, emesis, and diarrhea so frequently associated with administration of PGE_1 to dogs.

Misoprostol in many respects shows improved selectivity relative to 15-hydroxy prostaglandins. For example, the separation of gastric antisecretory and diarrheogenic properties in animals was much greater for misoprostol than for (15*S*)-15-methyl- and 16,16-dimethyl-PGE_2 (*23*) (Table 1). Further, no evidence of platelet, hypotensive, or hyperostotic effects has been found with misoprostol. One prostaglandin side effect that has not been eliminated, however, is stimulatory activity of uterine smooth muscle. Although

Table 1. Comparative Oral Gastric Antisecretory and Diarrheal Effects of (*E*)-Prostaglandin Analogues

Compound	*Gastric Antisecretory Effects in Dogs*	*Diarrheal Effects in Rats (mean ± SE)*	*Therapeutic Index: ED_{50} Diarrhea / ED_{50} Antisecretory*
(15*S*)-15-methyl-PGE_2 methyl ester	6.5–10.0	27 ± 6	4.2
16,16-dimethyl-PGE_2 methyl ester	2.4–3.0	10 ± 4.0	4.2
Misoprostol	5.5–10.0	366 ± 70	66.5

NOTE: Values are ED_{50} in micrograms per kilogram when drugs are given intragastrically.

uterotonic activity was neither observed nor suggested by preclinical investigations, clinical studies demonstrated that misoprostol can increase uterine contractile activity and may endanger pregnancy (*24*).

Development

Stability and Pharmaceutical Formulation

Misoprostol is a viscous oil with the consistency of honey. Its physical state, coupled with its inherent instability, created serious problems in pharmaceutical formulation. The various liquid systems developed were unstable and unacceptable to clinicians for human trials. Back in the laboratory we made several bulky aromatic esters (*25*) of misoprostol in an attempt to find a crystalline substitute, but to no avail. Further molecular modification to eliminate the cause of the instability also failed. For example, we prepared the 10,10-dimethyl and the 9-methylene analogues of misoprostol, but both were pharmacologically inferior. At one point, around 1977, the development of misoprostol was put on hold until a stable, solid dosage form could be found.

It was in this setting that the determination of researchers played a vital drug-saving role. Pappo and Dajani discussed the problem with D. Sandvordecker, a researcher in pharmaceutical development, who then initiated a study to determine the stability effects of dispersing misoprostol on various solid supports. The studies consisted simply of spraying an ethanol solution of misoprostol onto GRAS (generally recognized as safe) materials such as cellulose and lactose and then observing the rate of decomposition at various temperatures. One of these dispersions, a 1:99 mixture of misoprostol on hydroxypropyl methylcellulose (HPMC), was found to significantly stabilize the drug (*26*). Dajani then demonstrated that the dispersion did not alter the potency or profile of misoprostol's gastric antisecretory activity. Conventional tablets were prepared from the solid dispersion and found to have a shelf life of several years at room temperature. As a result of these efforts, the development of misoprostol was reinstated.

Synthesis and Stereochemistry

Misoprostol was originally synthesized by the process described earlier, but as advances in synthetic methodology occurred, the sequence was modified to reflect the improvements. In particular, we found that light-catalyzed hydrostannation of the acetylene side-chain precursor is a much superior method to hydroboration or hydroalumination.

Currently, misoprostol is synthesized as outlined in Figures 8 and 9. An undiluted mixture of the acetylenic side chain (**16**) and tri-*n*-butyl tin hydride is irradiated at room temperature with a sunlamp to provide the *E*-vinylstannane (**17**) (*27*). The stannane in tetrahydrofuran (THF) is cooled to −60 °C and treated with *n*-butyl lithium to generate the vinyl lithium species **18**, which is treated with a solution of copper pentyne solubilized with hexamethylphosphorus triamide in ether to give the cuprate reagent **19**. An ether solution of the cyclopentenone **20** is added to the cold (−60 °C) cuprate reagent dropwise, and the reaction mixture is worked up after about 1 h to give the protected prostaglandin **21** (Figure 9). The cuprate addition is very rapid and virtually quantitative. After removal of the protecting groups by mild acid hydrolysis and purification by column chromatography, misoprostol **22** is obtained as a pale yellow viscous oil in approximately 75% yield from **20**. The cyclopentenone **20** is synthesized by a multistep procedure developed in Searle laboratories (*14*).

As indicated in Figure 9, misoprostol is a mixture of two racemates or four stereoisomers. Even though the cuprate reaction is stereospecific, the use of racemic cyclopentenone and side chain produces two racemates. Preparation of a single stereoisomer or racemate is complicated by the fact that chromatographic separation of the misoprostol racemates, unlike the situation with the corresponding 15-methyl-15-hydroxy compound, is difficult and presently cannot be done on a practical scale. Thus the only effective way a single isomer can be produced is by resolution of both synthetic components or possibly by a combination of resolution and asymmetric induction. Although reasonably efficient methods exist to

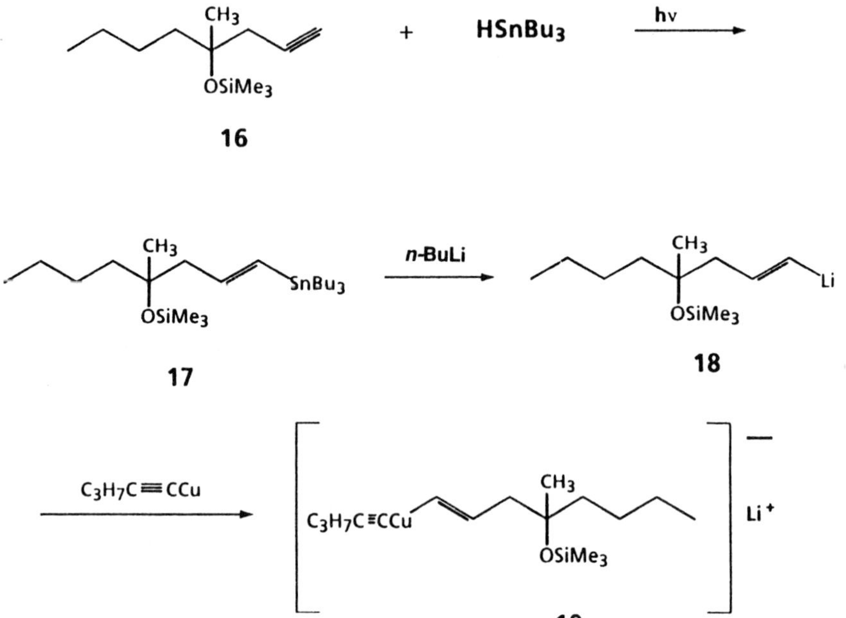

Figure 8. Synthesis of side-chain cuprate.

resolve the cyclopentenone (28, 29), no good method has been found to directly resolve tertiary alcohols in general or the particular alcohol **24** (Figure 10) required for misoprostol. Instead, rather circuitous routes to resolve **24** have been developed (30–33). At Searle, we resolved the hydroxy acid **23** via its naphthylethylamine salt and then converted **23** to (4*S*)- and (4*R*)-**24** by a series of chemical manipulations (Figure 10). Though not feasible on a commercial scale, this procedure has been utilized to prepare each of the four stereoisomers of misoprostol for pharmacological evaluation. Predictably, biological activity, both desired and undesired effects, resides almost entirely in the 11*R*, 16*S* isomer, which corresponds to the absolute configuration of natural PGE_1 (34, 35).

Vasodilatory Properties of Misoprostol

In the course of isolating and purifying PGE_1 and misoprostol, I frequently experienced reddening and discomfort of my face and eyes. These effects,

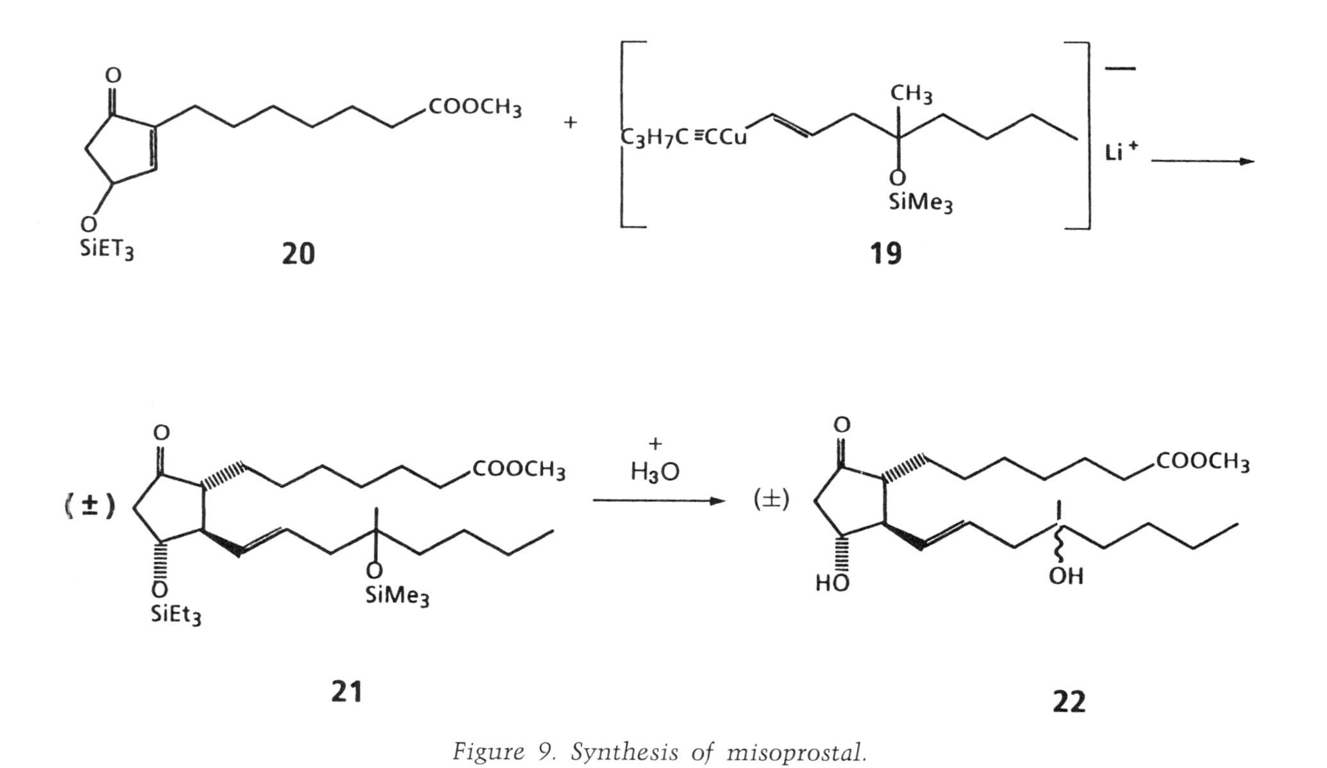

Figure 9. Synthesis of misoprostal.

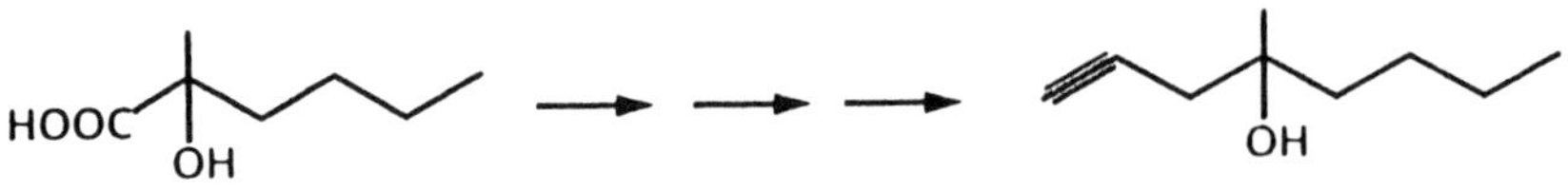

HOOC OH → → → OH

23 24

Figure 10. Resolution of side chain.

which were later attributed to the vasodilatory activity of the prostaglandins, would last about 3–4 h and then gradually disappear. I did not initially connect the effects with the prostaglandins themselves, but the relation soon became obvious, for every time I chromatographed the compounds, the facial events occurred. Apparently, in combining fractions, small amounts of prostaglandin solution were transferred to my fingers and then to my face by inadvertent touching. Although the fingers and hands were insensitive, the face and eyes were acutely sensitive, presumably because of their rich and shallow blood supply. Although the vasodilatory activity of misoprostol was a harmless and avoidable nuisance in the preparation and purification of small amounts of compound, it became more of a problem as larger amounts of material were prepared and more people became involved with its development. On several occasions, technicians were sent to the medical department for treatment after accidental exposure to the compound, and pessimists within the company questioned how a drug with these adverse effects could ever be a product. However, these concerns were laid to rest by institution of proper manufacturing procedures and handling precautions and the development of the HPMC dispersion, which is without effect.

Structure–Activity Relationships

Since the discovery of misoprostol, several hundred analogues have been prepared and evaluated for gastric antisecretory activity at Searle and other laboratories. In general, the modification strategies have been directed toward one or more of the three drawbacks of natural PGE_1, that is, toward preventing or impeding metabolism in order to increase the duration of action, toward reducing side effects, or toward improving chemical stability.

Carboxylate Group

Studies with isolated canine parietal cells suggested that the active antisecretory entity of misoprostol is its free acid (36). Esters, even hindered ones that can be enzymatically cleaved to the free acid in vivo, are active. In contrast, amides are either inactive or weakly active. Acid substitutes, such as tetrazoles and sulfonimides, are compatible with activity, as are groups that can be metabolized to carboxylic acids. Misoprostol is marketed as a methyl ester rather than as a free acid because of manufacturing and stability benefits. The free acid of misoprostol is significantly less stable because of autocatalytic acid dehydration.

Alpha Chain

Most of the work in this area has focused on blocking or impeding β-oxidation. With a few notable exceptions, modifications in the alpha chain have been unsuccessful, leading to poor antisecretory activity. For example, placement of one or two alkyl groups at C-2, substitution of oxygen for C-3, two methyl groups at C-4, a keto group at C-6, or a methyl group at C-7 greatly reduced antisecretory activity. Shortening or lengthening the chain by one or two carbons profoundly diminished activity.

Incorporation of double bonds into the alpha chain has been more fruitful. The $\Delta^5 Z$ compound (E_2 analogue) is approximately equipotent to misoprostol, while the $\Delta^3 Z$, $\Delta^2 Z$, $\Delta^3 E$, and $\Delta^2 E$ analogues are all about three to five times less active than misoprostol. In contrast, the $\Delta^4 Z$ analogue is about five times more active than misoprostol as a gastric antisecretory agent and is also longer acting (*23, 37*). This compound, enisoprost, was prepared based on literature information that introduction of a $\Delta^4 Z$ double bond in other prostaglandin compounds impeded β-oxidation (*38*) but did not alter gastric antisecretory activity (*39*). Interestingly, the $\Delta^4 E$ compound was about 30 times less active than enisoprost, and the corresponding acetylene was devoid of gastric antisecretory activity (*23*). In view of the favorable findings with enisoprost, we extended our studies to compounds having two double bonds in the alpha chain (*40*). The most active gastric antisecretory agent in a series of seven diene analogues was the 3*E*, 5*Z* compound (*41*), which was about three times more potent than misoprostol. Surprisingly, insertion of an *E* double bond at C-2 of enisoprost reduced activity. We also prepared allene analogues at C-3,4,5 and C-4,5,6, but both were less active than misoprostol.

Cyclopentane Ring

The changes in this portion of the molecule have been primarily aimed at increasing chemical stability while maintaining an adequate level of antisecretory activity. None has been successful to date. The 9-methylene, 9-chloro, 10,10-dimethyl, 11-deoxy, 11-epi, 11-deoxy-11-methyl, and the

PGD analogues of misoprostol are all either inactive or weakly active as gastric antisecretory agents.

Omega Chain

The $\Delta^{13}E$ double bond of misoprostol appears to be important, if not necessary, for good antisecretory activity. The $\Delta^{13}Z$ analogue, the saturated compound, the 13,14-cyclopropyl derivative, the 13,14-acetylene, and the 13,14,15-allene were all weak inhibitors of gastric secretion (10 to 30 times less active than misoprostol). Substitution of a methyl group on either C-13 or C-14 resulted in a nearly complete loss of activity.

At C-16, the methyl group is the only alkyl substituent compatible with good antisecretory activity. The ethyl, butyl, and vinyl analogues were much less active than misoprostol. Interestingly, the 16-vinyl $\Delta^{5}Z$ analogue of misoprostol was investigated as a potential antihypertensive agent (*42*). Blockage of the 16-hydroxy group as a methyl ether or as an acetate abolished activity, suggesting that the free hydroxy must be available for receptor interaction. Moving the C-16 hydroxy and methyl group to C-17 also abolished activity. An unexpected finding was that the 15,15-dimethyl and 17,17-dimethyl analogues of the 16-hydrogen parent of misoprostol were both devoid of gastric antisecretory activity. The latter compound is analogous to 16,16-dimethyl PGE_2, a very potent gastric antisecretory agent (*43*).

The $\Delta^{17}E$, $\Delta^{17}Z$, and 17-acetylene analogues (*30*) of misoprostol are each about five times less active than the parent compound. This finding is somewhat surprising, because the modifications reestablish the allylic nature of the hydroxy group common to natural prostaglandins. Lengthening the omega chain by one carbon has minimal impact on the activity of misoprostol, but activity decreases as the length of the chain is increased further. Unlike the case with some 15-hydroxy prostaglandins, the presence of a phenoxy or phenyl group in the omega chain of misoprostol is detrimental to antisecretory activity. No antisecretory activity was observed with 18- or 19-hydroxy analogues or with the 19-oxa analogue of misoprostol.

After the discovery of the cytoprotective properties of prostaglandins by Robert et al. (*44*) (described in the next section), the ethanol rat model was set up at Searle, and many of the analogues described earlier were assessed for cytoprotective activity. In general we have found that protective activity parallels gastric antisecretory activity. This generalization for the 16-hydroxy series is in contrast to the original finding by Robert that 15-hydroxy prostaglandins with varying structures and degrees of antisecretory activity are all cytoprotective. We also found protective activity to be stereospecific. The active antisecretory isomer of misoprostol is protective, while the other three stereoisomers are not (35).

Therapeutic Role of Misoprostol

Mucosal Protection

A milestone in the history of prostaglandins and in the development of misoprostol for the treatment of peptic ulcer disease was the discovery that prostaglandins can protect the gastrointestinal mucosa from injury caused by a variety of noxious agents. This intriguing phenomenon was first reported by Robert in 1975 (45) and termed "cytoprotection" by Jacobson (cited in ref. 44). The term "mucosal protection" has now supplanted the original term.

In his original work, Robert found that concomitant administration of prostaglandins with indomethacin to rats prevented or diminished the severity of intestinal lesions caused by this agent. Protection was greatest when the animals were treated with the prostaglandin 30 min before or up to 1 h after indomethacin. Robert suggested that the lesions were due to a deficiency of endogenous prostaglandins caused by indomethacin and that treatment with a prostaglandin prevented the deficiency and thus the damage. Later work extended the phenomenon to gastric protection. In a landmark publication (44), Robert et al. reported that pretreatment of rats with any one of a variety of prostaglandins protected the gastric mucosa from damage caused by administration of ethanol, strong base or acid, hypertonic solutions, or boiling water. This and other work strongly suggested that cytoprotection is independent of inhibition of gastric acid secretion. The basis for this hypothesis is fourfold: (1) Prostaglandins that are devoid of gastric antisecretory properties are cytoprotective. (2) Cytoprotection can be demonstrated at doses far below the acid inhibitory dose of antisecretory prostaglandins. (3) Cytoprotection occurs in acid-independent models such as ethanol injury to the stomach or indomethacin damage to the intestines. (4) Other antisecretory agents such as histamine H_2-receptor blockers and anticholinergic drugs are ineffective in acid-independent models.

Robert's work prompted a vigorous effort to determine the mechanism of cytoprotection and the specific cellular protective effects of prostaglandins. Despite the large amount of work done in this area, the mechanism of cytoprotection remains unclear. Among the prominent theories proposed and investigated are prevention of gastric mucosal barrier disruption, stimulation of gastroduodenal mucus and bicarbonate secretion, stimulation of mucosal blood flow, and modulation of endogenous sulfhydryl levels. None of these or other proposed mechanisms alone can adequately explain cytoprotection. The cytoprotective action of misoprostol and other prostaglandins may well involve an as yet unidentified mechanism rather than a combination of existing mechanisms. Current conceptual emphasis is on increased cellular resistance mechanisms (*46*) and improved mucosal regenerative capacity (*47, 48*) to explain the mucosal protective effects of prostaglandins. For example histological studies (*49, 50*) have demonstrated that, while prostaglandins did not prevent destruction of the superficial epithelium of the stomach during ethanol-induced damage, preservation of the deeper epithelium did occur, and moreover, restitution of the superficial epithelium was more rapid in prostaglandin-treated animals than in controls.

Natural prostaglandins play an important physiological role in maintaining the normal integrity of the gastrointestinal mucosa, and the actions of misoprostol on mucosal resistance factors generally mimic those of the natural prostaglandins (*47*). In fact, in healthy humans misoprostol elicits many of the mucosal defensive mechanisms that have been proposed to explain cytoprotection. For example, dose-related increases in gastric mucus production (*51*), stimulation of basal bicarbonate secretion in proximal and distal duodenal segments (*52*), and increases in gastric mucosal blood flow (*53*) have been reported.

Redirection of Primary Clinical Indication

Misoprostol in general has shown equal but not superior efficacy to the H_2-receptor antagonists in healing gastric and duodenal ulcers. Even though misoprostol is a new structural class of compound and has both antisecretory and mucosal protective properties, the H_2-receptor antagonists have a longstanding and widespread acceptance in the standard peptic ulcer market. Thus, a key positioning decision was made at Searle to redirect misoprostol away from the standard ulcer market to the prevention and treatment of NSAID-induced gastrointestinal damage, an area in which the mucosal protective properties of misoprostol would be uniquely beneficial.

NSAID-Induced Gastropathy

Nonsteroidal antiinflammatory drugs (NSAIDs) are widely used in arthritic diseases, and it is now widely acknowledged that NSAIDs cause gastroduodenal injury in a significant percentage of patients, especially those on long-term therapy and the elderly (*54*). In fact, NSAID use has been suggested as the cause of rising mortality rates among elderly patients with ulcer disease in the United Kingdom (*55*). Additionally, in many patients, NSAID damage to the gastrointestinal tract is often asymptomatic until a life-threatening complication such as bleeding or perforation occurs.

Two pivotal clinical studies demonstrated that misoprostol is effective in healing existing aspirin-induced gastric and duodenal ulcers when aspirin therapy is continued in rheumatoid arthritis patients (*56*), and is also effective in preventing gastric ulcers in osteoarthritis patients receiving either ibuprofen, peroxicam, or naproxen (*57*). Based on these studies, the FDA approved misoprostol in December 1988 for the prevention of NSAID-induced gastric ulcers and proclaimed it as "probably the most important drug we have approved all year".

Other Indications

Studies in animals and humans have shown that misoprostol exhibits protective properties in tissues other than the upper gastrointestinal tract. For instance, misoprostol reversed acute cyclosporin-induced renal vasoconstriction and renal dysfunction in rats (*58*), protected the intestinal mucosa of mice from radiation damage (*59*), and prevented injury to the colonic mucosa of rats following acetic acid instillation (*60*). Clinically, misoprostol improved renal function in cyclosporin-treated renal transplant recipients and also reduced the frequency of acute transplant rejection (*61*). Steatorrhea in cystic fibrosis patients was reduced by misoprostol therapy (*62*). Misoprostol has also been reported to ameliorate oral mucositis in cancer patients receiving chemotherapy and radiation therapy. Thus, the ultimate therapeutic role of misoprostol may be much broader than ulcer disease and may finally justify the early enthusiasm for prostaglandins as broad-spectrum drugs.

Acknowledgments

In addition to the key people mentioned in the text, many individuals played important and sometimes vital roles in the discovery and development of misoprostol. A partial list includes Frank Colton, Peter Cammarata, P. H. Jones, Raymond Bauer, Grant Schoenhard, Paul Klimstra, Dan Azarnoff, Alan Gasiecki, Steven Kramer, Chris Jung, Bob Bianchi, and J. Behling. In addition to his discovery contributions, Esam Dajani was instrumental in the clinical and marketing support areas.

Literature Cited

1. Watkinson, G.; Hopkins, A.; Akbar, F. A. *Postgrad. Med. J.* **1988,** *64* (Suppl. 1), 60.
2. Agrawal, N. M.; Dajani, E. Z. *J. Rheumatol.* **1990,** *17* (Suppl. 20), 7.
3. Kurzrok, R.; Lieb, C. C. *Proc. Soc. Exp. Biol. Med.* **1930,** *28,* 268.
4. von Euler, U. S. *J. Physiol. (London)* **1937,** *88,* 213–215.
5. Samuelsson, B. *Angew. Chem.* **1965,** *4,* 410–416.
6. Collins, P. W. *J. Med. Chem.* **1986,** *29,* 437–443.
7. Bindra, J. S.; Bindra, R. *Prostaglandin Synthesis*; Academic: New York, 1977.
8. Pappo, R.; Collins, P.; Jung, C. *Ann. N.Y. Acad. Sci.* **1971,** *180,* 64–75.
9. Collins, P.; Jung, C.; Pappo, R. *Isr. J. Chem.* **1968,** *6,* 839–841.
10. Kluge, A. F.; Untch, K. G.; Fried, J. H. *J. Am. Chem. Soc.* **1972,** *94,* 9256–9258.
11. Sih, C. J.; Solomon, R. G.; Price, P.; Peruzzoti, G.; Sood, R. *J. Chem. Soc. Chem. Commun.* **1972,** 240–241.
12. Floyd, M. B.; Weiss, J. J. *Prostaglandins* **1973,** *3,* 921–924.
13. Pappo, R.; Collins, P. W. *Tetrahedron Lett.* **1972,** 2627–2631.
14. Collins, P. W.; Dajani, E. Z.; Driskill, D. R.; Bruhn, M. S.; Jung, C. J.; Pappo, R. *J. Med. Chem.* **1977,** *20,* 1152–1159.
15. Corey, E. J.; Beames, D. J. *J. Am. Chem. Soc.* **1972,** *94,* 7210–7212.
16. Robert, A.; Nezamis, J. E.; Phillips, J. P. *Am. J. Dig. Dis.* **1967,** *12,* 1073–1076.
17. Dajani, E. Z.; Driskill, D. R.; Bianchi, R. G.; Collins, P. W.; Pappo, R. *Prostaglandins* **1975,** *10,* 733–745.
18. Bergstrom, S.; Carlson, L. A.; Ekelund, L.-G.; Oro, L. *Acta Physiol. Scand.* **1965,** *64,* 332–339.
19. Dingfelder, J. R.; Brenner, W. E. *Acta Obstet. Gynecol. Scand.* **1978,** *57,* 35–40.
20. Ueda, K.; Saito, A.; Nokano, H.; Aoskima, M.; Yokota, M.; Muraoka, R.; Iwaya, T. *J. Pediatr.* **1980,** *97,* 834–836.
21. Robert, A.; Magerlein, B. J. *Adv. Biosci.* **1973,** *9,* 247–253.
22. Dajani, E. Z.; Driskill, D. R.; Bianchi, R. G.; Collins, P. W.; Pappo, R. *Am. J. Dig. Dis.* **1976,** *21,* 1049–1057.

23. Collins, P. W.; Dajani, E. Z.; Pappo, R.; Gasiecki, A. F.; Bianchi, R. G.; Woods, E. M. *J. Med. Chem.* **1983,** *26,* 786–790.
24. Lewis, J. H. *Am. J. Gastroenterol.* **1985,** *80,* 743–745.
25. Morozowich, W.; Oesterling, T. O.; Miller, W. L.; Lawson, C. F. *J. Pharm. Sci.* **1979,** *68,* 833–836.
26. Sanvordeker, D. U.S. Patent 4,301,146, 1981.
27. Collins, P. W.; Jung, C. J.; Gasiecki, A. F.; Pappo, R. *Tetrahedron Lett.* **1978,** 3187–3190.
28. Pappo, R.; Collins, P.; Jung, C. *Tetrahedron Lett.* **1973,** 943–946.
29. Sih, C. J.; Heather, J. B.; Sood, R.; Price, P.; Peruzzotti, G.; Hsu Lee, L. F.; Lee, S. S. *J. Am. Chem. Soc.* **1975,** *97,* 865–874.
30. Pappo, R.; Collins, P. W.; Bruhn, M. S.; Gasiecki, A. F.; Jung, C. J.; Sause, H. W.; Schulz, J. A. In *Chemistry, Biochemistry and Pharmacological Activity of Prostanoids*; Roberts, S. M.; Scheinmann, F., Eds.; Pergamon: New York, 1979; p 17–26.
31. Fujimoto, Y.; Yadav, J. S.; Sih, C. J. *Tetrahedron Lett.* **1980,** 2481–2485.
32. Corey, P. F. *Tetrahedron Lett.* **1987,** 2801–2804.
33. Yadow, J. S.; Satyanarayana Reggy, P.; Jolly, R. S. *Ind. J. Chem.* **1986,** *25B,* 294–295.
34. Dajani, E. Z.; Driskill, D. R.; Bianchi, R. G.; Phillips, E. L.; Woods, E. M.; Colton, D. G.; Collins, P. W.; Pappo, R. *Drug Dev. Res.* **1983,** *3,* 339–347.
35. Bauer, R. F.; Bianchi, R. G.; Gullikson, G. W.; Tsai, B. S.; Collins, P. W. *Fed. Proc.* **1987,** *46,* 1085.
36. Tsai, B. S.; Kessler, L. K.; Stolzenbach, J.; Schoenhard, G.; Butchko, G. M.; Bauer, R. F. *Gastroenterology* **1986,** *90,* 1671.
37. Collins, P. W.; Gasiecki, A. F.; Weier, R. M.; Kramer, S. W.; Jones, P. H.; Gullikson, G. W.; Bianchi, R. G.; Bauer, R. F. *Prostaglandins* **1987,** *33* (Suppl.), 17–29.
38. Green, K.; Samuelson, B.; Magerlein, B. J. *Eur. J. Biochem.* **1976,** *62,* 527–537.
39. Walker, E. R. H. In *Chemistry, Biochemistry and Pharmacological Activity of Prostanoids*; Roberts, S. M.; Scheinmann, F., Eds.; Pergamon: New York, 1979; p 326–345.
40. Collins, P. W.; Gasiecki, A. F.; Jones, P. H.; Bauer, R. F.; Gullikson, G. W.; Woods, E. M.; Bianchi, R. G. *J. Med. Chem.* **1986,** *29,* 1195–1201.
41. Collins, P. W.; Kramer, S. W.; Gasiecki, A. F.; Weier, R. M.; Jones, P. H.; Gullikson, G. W.; Bianchi, R. G.; Bauer, R. F. *J. Med. Chem.* **1987,** *30,* 193–197.

42. Birnbaum, J. E.; Cervoni, P.; Chan, P. S.; Chen, S. M. L.; Floyd, M. B.; Grudzinskas, C. V.; Wiess, M. J. *J. Med. Chem.* **1982,** *25,* 492–494.
43. Robert, A.; Schultz, J. R.; Nezamis, J. E.; Lancaster, C. *Gastroenterology* **1976,** *70,* 359–370.
44. Robert, A.; Nezamis, J. E.; Lancaster, C. L.; Hanchar, A. J. *Gastroenterology* **1979,** *77,* 433–443.
45. Robert, A. *Gastroenterology* **1975,** *69,* 1045–1047.
46. Miller, T. A. *Surgery* **1988,** *103,* 389–397.
47. Wilson, D. W. *Postgrad. Med. J.* **1988,** *64* (Suppl. 1), 7–11.
48. Schmidt, K. L.; Miller, T. A. *Toxicol. Pathol.* **1988,** *16,* 223–236.
49. Lacy, E. R.; Ito, S. *Gastroenterology* **1982,** *83,* 619–625.
50. Tarnawski, A.; Hollander, D.; Stachura, J.; Krause, W. J.; Gergely, H. *Gastroenterology* **1985,** *88,* 334–352.
51. Wilson, D. E.; Quadros, E.; Rejapaska, T.; Adams, A.; Noar, M. *Dig. Dis. Sci.* **1986,** *31,* 126S–129S.
52. Isenberg, J. I.; Hogan, D. L.; Selling, J. A.; Koss, M. A. *Dig. Dis. Sci.* **1986,** *31,* 130S.
53. Sato, N.; Kawano, S.; Fukada, M.; Tsuji, S.; Kanada, T. *Am. J. Med.* **1987,** *83* (Suppl. 1A), 15–21.
54. Butt, J. H.; Barthel, J. S.; Moore, R. A. *Am. J. Med.* **1988,** *84* (Suppl. 2A), 5–14.
55. Langman, M. J. S. *Am. J. Med.* **1988,** *84* (Suppl. 2A), 15–19.
56. Roth, S.; Agrawal, N.; Mahowald, M. *Arch. Intern. Med.* **1989,** *149,* 775–779.
57. Graham, D. Y.; Agrawal, N. M.; Roth, S. H. *Lancet* **1988,** *2,* 1277–1280.
58. Paller, M. S. *Transplant Proc.* **1988,** *20,* 634–639.
59. Hanson, W. R.; DeLaurentiis, K. *Prostaglandins* **1987,** 33 (Suppl.), 93–104.
60. Fedorak, R. N.; Empey, L. R.; MacArthur, C.; Jewell, L. D. *Gastroenterology* **1990,** *98,* 615–625.
61. Moran, M.; Mozes, M. F.; Maddux, M. S.; Veremis, S.; Bartkus, C.; Ketel, B.; Pollak, R. *N. Engl. J. Med.* **1990,** *322,* 1183–1188.
62. Cleghorn, G. J.; Shepherd, R. W.; Holt, T. L. *Scand. J. Gastroenterol.* **1988,** *23* (Suppl. 143), 142–147.

Enalapril and Lisinopril

Arthur A. Patchett
Merck Research Laboratories

Enalapril and lisinopril are angiotensin-converting-enzyme (ACE) inhibitors now widely used in clinical practice for the treatment of essential hypertension and congestive heart failure. They were synthesized in the Merck Research Laboratories in a project whose goal was the de novo design of enzyme inhibitors. The biological rationale for this target was based on years of pioneering academic research that characterized the involvement of the renin-angiotensin system (RAS) in blood pressure control (for reviews see 1–3). The attention of drug designers was drawn in earnest to the RAS in the early 1970s. By that time angiotensin II (AII) had been established as an extremely potent pressor substance that also causes aldosterone release from the kidneys. In addition, AII has a direct effect on the proximal tubules of the kidneys to retain Na^+, which contributes to blood pressure elevation by an increase in blood volume. In the early 1970s it was by no means certain that modulating the biosynthesis or activity of AII would make a significant contribution to therapy because most hypertensive patients do not have a hyperactive RAS as measured by plasma renin activity. However, subsequent research produced investigational compounds whose efficacy in animals and in the clinic played a major role in establishing the RAS as an important target for hypertensive therapy.

2523–0/93/0125$10.50/0

The Renin-Angiotensin System

Figure 1 summarizes the major components of the RAS. Within the RAS there are several potential targets for possible therapeutic intervention. AII biosynthesis might be reduced by inhibiting either renin or ACE, or the actions of AII could be antagonized at its receptors. The first useful approach was to design peptide antagonists of AII (*4, 5*). These compounds, of which saralasin has been most studied, were effective in lowering blood pressure in high-renin models of hypertension, but saralasin's peptide structure, relatively short duration of action, and partial agonist properties limited its usefulness to diagnostic applications.

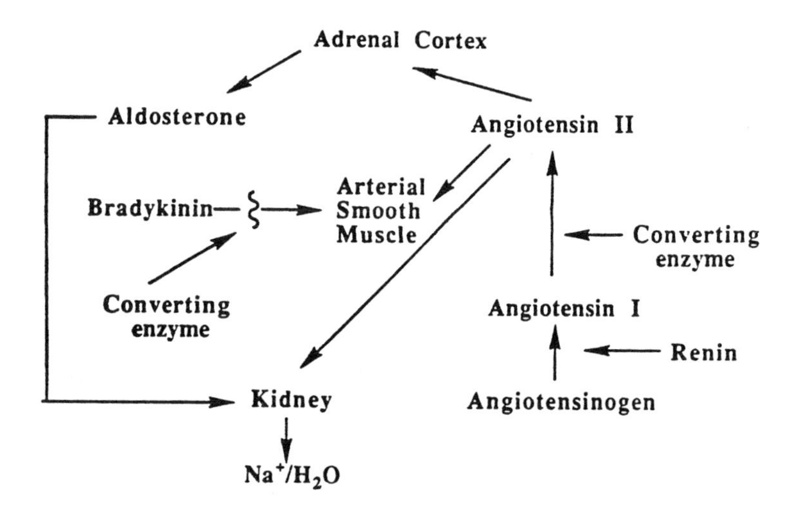

Figure 1. The renin-angiotensin system.

Merck research in the early 1970s focused on renin, although some ACE and AII antagonist screening was also conducted, without success. Renin is an attractive target because it is the rate-limiting enzyme in the biosynthesis of AII and because it is a very specific enzyme. Early research at Merck under the direction of Dr. Ralph Hirschmann focused on the design of peptide inhibitors of renin. Although no publications resulted from that work, the group's early attempts to utilize cyclic peptides to attain metabolic stability are worthy of mention. Subsequent extensive renin research in the Merck

Research Laboratories utilized pepstatin as a lead and has been reviewed several times (*6, 7*).

In 1974 Edward H. Ulm was asked by Dr. Clement A. Stone and his associates to set up a high-capacity ACE inhibitor assay in the Merck laboratories at West Point, Pennsylvania. In part the decision was influenced by results from the clinical testing of teprotide (*8, 9*), a nonapeptide ACE inhibitor developed in the Squibb laboratories (*10, 11*). The history of this compound's isolation from a snake venom and its development as an investigational drug have been reviewed (*12–14*). Unquestionably, teprotide's clinical efficacy in blocking the pressor effects of angiotensin I (*8*), in lowering blood pressure in patients with elevated renin levels (*9*), and later its efficacy in patients with normal renin levels (*15*) established major interest in ACE inhibitors as a drug target.

The main problem with teprotide was its lack of oral activity. Even so, questions remained concerning the ultimate usefulness of ACE inhibitors, because potentiation of the hypotensive, inflammatory mediator bradykinin could be expected. Indeed, teprotide and the other *Bothrops jararaca* peptides were originally detected in snake venom as bradykinin-potentiating substances. Further, ACE was known to be a widely distributed, nonspecific enzyme whose inhibition might have unforeseen consequences arising from effects on other hormonal or neuropeptide pathways. An additional argument was advanced, namely, that lowering AII levels could activate a compensatory release of renin that would lead to the increased production of angiotensin I, potentially overcoming the enzyme inhibition. This analysis seemed to favor the inhibition of renin over that of ACE. For these reasons, and because of the urgency of other Merck medicinal chemistry projects, including syntheses of cardioselective β-blockers, calcium channel antagonists, and a second-generation Aldomet, responsibility for the design of an ACE inhibitor was left with the New Lead Discovery Department. The project began with screening in Ulm's assay to find a nonpeptide lead using fermentation extracts and selected compounds. The latter included compounds synthesized in the department's novel compounds for screening project. These screening approaches failed to yield significant ACE inhibitor leads, nor were leads found in several other enzyme assays in progress at the time. Therefore, we decided to supplement random screening with an enzyme inhibitor design project. In 1975 Dr. Alan Maycock joined us from Prof. Robert H. Abeles' laboratory at Brandeis University to initiate some design activities.

The Enzyme Target and Initial Designs

ACE is a noncrystalline, membrane-bound glycoprotein whose molecular weight has been reported to range from 130,000 to 180,000 daltons (*16, 17*). It primarily acts as a dipeptidyl carboxypeptidase and is relatively nonspecific. It was known, however, not to hydrolyze peptide bonds containing a secondary amino group, nor are peptides containing dicarboxylic amino acids at the carboxyl terminus good substrates (*18*). Even blocked tripeptides are hydrolyzed well by the enzyme, and in fact we used an assay method adapted from Piquillod et al. (*19*) based on cleavage of the Phe–His bond of Cbz–Phe–His–[^{14}C–Leu]–OH to test potential inhibitors. Another frequently used assay employed hippuryl–His–Leu as a substrate (*20*). The fact that such small compounds are effective substrates of the enzyme encouraged us to believe that similarly small inhibitors might be discovered. Most important for design purposes, the enzyme had been inferred to be a metallopeptidase from early studies by Skeggs et al. (*21*) that described inhibition by EDTA (ethylenediaminetetraacetic acid). Subsequent reports described other chelating inhibitors, including *o*-phenanthroline, 2,2′-dipyridyl, 8-hydroxyquinoline, and, significantly, 2,3-dimercaptopropanol (*22*). In 1975 the cation essential for catalytic activity was shown to be Zn^{2+} (*23*).

Bi-product Designs

Dr. Richard Wolfenden was a consultant in the early days of the project, so we were quite familiar with transition state and bi-product (collected product) design strategies. The fact that ACE contains an essential Zn^{2+} atom places the enzyme in the class of metallopeptidases which includes carboxypeptidase A and thermolysin. We reasoned that mechanistic similarities within a class of enzymes should imply the broad usefulness of active-site inhibitor designs within that class, provided that specificity determinants were adjusted to fit the requirements of the individual enzymes. We therefore attempted to adapt to ACE the Byers and Wolfenden bi-product inhibitor design for carboxypeptidase A (*24, 25*). The Squibb group was also influenced by this important work in their design of captopril (see the section "The *N*-Carboxyalkyl Dipeptide Design"). Byers and Wolfenden had shown that DL-benzylsuccinic acid inhibited carboxypeptidase A with a K_i of 1.1×10^{-6} M, and that inhibition was better than that of DL-2-benzylglutaric acid (K_i of 5×10^{-6} M). To adapt this design to ACE, it was necessary to incorporate a succinic acid fragment within a dipeptide-like structure. Because the snake venom peptides, including teprotide, have proline at their carboxy termini, Maycock synthesized succinoylproline, but its IC_{50} of 3.3 ×

10^{-4} M was an extreme disappointment. He also made a derivative of this compound in which a chelating hydroxamic acid group replaced the succinoyl carboxylic acid. Unfortunately, its IC_{50} of 5×10^{-5} M was less than we expected, and the entire approach was set aside. In retrospect the extrapolation from the carboxypeptidase A inhibitors had been made too precisely in respect to the selection of a succinoyl rather than a glutaroyl group, and the importance of S_1' side-chain functionality had been underestimated.

Pentacovalent Phosphorus-Containing Inhibitors

Intrigued by the excellent inhibition of thermolysin by phosphoramidon (K_i of 2.8×10^{-8} M) (*26*), we moved next to pentacovalent phosphorous inhibitors. The negatively charged and tetrahedrally disposed oxygen atoms in such phosphorous inhibitors bear an easily imagined resemblance to a transition state for amide bond hydrolysis. Indeed, X-ray structure determination of the phosphoramidon-thermolysis complex provided support for transition-state inhibition (*26*). Furthermore, the rhamnose fragment in phosphoramidon seemed to be out of place in a peptidase inhibitor. Its replacement by an amino acid fragment and adjustment of the Leu–Trp portion to better reflect the substrate specificity of ACE were considered to be attractive ways to attain exceptionally potent ACE inhibitors.

Several effective phosphorus-containing inhibitors were synthesized, including $CH_3OP(O)(OH)NHAlaPro$, which showed intravenous (iv) activity as an antagonist of the angiotensin I pressor response in rats. Phosphorus-based inhibitor syntheses continued after captopril was announced in 1977 (*27, 28*), and the best of those synthesized have been reported (*29, 30*). Unfortunately, the highly active phosphoramides that we synthesized were quite unstable to hydrolysis, and the duration of action and oral activity in vivo were continuing problems that were not solved in the phosphate and phosphinic acid analogues that we synthesized.

The Captopril Achievement

The synthesis of captopril was a superb achievement in the history of drug research. Its design and development have been reviewed a number of times (see refs. 12, 31, and 32). Early clinical reports, however, did indicate some shortcomings of captopril. These shortcomings included the necessity for multiple dosing and the occurrence of side effects, the most common of which were rash and diminution or loss of taste perception (*33*). Similar side effects are observed with penicillamine, which is marketed by Merck for the treatment of Wilson's disease. Thus, we hypothesized that designing an ACE inhibitor without a thiol group might reduce the incidence of these side effects. In addition, metabolic stability and duration of action might be improved, because thiol groups provide a sensitive site for metabolism (*34*). Recognizing captopril to have been a major clinical breakthrough, we increased our synthetic manpower to seven chemists. We continued the phosphorus-based designs, searched for new inhibitor designs, and briefly examined possible replacement of the thiol group by other Zn^{2+} liganding functionality. A selection of compounds in the latter category along with their poor ACE inhibitory potencies is shown in Table 1.

The *N*-Carboxyalkyl Dipeptide Design

When captopril was announced, the Squibb group summarized its attempts to adapt the Byers and Wolfenden carboxypeptidase A collected product design to ACE (*27, 28*). Several of the compounds they reported at that time are shown in Table 2. The breakthrough by Ondetti and Cushman in the design of captopril was achieved by replacing the putative zinc-binding carboxyl in compound **10** by an SH group. This much more potent Zn^{2+} ligand increased the intrinsic potency of **10** nearly three orders of magnitude.

In our resumed interest in these acyl proline inhibitors, we returned to the idea that compounds such as **11** are collected product inhibitors. If so, it would seem to follow that introducing a properly placed NH group in compound **11** should enhance its resemblance to an Ala–Pro product fragment and, thereby, its potency. To our chagrin the resulting compound **13** was no more potent than the parent methylene analogue **11**. We turned our attention to other approaches for 3 months and fortunately had forgotten the reduced activity reported by the Squibb group for the isomerically methylated analogue **12**. In rethinking the apparently unchanged activity of compound **13** despite the introduction of an NH group, it seemed possible that counterbalancing factors had nullified the expected potency enhancement. The added NH group might have achieved a desirable hydrogen bond donor or acceptor interaction with the enzyme, but its added polarity could have had a negative effect in the hydrophobic enzyme active site. Therefore, we decided to add the second methyl group, which is present in structure **14**. To our delight, inhibitory potency now increased approximately 25 times, even in the unseparated mixture of diastereomers, and iv activity to block angiotensin I's pressor effect in rats was realized. After 3

Table 1. The ACE Inhibition of Compounds Bearing Possible Zn^{2+} Ligands Other Than a Thiol Group

No.	*Structure*[a]	*ACE Inhibition at 3.3 × 10^{-4}M*[b]
1	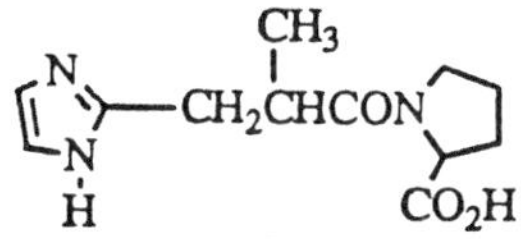 N, N, H; CH_3; $-CH_2CHCON$; CO_2H	61%
2	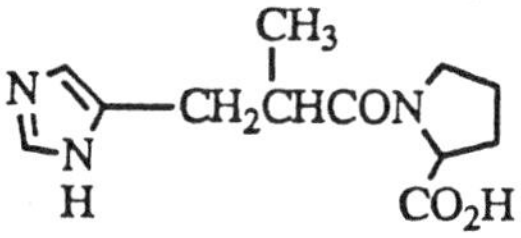 N, N, H; CH_3; $-CH_2CHCON$; CO_2H	5%
3	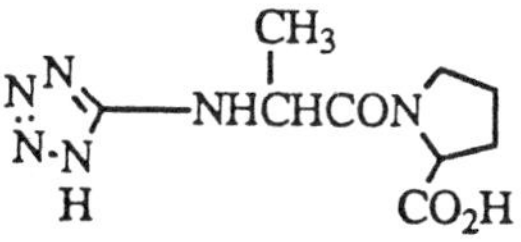 N, N, N, N, H; CH_3; $-NHCHCON$; CO_2H	0%
4	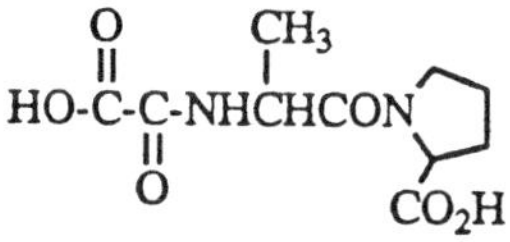 O; CH_3; HO-C-C-NHCHCON; O; CO_2H	38%
5	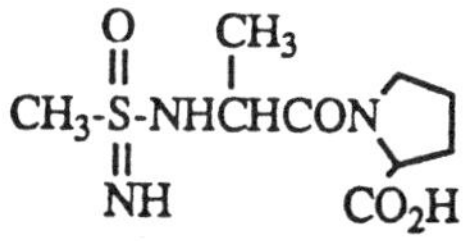 O; CH_3; CH_3-S-NHCHCON; NH; CO_2H	61%
6	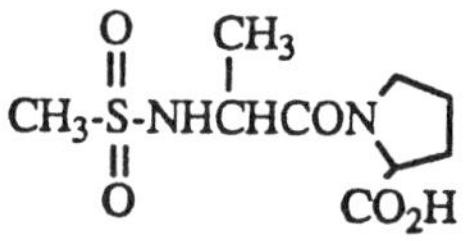 O; CH_3; CH_3-S-NHCHCON; O; CO_2H	14%
7	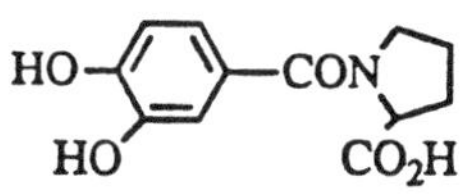 HO; -CON; HO; CO_2H	12%

[a] Compounds are from the unpublished work of E. E. Harris, E. D. Thorsett, J. Toth, E. R. Peterson, N. Grodin, and Y. Cheung.

[b] Assay results in this and succeeding tables were determined with angiotensin-converting enzyme prepared from fresh citrated hog plasma using Cbz–Phe–His–[^{14}C–Leu]–OH as described in ref. 34.

Table 2. Carboxyalkanoylproline Analogues as ACE Inhibitors

No.	*Structure*	*ACE Inhibition* *(IC_{50})*
8	$HO\text{-}\overset{O}{\overset{\|}{C}}\text{-}CH_2CH_2CON$ (pyrrolidine ring) CO_2H	330 μM[a]
9	$HO\text{-}\overset{O}{\overset{\|}{C}}\text{-}CH_2CH_2CH_2CON$ (pyrrolidine ring) CO_2H	70 μM[a]
10	$HO\text{-}\overset{O}{\overset{\|}{C}}\text{-}CH_2\overset{CH_3}{CH}CON$ (pyrrolidine ring) CO_2H	22 μM[a]
11	$HO\text{-}\overset{O}{\overset{\|}{C}}\text{-}CH_2CH_2\overset{CH_3}{CH}CON$ (pyrrolidine ring) CO_2H	4.9 μM[a]
12	$HO\text{-}\overset{O}{\overset{\|}{C}}\text{-}\overset{CH_3}{CH}CH_2CH_2CON$ (pyrrolidine ring) CO_2H	260 μM[a]
13	$HO\text{-}\overset{O}{\overset{\|}{C}}\text{-}CH_2NH\overset{CH_3}{CH}CON$ (pyrrolidine ring) CO_2H	2.4 μM[b]
14	$HO\text{-}\overset{O}{\overset{\|}{C}}\text{-}\overset{CH_3}{CH}NH\overset{CH_3}{CH}CON$ (pyrrolidine ring) CO_2H	0.09 μM[b]

[a] Data are from ref. 28.
[b] Data are from ref. 34.

years of frustrating failures we had made the breakthrough that led to enalapril and lisinopril. We were within a factor of 4 of captopril's potency, and the achievement had been made with a novel design that did not rely on potent chelation to Zn^{2+}. Moreover, it seemed likely that the methyl group was pointed toward the S_1 pocket of the enzyme and that analogues with much greater potency could be synthesized. Subsequent elaboration at this position quickly justified our expectations, and later modeling studies were consistent with the S_1 subsite binding of homologues of the methyl group.

We were left with a puzzle in respect to the methyl group's potency enhancement as expressed in compound **14** but not in the all-carbon compound **12**. Possibly the NH group in compound **14** has a directive effect on the enzyme-bound conformation of this inhibitor to position the methyl group for an effective interaction. If so, that conformation apparently is not achieved when compound **12** binds to ACE, since it may be energetically unfavorable to place a CH_2 group where the NH group binds to the enzyme. In any event, it was fortunate that we had not erroneously used published data to conclude that a methyl group next to the Zn^{2+} liganding carboxyl would cause a loss in activity.

Elaboration at the P_1 Subsite

The breakthrough compound **14** was synthesized by Dr. Matthew Wyvratt in September 1978. As soon as its remarkable activity was known, he and his associates, in particular Dr. Edward Tristram, Mr. Ted Ikeler, Dr. Elbert Harris, and Mr. Elwood Peterson, focused their efforts on substitutions at the *N*-carboxyalkyl position. By late 1978 they had defined in broad outline the structure–activity relationship requirements of this position and had synthesized the exceptionally potent compound **20**, later to be named enalaprilat.

Some of the more important compounds from this and subsequent elaborations at the *N*-carboxyalkyl position are listed in Table 3. In agreement with what was known of the enzyme's specifity, side chains of hydrophobic amino acids conferred high levels of potency as P_1 substituents. Branching at the α-position of these substituents (cf. **15** and **16**) reduced activity, but phenylalkyl substituents were consistent with excellent activity, provided that the *n*-alkyl group contained at least two carbon atoms. All of these compounds were tested iv as potential inhibitors of angiotensin I's pressor effect in anesthetized rats. Special attention was drawn to compound **19** because of its outstanding potency and duration of action in this assay. It and the other analogues listed in Table 3 were tested initially as the mixtures

O R CH_3

HO—C—CHNHCHCON

CO_2H

Table 3. *N*-Carboxyalkyl Derivatives of Ala–Pro

No.	R	*ACE Inhibition* *(IC_{50})*
13	–H	2,400 nM
14	$-CH_3$	90 nM
15	$-CH_2CH_2CH_3$	5 nM
16	$-CH(CH_3)CH_2CH_3$	28 nM
17	$-CH_2CH_2CH(CH_3)_2$	2.6 nM
18	$-CH_2Ph$[a]	39 nM
19	$-CH_2CH_2Ph$	3.8 nM
20	$-CH_2CH_2Ph$ (S-)	1.2 nM
21	$-CH_2CH_2Ph$ (R-)	820 nM
22	$-CH_2CH_2CH_2Ph$	8.3 nM
23	$-CH_2CH_2NH_2$	320 nM
24	$-CH_2CH_2CH_2NH_2$	99 nM
25	$-CH_2CH_2CH_2CH_2NH_2$	2.2 nM
26	$-CH_2CH_2CH_2CH_2NH$–Phth[b]	5.1 nM
27	$-CH_2CH_2CO_2H$	500 nM
28	$-CH_2CH_2CH_2CO_2H$	76 nM

NOTE: All compounds were tested as mixtures of diastereomers except where noted.
[a] Data are taken from ref. 34.
[b] Phth is *N*-phthalimide.
SOURCE: Data are taken from ref. 30 except where noted.

of diastereomers resulting from the sodium cyanoborohydride reductive alkylation of Ala–Pro with an α-ketoacid. The diastereomers of compound **19** were separated by Dr. Henry Joshua by chromatography on XAD-2 (styrene–divinylbenzene resin) and were crystallized from water by him. As expected, most activity resided in one diastereomer (**20**), whose potency was over 2,000-fold greater than that of the unsubstituted analogue, **13**. The importance of hydrophobic binding in enhancing the efficacy of an enzyme inhibitor is strikingly illustrated in this series of compounds.

A surprising result was the exceptional activity of the aminobutyl-substituted compound **25**, whose potency was nearly identical to that of the phthalimidoyl intermediate from which it was derived. Evidently the four methylene groups in these compounds provide hydrophobic binding, but functionality beyond the fourth carbon atom no longer effectively interacts with the enzyme. A similar situation exists in the P_1' position beyond about five atoms from the peptide backbone. Such bulk tolerance positions at the periphery of inhibitors provide useful attachment sites to synthesize dual-action drugs, to modify physical properties, to append radioactive ligands, and to provide linking groups to affinity columns. Most of these opportunities were investigated by us from the P_1' position, although undoubtedly P_1 substituents could also have been used for these purposes.

The in vitro and in vivo potencies in rats of compound **20** in comparison with captopril are summarized in Table 4. As expected, the iv potency of **20** as measured by its ED_{50} was significantly greater than that of captopril, as was its duration of action as measured by its area under the curve (AUC; percent inhibition vs. time). However, compound **20** did not show good oral bioavailability. In fact, its oral ED_{50}, instead of being roughly 10 times better than that of captopril, was about 10 times worse. If *N*-carboxyalkyl dipeptides were to become candidates for safety assessment studies, we would need to improve their oral activity. Everyone on the project was now making analogues of compound **20** in an effort to reach this goal.

Table 4. In Vitro and In Vivo Oral Activity Comparisons of Captopril and Compound 20 in the Blockade of Angiotensin I

Compound	*Enzyme Inhibition*	*ED_{50}[a] in rats*		*ACU[b] at ED_{50}*
		Intravenous	*Oral*	
Captopril	2.3×10^{-8}M	26 μg/kg	0.33 mg/kg	190
20	1.2×10^{-9}M	8.2 μg/kg	2.29 mg/kg	266

[a] Dose that inhibited the maximum pressor response to angiotensin I by 50%.
[b] AUC is the area under the percent inhibition vs. time curve during a 7-h period following oral administration.
SOURCE: Data are taken from ref. 35.

Achieving Good Oral Activity

Excellent oral bioavailability for enalaprilat was realized remarkably quickly using a prodrug approach. When both carboxylate groups of compound **20** were esterified, the resulting weakly basic compound was orally active in rats, although relatively facile diketopiperazine formation involving the alanine NH and the proline carboalkoxy group vitiated this approach.

Fortunately, however, monoesterification of the *N*-carboxyalkyl group was a very effective way of obtaining good oral activity. Because esterification increases the lipophilicity of carboxylic acids, we synthesized methyl, ethyl, butyl, and benzyl esters but saw no major differences among them with limited testing in the oral rat angiotensin I pressor assay. Possibly counterbalancing factors affect the bioavailability of these compounds if both physical properties and the ease with which they are deesterified are important. In any event, with little basis to choose among them, we elected to go forward with the ethyl ester, relying on the precedent of ethyl esters being used in clinical medicine and being metabolized to varying degrees by human esterases. The monoethyl ester (**21**) of compound **20** is indeed a prodrug requiring deesterification because its IC_{50} in the inhibition of ACE was at least 1,000-fold poorer than that of compound **20** itself (*34*).

Studies of the ethyl ester (**21**) in rats are summarized in Table 5 (*35*). It was clear that in this species the goal had been reached of obtaining a compound that was at least as potent as captopril both iv and orally.

Our group sought now to obtain a useful synthesis of compound **21** and to obtain a crystalline salt of it. The initial synthesis involving the sodium cyanoborohydride reduction of ethyl 2-oxo-4-phenylbutanoate with Ala–Pro in the presence of molecular sieves afforded only a slight diastereomeric excess of the more active compound. However, reduction using 10% palladium-on-carbon in the presence of molecular sieves favored compound **21** over its less active diastereomer by a 60:40 margin (*36*). Still, the purification of compound **20** required chromatography. A breakthrough in its synthesis was the preparation of a crystalline, nonhygroscopic maleate salt that could be preferentially precipitated by seeding crystals of it into an acetonitrile solution of the 60:40 diastereomeric mixture of maleate salts. In this way the maleate salt of compound **21** could easily be prepared in quantity for extensive pharmacology studies (*36*).

Oral activity studies to block the pressor activity of angiotensin I in dogs showed good bioavailability of compound **21** maleate at 0.3 mg/kg, with a duration of effect that exceeded 6 h (*34*). In addition, Dr. Charles Sweet and co-workers established that this compound, later to be called enalapril maleate, was effective in animal models of hypertension, including renovascular and spontaneously hypertensive rats (*37, 38*).

The absorption of enalaprilat (compound **20**) was estimated to be 5% in rats and 12% in dogs, whereas for enalapril maleate (**21**) the estimated absorption was 39% in rats and 64% in dogs (*39*). However, both compounds were selected for safety assessment studies because enalaprilat would likely be preferred as an injectable antihypertensive agent and there was a

Table 5. In Vivo Comparisons of Enalaprilat (20), Enalapril (21), and Captopril (22) in the Blockade of Angiotensin I

No.	Structure	ED_{50} in Rats[a] Intravenous	ED_{50} in Rats[a] Oral	AUC at ED_{50}
20	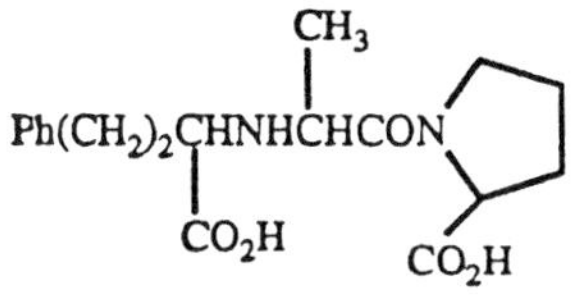 $Ph(CH_2)_2CH(CO_2H)NHCH(CH_3)CON$ (pyrrolidine ring with CO_2H)	8.2 μg/kg	2.29 mg/kg	266
21	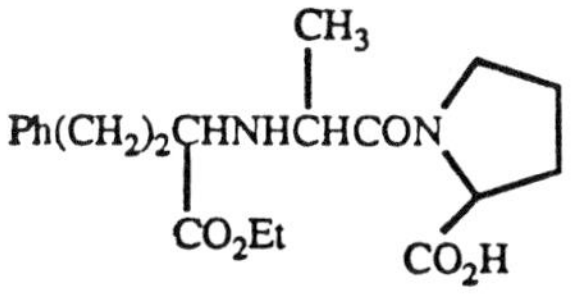 $Ph(CH_2)_2CH(CO_2Et)NHCH(CH_3)CON$ (pyrrolidine ring with CO_2H)	14 μg/kg	0.29 mg/kg	219
22	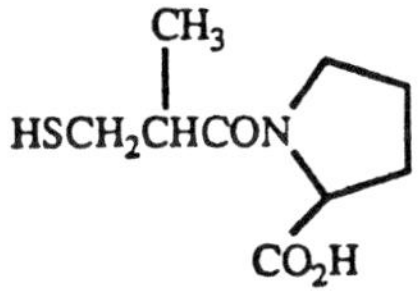 $HSCH_2CH(CH_3)CON$ (pyrrolidine ring with CO_2H)	26 μg/kg	0.33 mg/kg	190

[a] Anesthetized rats were used in the intravenous studies and conscious rats in the oral studies.
SOURCE: Data are taken from ref. 35.

possibility that humans might absorb it adequately even though rats and dogs did not.

At the time we had no completely satisfactory explanation for the oral activity of enalapril. The pK_a values of enalaprilat were determined to be 2.8, 3.5, and 7.6, and of enalapril, 3.0 and 5.4. Therefore, one possibility was that at near neutral pH in the gastrointestinal tract, enalapril was predominantly ionized as a monocarboxylate, because esterification had both removed a negative charge and reduced the basicity of the alanine NH to a pK_a of 5.4. In effect, enalapril might be absorbed as are other carboxyl-containing drugs, including captopril. Friedman and Amidon (*40*) recently reported that in the rat jejunum, enalapril is absorbed largely by carrier-mediated transport, which is inhibitable by the dipeptide Tyr–Gly and by cephradine. These findings help explain why changes in ester lipophilicity that were designed to alter passive absorption apparently had little effect.

Analogues of the Alanine–Proline Fragment

The Ala–Pro part–structure of enalapril was derived from the functionality of captopril and ultimately from the carboxy terminal sequence of the snake venom peptide BPP_{5a} (*11*). We now needed to determine if this part–structure was optimal. Because ACE is a relatively nonspecific enzyme, there was no assurance that dipeptides taken from the carboxy termini of angiotensin I (His–Leu) or of bradykinin (Phe–Arg) would afford inhibitors of maximal intrinsic potency. Indeed, *N*-carboxyalkyl derivatives of these dipeptides were not as potent as corresponding ones derived from Ala Pro. We therefore relied on systematic variation of the Ala–Pro fragment in which we held Ala constant and varied the terminal amino acid, and held Pro constant and varied the penultimate position. The results of some of these changes at the P_2' position are shown in Tables 6 and 7 (*41*).

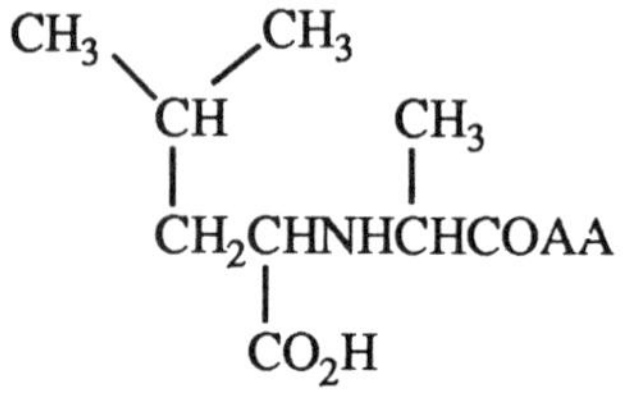

$$(CH_3)_2CH\text{-}CH_2CH(CO_2H)NHCH(CH_3)COAA$$

Table 6. ACE Inhibition of *N*-Carboxyalkyl Dipeptides Varied at the P_2' Position

No.	*Amino Acid (AA)*	*IC_{50}*
23	Proline	5.6×10^{-9} M
24	Tryptophan	3×10^{-8} M
25	Tyrosine	4×10^{-8} M
26	Phenylalanine	8.6×10^{-8} M
27	Alanine	1.3×10^{-7} M
28	Valine	8.7×10^{-7} M

SOURCE: Data are taken from ref. 41.

Proline among the natural amino acids was found to be preferred at the carboxy terminus. The negative charge of the proline carboxylate is an important contributor to potency, but the proline ring itself was tolerant of extensive variations both in size and lipophilicity. Furthermore, imino acids could be used in place of proline to produce potent inhibitors, as exemplified by compound **33**. The relative insensitivity of intrinsic potency to structural changes in the proline fragment discouraged extensive investigation of this

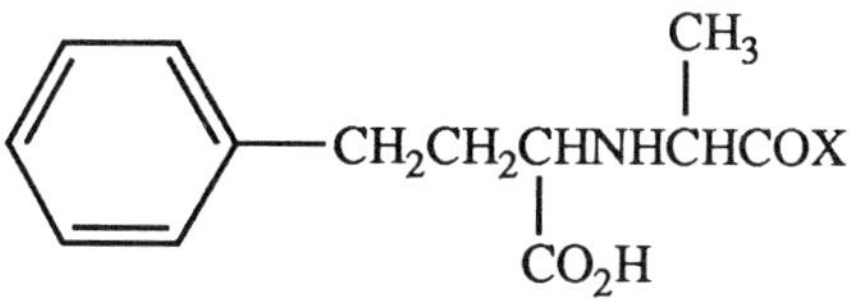

C_6H_5-$CH_2CH_2CH(CO_2H)NHCH(CH_3)COX$

Table 7. ACE Inhibition of Proline Analogues in the Enalaprilat Design

No.	X	IC_{50}
29	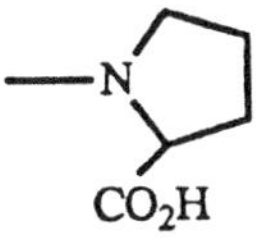 —N (pyrrolidine), CO_2H	3.8×10^{-9} M
30	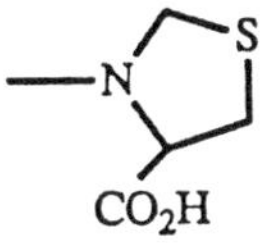 —N (thiazolidine, S), CO_2H	7.6×10^{-9} M
31	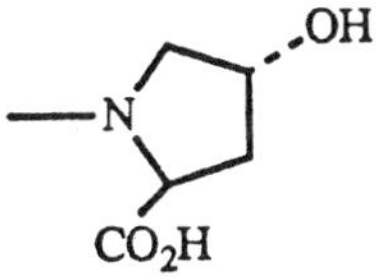 —N (pyrrolidine, OH), CO_2H	1.2×10^{-9} M
32	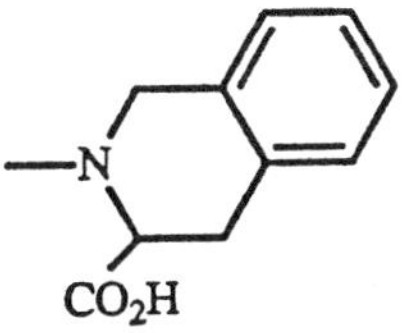 —N (tetrahydroisoquinoline), CO_2H	4.4×10^{-9} M
33	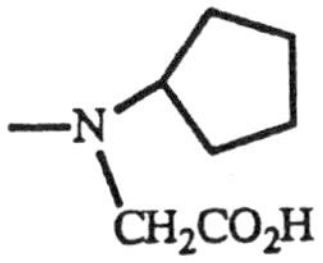 —N(cyclopentyl), CH_2CO_2H	4.4×10^{-9} M
34	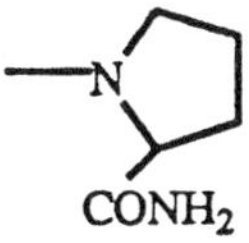 —N (pyrrolidine), $CONH_2$	$>1.67 \times 10^{-7}$ M[a]

[a] Data are taken from ref. 30.
SOURCE: Data are taken from ref. 41 except where noted.

position by us, and we felt we would be behind other groups already modifying this fragment in the search for better captopril analogues. Indeed, structural latitude at this position has allowed other research groups to generate many novel variants of enalapril and lisinopril. Variations in time to onset of action, tissue penetration, and in the major route of excretion have been ascribed to some of these newer analogues (*42, 43*), although it is not clear that differences of major importance in the clinical usefulness of *N*-carboxyalkyl inhibitors have been achieved.

The effects of changes at the P_1' position of enalaprilat analogues are summarized in Table 8. As expected from the known specificity of the enzyme, imino acids at P_1', such as proline (compound **37**) and *N*-methyl-L-alanine (compound **38**), afforded poor inhibitors. Alanine in this position provided an inhibitor much more active than the glycine analogue, perhaps in part for conformational reasons, as has been suggested to explain the importance of the methyl group in captopril (*28*). Larger alkyl groups, especially if they are branched, decreased activity as did arylalkyl groups, and a glutamic acid residue at P_1' afforded a very poorly active compound. It was therefore surprising to find that the lysine analogue **45** had extremely good intrinsic potency. Apparently an additional binding site had been picked up on the enzyme that compensated for the presence of a substituted higher alkyl group. We were even more surprised to learn that compound **45** itself was orally active in rats. Analogues of **45** and conformationally constrained lactam analogues of enalaprilat then became our main approaches to the design of orally active ACE inhibitors that were not prodrugs.

We also examined the essentiality of the carboxy group in the *N*-carboxyalkyl fragment of enalaprilat. The results of some of those variations are shown in Table 9. Removing the carboxyl group (compound **46**) caused a profound drop in potency, as did conversion to an amide group (compound **47**) or reduction to an alcohol (compound **48**). Such losses in potency are consistent with an electrostatic interaction of the carboxyl group as a ligand to zinc. We did not synthesize a thiolmethyl group in this position, because of our determination to synthesize potent inhibitors devoid of the thiol functionality. Nonetheless we did synthesize hydroxamate-, phosphonic acid-, and imidazole-containing analogues (compounds **49**, **50**, and **51**) because they are also known to be potent Zn^{2+} ligands. The former two had reasonable but reduced potencies compared with compound **20**, and the uncharged or positively charged imidazole analogue was poorly active. Apparently a carboxyl group makes better hydrogen bond interactions or makes a better steric fit to the active site of ACE than does the hydroxamate or phosphonic acid group. The carboxyl group in an *N*-carboxyalkyl

C_6H_5–$CH_2CH_2CH(CO_2H)$—X—Pro

Table 8. ACE Inhibition of *N*-Carboxyalkyl Dipeptides Varied at the P1′ Position

No.	*X*	IC_{50}
35	Glycine	2.3×10^{-7} M
19	L-Alanine	3.8×10^{-9} M
36	α-Methylalanine	2.5×10^{-6} M
37	Proline	2.7×10^{-6} M
38	*N*-Methyl-L-alanine	1.0×10^{-7} M
39	L-α-Amino-*n*-pentanoic acid	8.8×10^{-9} M[a]
40	Valine	7.8×10^{-8} M[b]
41	Phenylalanine	5.2×10^{-8} M
42	Leucine	2.3×10^{-7} M[b]
43	Histidine	7.4×10^{-8} M
44	Glutamic acid	1.3×10^{-6} M
45	Lysine	1.2×10^{-9} M[b]

[a] Data are taken from ref. 30.
[b] Data were obtained on a single diastereomer.
SOURCE: Data are taken from ref. 41 except where noted.

inhibitor bound to thermolysin makes a tight, bidentate fit to Zn^{2+} with multiple hydrogen bond interactions to that enzyme, as described in the section "The Binding of *N*-Carboxyalkyl Peptides to Zinc Metallopeptidases." If these same geometries and interactions apply to ACE, then hydroxamate and phosphonic acid analogues can be rationalized to bind well but slightly less effectively than the carboxy parent **20**.

Finally, it was of interest to learn if the potency of *N*-carboxyalkyl dipeptides paralleled the much weaker end-product potency of the dipeptides from which they were synthesized. If so, support would be given for a collected product inhibitor designation for *N*-carboxyalkyl dipeptides. More important for this and subsequent metallopeptidase design projects, optimum *N*-carboxyalkyl peptides might be designed if we knew the K_m of peptide products. The Squibb group determined the ACE inhibitory potency of a large number of dipeptides, and in general the potency of captopril analogues paralleled the corresponding dipeptides (*44*). In our case this conclusion was less clear-cut (*41*). There were notable exceptions such as *N*-

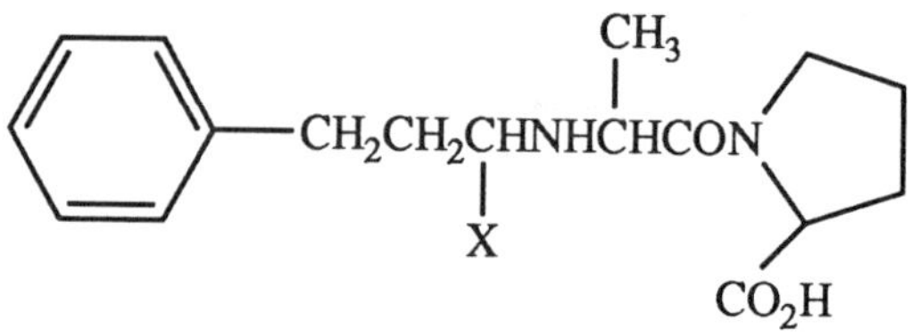

Table 9. Carboxy Replacements in the N-Carboxyalkyl Fragment

No.	*X*	IC_{50}
20	$-CO_2H$	1.2×10^{-9} M[a]
46	$-H$	5.8×10^{-6} M
47	$-CONH_2$	2.6×10^{-5} M
48	$-CH_2OH$	5.7×10^{-6} M
49	$-CONHOH$	9.8×10^{-9} M[a]
50	$-PO(OH)_2$	5.1×10^{-8} M
51	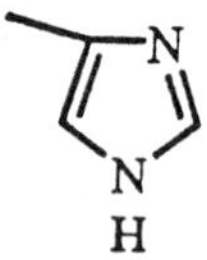 N N H	1.5×10^{-6} M

[a] Data obtained with a single diastereomer.
SOURCE: Data are taken from ref. 30.

(1-carboxyl-3-phenylpropyl)–Val–Trp, whose IC_{50} of 7×10^{-9} M was less than that of the Ala–Pro analogue, even though Val–Trp is a much better inhibitor of ACE than is Ala–Pro (*44*). No definitive position was reached, although it seems likely that peptide data can provide a useful starting point for the design of *N*-carboxyalkyl peptide inhibitors.

The Properties of Lisinopril

The good oral activity of lisinopril (**45**) was unexpected. It was detected in a high-capacity assay that Dr. Charles Sweet and Dr. Dennis Gross had developed to evaluate the oral activity of ACE inhibitors in rats. Evaluations were based on the measurement of angiotensin I inhibitory activity integrated over time, with AUCs established for each compound after the iv and oral administration of compounds. Gross and his associates tested all of our potent compounds in this assay without regard for rationales based on structure or duration of action following iv administration. If we had had to justify the oral testing of lisinopril, its good absorption may never have been discovered, for enalaprilat had poor oral activity and lisinopril was an even more polar compound. Despite attempts to achieve oral activity with nonpeptide designs, lisinopril and its congeneric designs even today provide the only non-prodrug means of attaining useful oral bioavailability in (*N*-carboxylalkyldipeptide) ACE inhibitors. We were extremely pleased with the lisinopril discovery because it provided us with a backup compound for enalapril. Prodrugs can lose some effectiveness when there is hepatic insufficiency; and when there is tissue-selective activation of drugs, potential effects or side effects may be accentuated in the organs involved. Such concerns proved to be unimportant in the case of enalapril, but when we began safety assessment and clinical testing, it was reassuring that both types of compounds were being evaluated.

Analogues of Lisinopril

Oral testing of the monoethyl ester of lisinopril was of great interest. Surprisingly, this compound was not better absorbed in rats than lisinopril, even though esterification had immensely improved the oral activity of enalaprilat. Additional analogues of lisinopril showed that it was relatively easy to retain high in vitro ACE inhibitory activities (Tables 10 and 11). The ornithine (**54**) analogue was well absorbed orally in rats, but the other analogues in Table 1 were not. Apparently a basic group is important because of the poor oral activity of the *N*-acetyl analogue (**59**). We surmised that bis-zwitterionic properties were important contributors to the absorption of lisinopril, but the poor oral activity of the *N*-dimethyl analogue (**58**) and of the guanidine analogues (**56**) and (**57**) was not consistent with that interpretation. We then hypothesized that lisinopril might be absorbed by a facilitated process. One possibility was absorption via a tripeptide transport system (*45*), which would be consistent with the apparent requirement for a

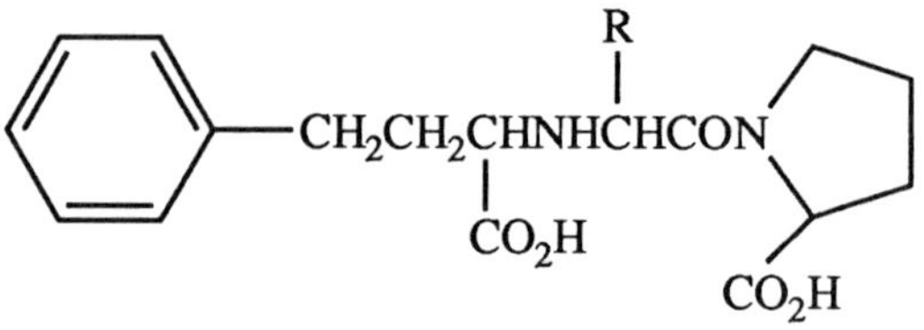

R

$CH_2CH_2CHNHCHCON$

CO_2H

CO_2H

Table 10. Lisinopril Analogues Varied at the Amino Group and in the Length of the Alkyl Group

No.	*R*[a]	*IC*$_{50}$
45	$-CH_2CH_2CH_2CH_2NH_2$	1.2×10^{-9} M[b]
52	$-(CH_2)_5NH_2$	5.2×10^{-9} M[b]
53	$-CH_2NH_2$	4.8×10^{-7} M[b]
54	$-(CH_2)_3NH_2$	4.5×10^{-9} M[c]
55	$-(CH_2)_2NH_2$ (R, S)	3.8×10^{-8} M[d]
56	NH ‖ $-(CH_2)_3NHCNH_2$	6.4×10^{-9} M[c]
57	NH ‖ $-(CH_2)_4NHCNH_2$	9.9×10^{-10} M[c]
58	$-(CH_2)_4N(CH_3)_2$	9.1×10^{-9} M[c]
59	$-(CH_2)_4NHCOCH_3$	1.6×10^{-8} M[c]

[a] Dipeptide amino acids were all of the natural (S-) configuration except where indicated. Compounds were synthesized by the $NaCNBH_3$ reductive condensation of 4-phenyl-2-oxobutyric acid with the dipeptides and were assayed without separation of diastereomers, with the exception of **45**, which is the most active (S, S, S-) diastereomer. Unpublished syntheses of M. T. Wu, L. G. Payne, D. L. Ondeyka, and T. J. Ikeler.
[b] Data are taken from ref. 34.
[c] Data are the unpublished results of E. H. Ulm and T. C. Vassil.
[d] Data are taken from ref. 35.

free carboxyl separated at tripeptide-like spacing from a basic primary or secondary amino group. To test this hypothesis, Dr. Wu and his associates synthesized several analogues in which peptide functionality was introduced into the lysine side chain; several of these compounds are shown in Table 11. Dr. Wu also included thiomethylene groups, because they have been used as amide surrogates (*46*). The best of these compounds in terms of oral activity was **64**; however, superiority to lisinopril was not demonstrated. Nonetheless a 1989 publication by Friedman and Amidon (*47*) showed that nonpassive, peptide-carrier-mediated transport is likely for lisinopril because its absorption in perfused rat intestine was decreased by Tyr–Gly and by cephradine. There may be some generality to the lisinopril "design" for oral absorption insofar as the orally active ACE inhibitors ceranapril (SQ 29,852) (*48*) and CGS 16617 (*49*), whose structures are shown in Figure 2, incorporate lysine-related part–structures.

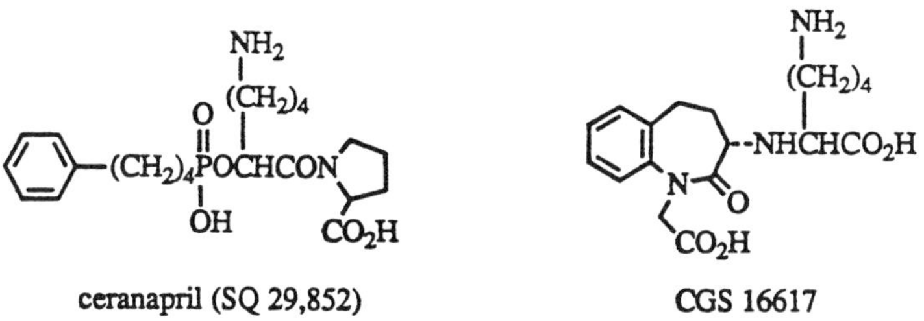

ceranapril (SQ 29,852) **CGS 16617**

Figure 2. Orally active angiotensin-converting-enzyme inhibitors that are not prodrugs.

It is also evident from the analogues shown in Table 11 that heteroatom substitution at the β-carbon atom of the lysine side chain lowers intrinsic activity but derivatization of the terminal amino group is quite tolerant of variations in size and lipophilicity. Dr. Martin Hichens derivatized lisinopril to afford a *para*-hydroxybenzamidine group at this terminal amino position (compound **69**). This derivative is highly potent and has become known in the literature as compound 351A. When it is reacted with $^{125}I_2$, nanomolar potency is retained, and this labeled analogue has been of considerable utility in autoradiography studies to visualize the localization of ACE in tissues (*50–54*). Compound 351A has also been used in a sensitive inhibitor binding assay of serum ACE (*55*).

In addition, the lysine amino group of lisinopril has been reacted with epoxy-activated Sepharose CL-4B (cross-linked agarose beads) to afford an extremely selective affinity column for the purification of ACE (*56*). With its use, electrophoretically homogeneous ACE was directly obtained from crude homogenates of rabbit lung enzyme, which is a 1,000-fold purification in a single step. This affinity column method was also used in the isolation of rabbit testicular ACE, which has catalytic properties similar to the lung enzyme but whose molecular weight is lower by about one-third (*57*). These experiments are beautiful examples of the power of affinity chromatography

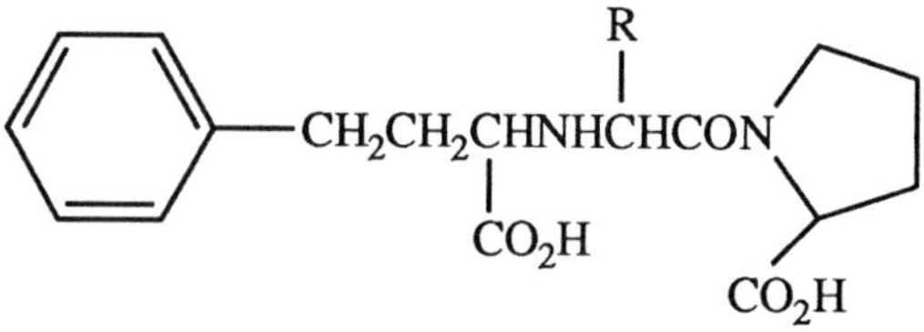

$$C_6H_5\text{-}CH_2CH_2CH(CO_2H)NHCH(R)CON\text{(proline-}CO_2H)$$

Table 11. Modifications in the Lysine Side Chain of Lisinopril

No.	R[a]	Stereochemistry	IC_{50}
60	$-CH_2NHCOCH_2NH_2$	(R,S; R,S; S)	1.5×10^{-7} M
61	$-CH_2CH_2NHCOCH_2NH_2$	(R,S; S; S)	6.8×10^{-9} M
62	$-CH_2CH_2NHCOC(CH_3)_2NH_2$	(R,S; R,S; S)	4.7×10^{-8} M
62	$-CH_2SCH_2CH_2NH_2$	(R,S; S; S)	1.3×10^{-7} M
64	$-CH_2CH_2SCH_2CH_2NH_2$	(S, S, S)	5.2×10^{-9} M
65	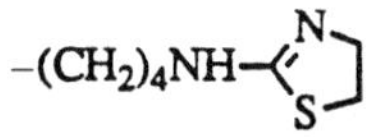 $-(CH_2)_4NH-$(thiazolin-2-yl)	(S, S, S)	1.9×10^{-9} M
66	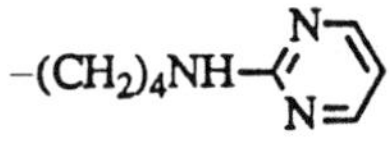 $-(CH_2)_4NH-$(pyrimidin-2-yl)	(S, S, S)	4.1×10^{-9} M
67	$-(CH_2)_4NH(CH_2)_3NH_2$	(S, S, S)	3.7×10^{-9} M
68	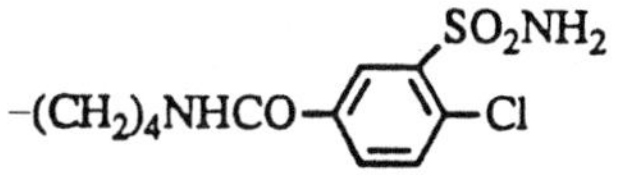 $-(CH_2)_4NHCO-C_6H_3(SO_2NH_2)(Cl)$	(S, S, S)	2.8×10^{-9} M
69[b]	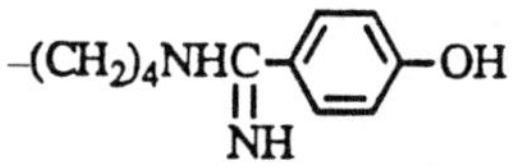 $-(CH_2)_4NHC(=NH)-C_6H_4-OH$	(S, S, S)	2.1×10^{-10} M
70[b]	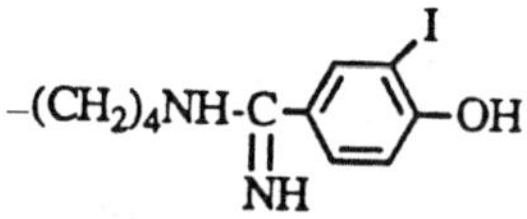 $-(CH_2)_4NH-C(=NH)-C_6H_3(I)-OH$	(S, S, S)	2.6×10^{-9} M

[a] Compounds except where noted are from the unpublished work of M. T. Wu, D. L. Ondeyka, L. G. Payne, R. D. Hoffsommer, and M. Tischler.
[b] Compound synthesized by Dr. M. Hichens.

in the purification of enzymes, and their success underlines the specificity of lisinopril as an ACE inhibitor.

Lactam Inhibitors

A third approach to synthesizing orally active ACE inhibitors had been under way since the announcement of captopril. The goal was to synthesize conformationally constrained dipeptide surrogates onto which enzyme inhibitory functionality could be appended. It was hoped that entropy reduction would increase potency and that the nonpeptide structure would promote better oral bioavailability. At the very least, stability to possible peptidase cleavage was expected.

Monocyclic Lactams

The early work of Thorsett et al. (58) on cyclic captopril analogues involved linking the methyl group of this compound to the 5-position of proline. This arrangement fixed the amide bond in a trans configuration, which by calculation was of lower energy than a cis conformer. By varying the number of linking atoms to the proline ring, it was possible to achieve variations in the psi angle within the lactam ring and thereby vary the SH position relative to the other functionality of the molecule. For ease of synthesis in these initial inhibitors, a complete proline ring was not synthesized, and thus some lipophilicity and the conformational restriction of the proline carboxyl were lost. Nonetheless, compound **72** (Table 12), corrected for its racemic character, is equivalent in potency to captopril. Thus we knew that a *trans*-amide bond was preferred in captopril and undoubtedly also in the *N*-carboxyalkyl dipeptides. We were also encouraged that even monocyclic lactams of high potency could be generated. Evidently the additional atoms used in these lactams to restrict conformational flexibility at the psi angle did not interfere with binding to the enzyme surface.

With the discovery of the enalaprilat lead, efforts now shifted to the synthesis of conformationally accurate replacements for the Ala–Pro dipeptide. The activities of some of the monocyclic lactams bearing a carboxyalkylamino group are shown in Table 13 (59). As expected from the lactams related to captopril, the highest activity was found with the eight-membered ring **77**, whose potency, even lacking the proline, was essentially equivalent to that of enalaprilat. The psi angle in this inhibitor, 145°, was virtually identical to that observed for enalaprilat itself by X-ray crystallography ($\psi =$

Table 12. Mercaptomethyl Lactams

No.	*Compound*	*IC₅₀ (μM)*
71	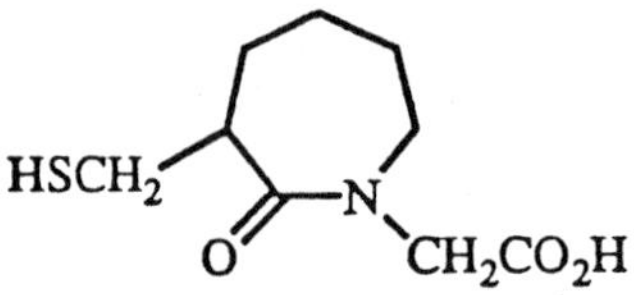 $HSCH_2$–(2-oxo-azepan-1-yl)–CH_2CO_2H	0.17
72	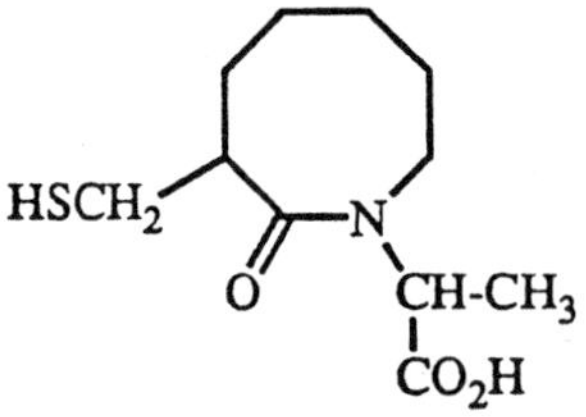 $HSCH_2$–(2-oxo-azocan-1-yl)–$CH(CH_3)CO_2H$	0.053[a]
73	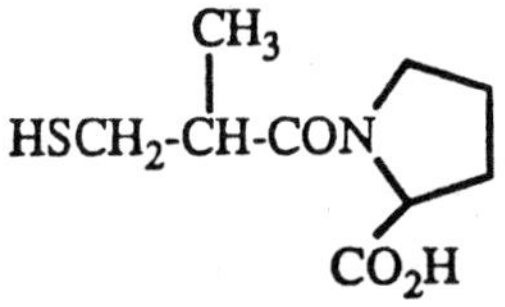 $HSCH_2$-$CH(CH_3)$-CON(pyrrolidine)–CO_2H	0.023

[a] Most active racemate.
SOURCE: Data are from ref. 58.

143°) (*36*). Furthermore, the calculated minimum energy conformation of enalaprilat was also characterized by a psi angle of 140° (*60*). Evidently the dipeptide backbone of enalaprilat does not need to undergo major conformational changes from its minimum energy conformer on binding to the enzyme. Unfortunately, if so, potency enhancements by restricting the conformational freedom of this part of the molecule would not be large.

Nonetheless, interest remained high to add on to these lactams elements of the proline ring. Initially methyl groups were added to the seven-membered lactam to mimic either carbon atoms 3 or 4 of proline and in the former case to provide rotational restrictions on the carboxyl group. The results of these changes are shown in Table 14 (*59*), and in both cases these added methyl groups enhanced the activity of the parent lactam when they were present in their preferred stereochemistry.

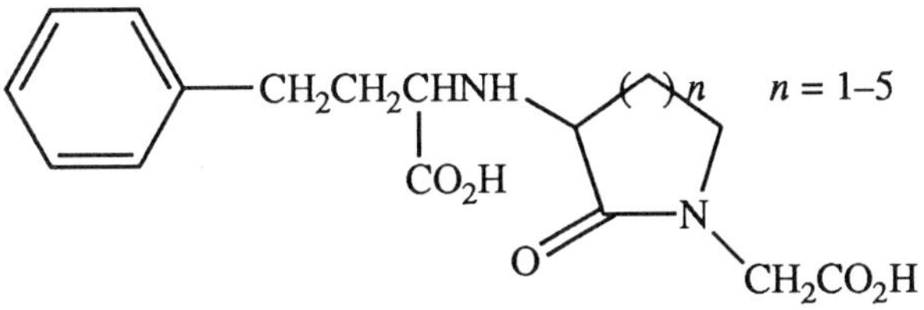

Table 13. Monocyclic Lactam Analogs of Enalaprilat

No.	Ring Size	IC_{50}
74	5	5.3×10^{-6} M
75	6	4.3×10^{-7} M
76	7	1.9×10^{-8} M
77	8	1.9×10^{-9} M[a]
78	9	8.1×10^{-9} M

[a] More active diastereomer.
SOURCE: Data are taken from ref. 59.

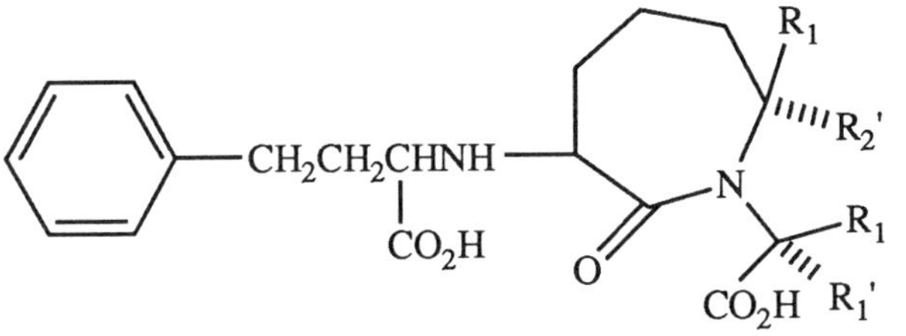

Table 14. Methyl-Substituted Caprolactam Inhibitors

No.	R_1	R_1'	R_2	R_2'	IC_{50}
79	H	H	H	H	1.9×10^{-8} M
80	CH_3	H	H	H	8×10^{-9} M
81	H	CH_3	H	H	5.7×10^{-8} M
82	H	H	CH_3	H	3×10^{-9} M
83	H	H	H	CH_3	7×10^{-9} M

SOURCE: Data taken from ref. 59.

Bicyclic Lactam Inhibitors

The next step was to construct lactams that contained a complete proline ring. Several of the structures that were synthesized are shown in Table 15. The most active of these compounds (**84**) inhibited ACE with an IC_{50} of 0.6 nM and with a K_i of 7.6×10^{-11} M, which makes it one of the most potent *N*-carboxyalkyl dipeptide inhibitors of ACE to have been synthesized. Oxidation of **84** to the corresponding sulfoxide reduced activity about 40-fold, which is consistent with the hydrophobic character of highly active proline surrogates (30).

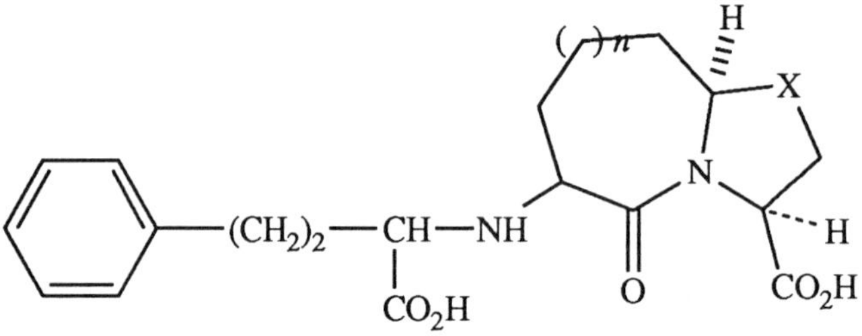

Table 15. Bicyclic Inhibitors of ACE

No.	*X*	*n*	IC_{50}
84	S	1	6×10^{-10} M[a]
85	CH_2	1	2.9×10^{-9} M[b]
86	CH_2	2	3.4×10^{-9} M[b]

[a] Data from ref. 61.
[b] Data from ref. 62.

In Vivo Testing of Lactam Inhibitors

Most of the lactam inhibitors as diacids had no better oral bioavailability than enalaprilat. In retrospect this result might have been anticipated from the metabolic stability of enalaprilat. In rats and dogs the only metabolite formed from enalapril was enalaprilat itself. In monkeys peptidase cleavage between alanine and proline occurred as only a minor metabolic event (39). Thus, very little advantage could be realized from the anticipated peptidase stability of lactam inhibitors. The most promising diacid inhibitor was **84**, which blocked the pressor effect of iv AII in rats with an oral ED_{50} of about 0.4 mg/kg, compared with an oral ED_{50} for enalaprilat of about 2.3 mg/kg (C. S. Sweet, Merck Sharp & Dohme Research Laboratories, pers. commun.). These results are nearly in accord with differences in their intrinsic potencies. Consequently, monoethyl esters were used in most oral studies of this and other lactam inhibitors. Apparently diacidic *N*-carboxyalkyl inhibitors are too polar to be absorbed from the gastrointestinal tract, even though some of the lactam structures are more lipophilic than enalaprilat.

The Binding of *N*-Carboxyalkyl Peptides to Zinc Metallopeptidases

The high potency of enalaprilat and lisinopril as ACE inhibitors encouraged us to test the generality of the *N*-carboxyalkyl design for the inhibition of other metallopeptidases. To perfect the design, we also wished to determine by X-ray crystallography how these inhibitors might bind to metallopeptidases. Early concerns that these compounds might also inhibit serine, cysteine, or aspartic acid proteases because they were being called collected product inhibitors were quickly dispelled. Because the design concept had arisen from Wolfenden's carboxypeptidase A inhibitors, and because the structure of that enzyme had been determined by X-ray crystallography (*63*), we first probed the possible inhibition of that enzyme by *N*-carboxyalkyl inhibitors. ACE had not been crystallized, nor has it been obtained in crystalline form to this date, although the amino acid sequences of human (*64*) and mouse enzyme (*65*) have now been determined. Presumably the fact that ACE contains up to 30% of its weight as carbohydrate residues makes crystallization difficult. Ironically, the *N*-carboxyalkyl design did not work well to inhibit carboxypeptidase A (Table 16), although a captopril design (compound **91**) was highly active (*66*). If we had tested new designs first as carboxypeptidase A inhibitors and only carried over to ACE the promising ones, or if Wolfenden had already tested the effect of nitrogen substitution in his benzyl glutaric acid lead, we may never have attempted *N*-carboxyalkyl inhibitors for ACE.

Binding Studies with Thermolysin

Thermolysin is the other zinc metallopeptidase that had been extensively studied and whose structure was known since the early 1970s by X-ray crystallography (*67*). Fortunately, *N*-carboxyalkyl dipeptides are good inhibitors of thermolysin, providing their structures reflect the specificity requirements of the enzyme (*68*). The best of our compounds was *N*-(1-carboxy-3-phenylpropyl) Leu–Trp (**92**), whose K_i is approximately 5×10^{-8} M. Because Prof. Brian Matthews and his group at the University of Oregon had determined the structures of thermolysin and its inhibitor complexes, we turned to them in late 1980 in the hope that the structure of compound **92** bound to thermolysin could be solved by X-ray crystallography. If not, we planned to study its binding to the enzyme by molecular modeling using the coordinates of phosphoramidon bound to thermolysin (*69*) to fix the Leu–Trp portion of compound **92** within the active site. Fortunately, Monzingo and Matthews were able to diffuse **92** into pregrown crystals of thermolysin and to determine and refine the structure of the enzyme-inhibitor complex to an R value of 17.1% at 1.9 Å resolution (*70*).

The active-site interactions of this inhibitor with thermolysin are shown schematically in Figure 3. Of special interest was the bidentate binding of the carboxylate oxygens to Zn^{2+} with displacement of a water molecule that is the fourth ligand bound to Zn^{2+} in the native enzyme. We wondered if these oxygen atoms were occupying positions corresponding to those of an amide carbonyl of a thermolysin substrate hydrated with this inhibitor-displaceable water molecule. Interactive computer graphics studies by Dr. David Hangauer in our laboratories were consistent with that possibility, using Cbz–Phe–Phe–Leu–Trp as a model substrate (*71*). A mechanism of action of thermolysin was proposed that involved Glu 143 activating the Zn^{2+}-bound water molecule to add it to the scissile carbonyl group. The latter's negatively charged oxygen in the transition state can be stabilized by hydrogen bonds from Tyr 157 and His 231, in addition to an electrostatic interaction with Zn^{2+}. Glu 143 then immediately shuttles a proton to the departing nitrogen atom to facilitate breakdown of the transition state. Interactions analogous to those seen with compound **92** have been observed with a phosphoramide inhibitor Cbz–Phe^{P} Leu–Ala, which has strengthened this proposed mechanism for thermolysin and supported extension of the mechanism to carboxypeptidase A as well (*72*).

Tyrosine, arginine, glutamic acid, and lysine residues have been shown to be essential to the catalytic process of ACE, as summarized in ref. 73. In addition, ACE contains the sequence HEMGH, which is analogous to that found in the zinc metallopeptidases and is thought to be involved in binding Zn^{2+} to these enzymes (*74*). Therefore, we believe a reasonable extrapolation can be suggested for the binding of enalaprilat to ACE (Figure 4).

Table 16. Inhibitors of Carboxypeptidase A

No.	*Structure*	K_i
87	$C_6H_5CH_2$ on C: $HOCOCH_2CH_2CH(CH_2C_6H_5)CO_2H$ — drawn as $HOCOCH_2CH_2\overset{CH_2C_6H_5}{C}H_2CO_2H$ (R,S)	5×10^{-6} M[a]
88	$HOCOCH_2NH\overset{CH_2C_6H_5}{C}H_2CO_2H$ (S)	2.5×10^{-4} M[b]
89	$HOCOCH(CH_2CH_2C_6H_5)NHCH(CH_2C_6H_5)CO_2H$ (R,S; S)	5×10^{-5} M[b]
90	$HOCOCH((CH_2)_5NH\text{-}t\text{-}Boc)NHCH(CH_2C_6H_5)CO_2H$	1.3×10^{-6} M[b]
91	$HSCH_2CH(CH_2C_6H_5)CO_2H$ (R,S)	1.1×10^{-8} M[c]

[a] Data taken from refs. 24 and 25.
[b] Data taken from ref. 41.
[c] Data taken from ref. 68.

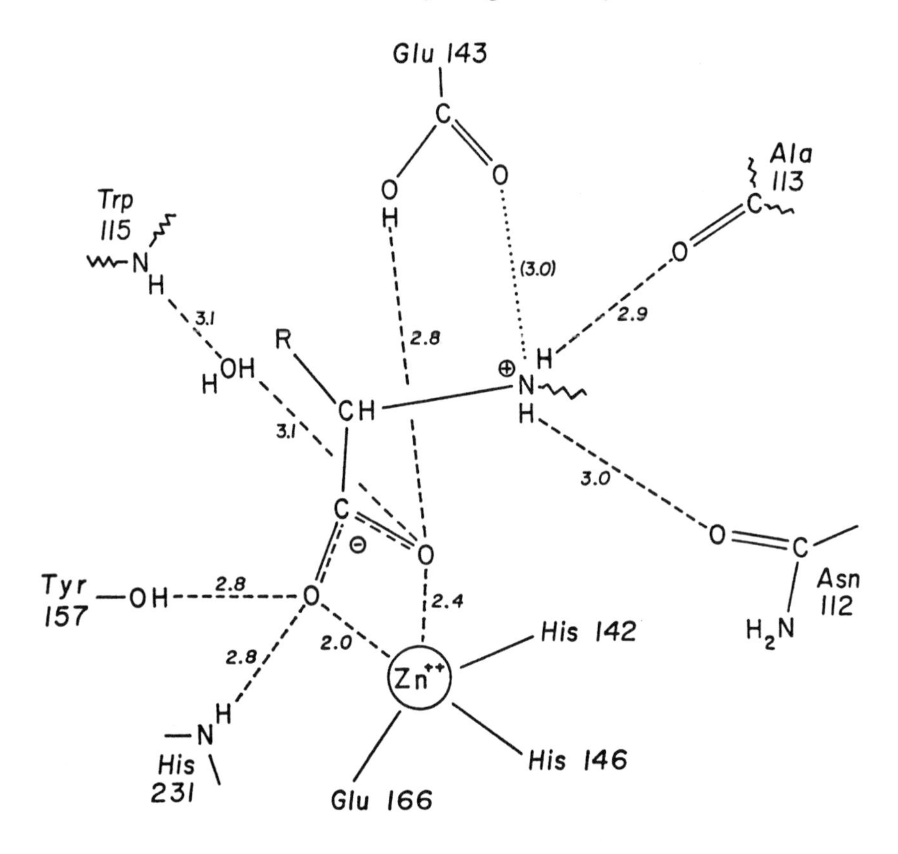

Figure 3. The active-site binding interactions of compound **92** *with thermolysin. Reproduced from ref. 72. Copyright 1988 American Chemical Society.*

Extensions to Other Enzymes

We were interested in the structure of an *N*-carboxyalkyl inhibitor bound to thermolysin, with the anticipation that the active site of this enzyme might be similar to that of ACE. If so, we might be able to use our model for the binding of enalaprilat and lisinopril to ACE to provide some approximate guidance for our synthetic agenda. For example, because the Zn^{2+}-bound carboxylate in these inhibitors already occupies all of the accessible ligand valencies of Zn^{2+}, there seemed to be no point in trying to add additional chelating functionality such as a thiol group alpha to the carboxylate. Also, the hydrogen-bond interactions to nitrogen shown in Figure 3 made it seem

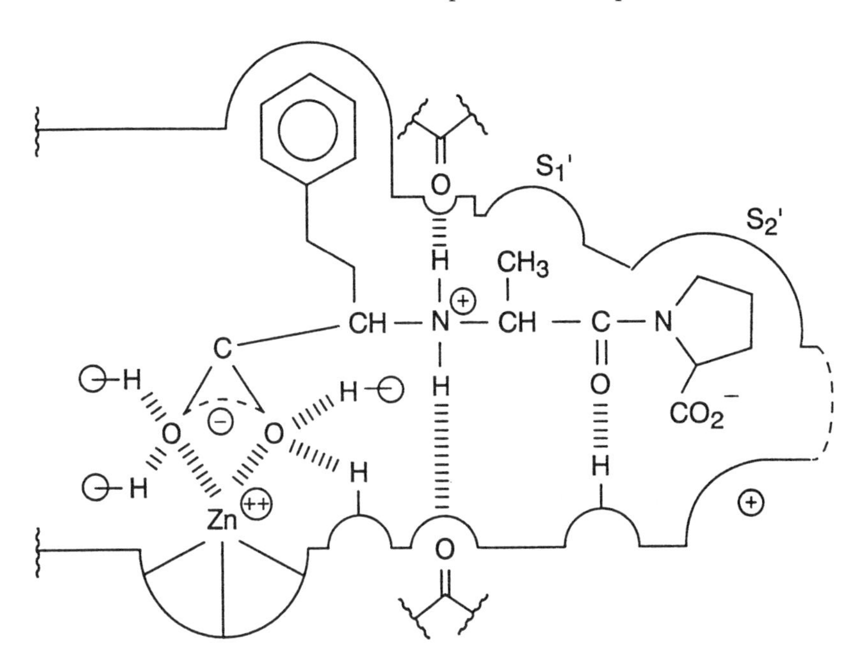

*Figure 4. Proposed binding interactions of enalaprit (**2**) with the angiotensin-converting enzyme. Adapted from ref. 90.*

unlikely that either –O–, –S–, or $-CH_2-$ would be important as NH replacements in the enalaprilat design. Indeed, the explanation offered for the poor activity of the $-CH_2-$ analogue **12** finds support in the bound environment of the NH group, and others have reported that –O– and –S– substitutions for NH in enalaprilat are only active at the 10^{-7} M level (75). We had hoped to extend the enalaprilat design to pick up binding interactions in the S_2 subsite of ACE with inhibitors such as the one shown in Figure 5. Unfortunately, this compound's IC_{50} of 7.7×10^{-9} M was less than that of enalaprilat (76), and modeling an analogous structure to thermolysin suggested that the extra carbon atom that the zinc coordinating carboxyl introduces between the zinc ligands and the main chain of the inhibitor causes an upward displacement on the nonprimed side of these inhibitors (77). Quite possibly the benzoyl Phe part–structure of the inhibitor shown in Figure 5 binds to ACE imperfectly only in the S_1 subsite of the enzyme. Efforts to achieve an ACE inhibitor design that incorporated interactions in both S_1 and S_2 were continued with thermolysin, but the results were inconclusive and thus were not incorporated into preferred sequences for ACE (77).

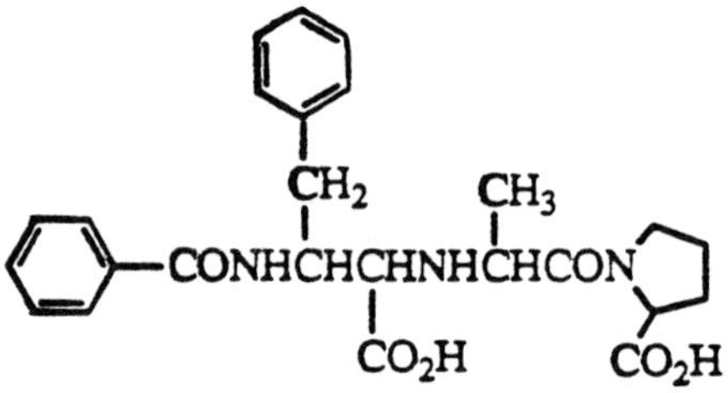

Figure 5. An angiotensin-converting-enzyme inhibitor designed to interact with the enzyme's S_2 subsite.

Extensions of the *N*-carboxyalkyl design to inhibit zinc metallopeptidases in addition to carboxypeptidase A and thermolysin have been made by ourselves and others. In most cases this design appears to be the method of choice, affording stability advantages in respect to sulfhydryl compounds, oral activity advantages in comparison with pentacovalent phosphorus compounds, and better prospects in safety assessment studies relative to hydroxamic acids.

While our studies of ACE inhibitors were being concluded, we briefly investigated *N*-carboxyalkyl dipeptides as inhibitors of neutral endopeptidase 24.11, also called enkephalinase and more recently atriopeptidase. Our best compounds inhibited the enzyme with IC_{50} values at the 10^{-7} M level (*78*). Subsequently *N*-carboxyalkyl peptide inhibitors of this enzyme with IC_{50} values around 10^{-8} M have been designed by groups at Pfizer (*79*) and Schering (*80*). Several of these compounds are currently under study clinically to potentiate the antihypertensive and natriuretic properties of atrionatriuretic peptide (*81*). *N*-Carboxyalkyl peptides have also been synthesized that are inhibitors of collagenase (*82*), of endopeptidase 24.15 (*83*), of the Pz-peptidase (*84*) and of *Pseudomonas* elastase (W. J. Greenlee, Merck Research Laboratories, pers. commun.). Additional applications of the *N*-carboxyalkyl peptide design can be anticipated as new metallopeptidase targets are identified.

Clinical Results

The Phase I clinical testing of enalapril maleate (MK-421, compound **21**), enalaprilat (MK-422, compound **20**) and lisinopril (MK-521, compound **45**) began in 1980. The first study was conducted by H. R. Brunner and his associates in Lausanne, Switzerland. In that study, conducted in healthy volunteers, oral doses of enalapril maleate as low as 2.5 mg inhibited the pressor response of iv injected antiotensin I, whereas with enalaprilat 40-mg doses were required to obtain significant inhibition (*85*). Lisinopril was also effective when given orally, although its onset of action was somewhat slower. Both enalapril maleate and lisinopril reduced angiotensin I's pressor effect 24 hours after a 20-mg dose. These long durations of action and the poor absorption of enalaprilat confirmed the results of animal studies. In addition, the activity of enalapril maleate assured us that prodrug deesterification was also proceeding as it had in animal models. Later studies showed that the absorption of enalapril maleate in humans, based on urinary recovery, was at least 61%, whereas the absorption of lisinopril was at least 29% (*86*). The superior bioavailability of enalapril maleate and the fact that it was ahead of lisinopril in the development process led to its selection as Merck's front-running clinical candidate. Phase III trials of enalapril maleate were initiated as soon as a high degree of efficacy and safety had been established. By the time these trials were completed, in March 1983, more than 40 Phase I and Phase II trials and eight Phase III multicenter trials had been completed, involving over 3,000 subjects (*87*). NDA approval for enalapril maleate (Vasotec) was received in December 1985. The approval of lisinopril (Prinivil, Zestril) came 2 years later, in December 1987, after it had been tested clinically in more than 2,000 patients, including 420 who were also treated with digitalis and/or diuretic therapy for congestive heart failure (*88, 89*). In the United States, both drugs are marketed alone and in combination with the diuretic hydrochlorothiazide for the treatment of hypertension, and Vasotec is indicated in the management of congestive heart failure. Enalaprilat has been approved for iv use when oral therapy for hypertension is not practical and is marketed in the United States as Vasotec IV.

Acknowledgments

The author thanks Drs. M. J. Wyvratt, C. S. Sweet, and R. F. Hirschmann for reviewing this manuscript and for their suggestions concerning it. Barbara Barsnica typed the manuscript and prepared the tables.

Literature Cited

1. Skeggs, L. T.; Dorer, F. E.; Kahn, J. R.; Lentz, K. E.; Levine, M. In *Biochemical Regulation of Blood Pressure*; Soffer, R. L., Ed.; John Wiley & Sons: New York, 1981; pp 3–38.

2. Skeggs, L. T. In *Hypertension and the Angiotensin System*; Doyle, A. E.; Bearn, A. G., Eds.; Raven: New York, 1984; pp 31–45.
3. Fasciolo, J. C. In *Hypertension*; Genest, J.; Koiw, E.; Kuchel, O., Eds.; McGraw-Hill: New York, 1977; pp 134–139.
4. Marshall, G. R.; Vine, W.; Needleman, P. *Proc. Natl. Acad. Sci. U.S.A.* **1970,** *67,* 1624–1630.
5. Pals, D. T.; Masucci, F. D.; Sipos, F.; Denning, D. S., Jr. *Circ. Res.* **1971,** *29,* 664–672.
6. Greenlee, W. J. *Med. Res. Rev.* **1990,** *10,* 173–236.
7. Veber, D. F.; Payne, L. S.; Williams, P. D.; Perlow, D. S.; Lundell, G. F.; Gould, N. P.; Siegl, P. K. S.; Sweet, C. S.; Freidinger, R. M. *Biochem. Soc. Trans.* **1990,** *18,* 1291–1294.
8. Collier, J. G.; Robinson, B. F.; Vane, J. R. *Lancet* **1973,** *1,* 72–74.
9. Gavras, H.; Brunner, H. R.; Laragh, J. H.; Sealey, J. E.; Gavras, I.; Vukovich, R. A. *N. Engl. J. Med.* **1974,** *291,* 817–821.
10. Ondetti, M. A.; Williams, N. J.; Sabo, E. F.; Pluscec, J.; Cleaver, E. R.; Kocy, O. *Biochemistry* **1971,** *10,* 4033–4039.
11. Cheung, H. S.; Cushman, D. W. *Biochim. Biophys. Acta* **1973,** *293,* 451–463.
12. Ondetti, M. A.; Cushman, D. W.; Rubin, B. In *Chronicles of Drug Discovery*; Bindra, J. S.; Lednicer, D., Eds.; John Wiley & Sons: New York, 1983; Vol. 2, pp 1–31.
13. Cushman, D. W.; Ondetti, M. A. In *Progress in Medicinal Chemistry*; Ellis, G. P.; West, G. B., Eds.; Elsevier/North-Holland Biomedical Press: Amsterdam, 1980; Vol. 17, pp 41–104.
14. Ferreira, S. H. *Drugs* **1985,** *30* (Suppl. 1), 1–5.
15. Case, D. B.; Wallace, J. M.; Keim, J. H.; Weber, M. A.; Sealey, J. E.; Laragh, J. H. *N. Engl. J. Med.* **1977,** *296,* 641–646.
16. Erdos, E. G. *Circ. Res.* **1975,** *36,* 247–255.
17. Soffer, R. L. *Annu. Rev. Biochem.* **1976,** *45,* 73–94.
18. Eliseeva, Y. E.; Orekhovich, V. N.; Pavlikhina, L. V.; Alekscenko, L. P. *Clin. Chim. Acta* **1971,** *31,* 413–419.
19. Piquilloud, Y., Reinkarz, A.; Roth, M. *Biochim. Biophys. Acta* **1970,** *206,* 136–142.
20. Cushman, D. W.; Cheung, H. S. *Biochim. Biophys. Acta* **1971,** *250,* 261–265.
21. Skeggs, L. T.; Kahn, J. R.; Shumway, N. P. *J. Exp. Med.* **1956,** *103,* 295–299.
22. Bakhle, Y. S. In *Angiotensin*; Page, I. K.; Bumpus, F. J., Eds.; Springer-Verlag: New York, 1974; pp 413–480.

23. Das, J.; Soffer, R. L. *J. Biol. Chem.* **1975,** *250,* 6762–6768.
24. Byers, L. D.; Wolfenden, R. *J. Biol. Chem.* **1972,** *247,* 606–608.
25. Byers, L. D.; Wolfenden, R. *Biochemistry* **1973,** 2070–2078.
26. Weaver, L. H.; Kester, W. R.; Matthews, B. W. *J. Mol. Biol.* **1977,** *114,* 119–132.
27. Ondetti, M. A.; Rubin, B.; Cushman, D. W. *Science (Washington, DC)* **1977,** *196,* 441–444.
28. Cushman, D. W.; Cheung, H. S.; Sabo, E. F.; Ondetti, M. A.*Biochemistry* **1977,** *16,* 5484–5491.
29. Thorsett, E. D.; Harris, E. E.; Peterson, E. R.; Greenlee, W. J.; Patchett, A. A.; Ulm, E. H.; Vassil, T. C. *Proc. Natl. Acad. Sci. U.S.A.* **1982,** *79,* 2176–2180.
30. Wyvratt, M. J.; Patchett, A. A. *Med. Res. Rev.* **1985,** *5,* 483–531.
31. Ondetti, M. A.; Cushman, D. W. In *Biochemical Regulation of Blood Pressure*; Soffer, R. L., Ed.; John Wiley & Sons: New York, 1981; pp 165–204.
32. Petrillo, E. W.; Ondetti, M. A. *Med. Res. Rev.* **1982,** *2,* 1–41.
33. Atkinson, A. B.; Robertson, J. I. S. *Lancet* **1979,** *ii,* 836–839.
34. Patchett, A. A. et al. *Nature (London)* **1980,** *288,* 280–283.
35. Sweet, C. S.; Patchett, A. A.; Ulm, E. H.; Gross, D. M. In *Proceedings of the Second SCI-RSC Medicinal Chemistry Symposium*; Emmett, J. C., Ed.; Royal Society of Chemists: London, 1984; Special Publication No. 50, pp 36–46.
36. Wyvratt, M. J.; Tristram, E. W.; Ikeler, T. J.; Lohr, N. D.; Joshua, H.; Springer, J. P.; Arison, B. H.; Patchett, A. A. *J. Org. Chem.* **1984,** *49,* 2816–2819.
37. Sweet, C. S.; Gross, D. M.; Arbegast, P. T.; Gaul, S. L.; Britt, P. M.; Ludden, C. T.; Weitz, D.; Stone, C. A. *J. Pharmacol. Exp. Ther.* **1981,** *216,* 558–566.
38. Sweet, C. S.; Arbegast, P. T.; Gaul, S. L.; Blaine, E. H.; Gross, D. M. *Eur. J. Pharmacol.* **1981,** *76,* 167–176.
39. Ulm, E. H. *Drug Metab. Rev.* **1983,** *14,* 99–110.
40. Friedman, D. I.; Amidon, G. L. *Pharm. Res.* **1989,** *6,* 1043–1047.
41. Patchett, A. A.; Cordes, E. H. In *Advances in Enzymology and Related Areas of Molecular Biology*; Meister, A., Ed.; John Wiley & Sons: New York, 1985; pp 1–84.
42. Wyvratt, M. J. *Clin. Physiol. Biochem.* **1988,** *6,* 217–229.
43. Salvetti, A. *Drugs* **1990,** *40,* 800–828.
44. Cheung, H. S.; Wang, F.-L.; Ondetti, M. A.; Sabo, E. F.; Cushman, D. W. *J. Biol. Chem.* **1980,** *255,* 401–407.

45. Matthews, D. M. *Physiol. Rev.* **1975,** *55,* 537–608.
46. Fok, K-F.; Yankeelov, J. A., Jr. *Biochem. Biophys. Res. Commun.* **1977,** *74,* 273–278.
47. Friedman, D. I.; Amidon, G. L. *J. Pharm Sci.* **1989,** *78,* 995–998.
48. Karanewski, D. S.; Badia, M. C.; Cushman, D. W.; DeForrest, J. M.; Dejneka, T.; Loots, M. J.; Perri, M. G.; Petrillo, E. W., Jr.; Powell, J. R. *J. Med. Chem.* **1988,** *31,* 204–212.
49. Zimmerman, M. B.; Barclay, B. W.; Lehman, M.; Norman, J. A. *Clin. Exp. Hypertens.* **1987,** *A9 283,* 461–468.
50. Mendelsohn, F. A. O. *Clin. Exp. Pharmacol. Physiol.* **1984,** *11,* 431–436.
51. Correa, F. M. A.; Plunkett, L. M.; Saavedra, J. M.; Hichens, M. *Brain Res.* **1985,** *347,* 192–195.
52. Correa, F. M. A.; Plunkett, L. M.; Saavedra, J. M. *Brain Res.* **1986,** *375,* 259–266.
53. Sakaguchi, K.; Chai, S-Y.; Jackson, B.; Johnston, C. I.; Mendelsohn, F. A. O. *Hypertension* **1988,** *11,* 230–238.
54. Sakaguchi, K.; Chai, S-Y.; Jackson, B.; Johnston, C. I.; Mendelsohn, F. A. O. *Neuroendocrinology* **1988,** *48,* 223–228.
55. Fyhrquist, F.; Tikkanen, I.; Gronhagen-Riska, C.; Hortling, L.; Hichens, M. *Clin. Chem.* **1984,** *30,* 696–700.
56. Bull, H. G.; Thornberry, N. A.; Cordes, E. H. *J. Biol. Chem.* **1985,** *260,* 2963–2972.
57. El-Dorry, H. A.; Bull, H. G.; Iwata, K.; Thornberry, N. A.; Cordes, E. H.; Soffer, R. L. *J. Biol. Chem.* **1982,** *257,* 14128–14133.
58. Thorsett, E. D.; Harris, E. E.; Aster, S.; Peterson, E. R.; Taub, D.; Patchett, A. A.; Ulm, E. H.; Vassil, T. C. *Biochem. Biophys. Res. Commun.* **1983,** *111,* 166–171.
59. Thorsett E. D.; Harris, E. E.; Aster. S. D.; Peterson, E. R.; Tristram, E. W.; Snyder, J. P.; Springer, J. P.; Patchett, A. A. In *Peptides: Structure and Function,* Proceedings of the 8th American Peptide Symposium; Hruby, V. J.; Rich, D. H., Eds.; Pierce Chemical Co.: Rockford, IL, 1983; pp 555–558.
60. Thorsett, E. D.; Harris, E. E.; Aster, S. D.; Peterson, E. R.; Snyder, J. P.; Springer, J. P.; Hirshfield, J.; Tristram, E. W.; Patchett, A. A.; Ulm, E. H.; Vassil, T. C. *J. Med. Chem.* **1986,** *29,* 251–260.
61. Wyvratt, M. J.; Tischler, M. H.; Ikeler, T. J.; Springer, J. P.; Tristram, E. W.; Patchett A. A. In *Peptides: Structure and Function,* Proceedings of the 8th American Peptide Symposium; Hruby, V. J.; Rich, D. H., Eds.; Pierce Chemical Co.: Rockford, IL, 1983; pp 551–554.

62. Thorsett, E. D. *Actual. Chim. Ther.* **1986,** *13,* 257–268.
63. Lipscomb, W. N. *Acc. Chem. Res.* **1982,** *15,* 232–238.
64. Soubrier, F.; Alhenc-Gelas, F.; Hubert, C.; Allegrini, J.; John, M.; Tregear, G.; Corvol, P. *Proc. Natl. Acad. Sci. U.S.A.* **1988,** *85,* 9386–9390.
65. Bernstein, K. E.; Martin, B. M.; Edwards, A. S.; Bernstein, E. A. *J. Biol. Chem.* **1989,** *264,* 11945–11951.
66. Ondetti, M. A.; Condon, M. E.; Reid, J.; Sabo, E. F.; Cheung, H. S.; Cushman, D. W. *Biochemistry* **1979, 18,** 1427–1430.
67. Colman, P. M.; Jansonius, J. N.; Matthews, B. W. *J. Mol. Biol.* **1972,** *70,* 701–724.
68. Maycock, A. L.; DeSousa, D. M.; Payne, L. G.; tenBroeke, J.; Wu, M. T.; Patchett, A. A. *Biochem. Biophys. Res. Commun.* **1981,** *102,* 963–969.
69. Weaver, L. H.; Kester, W. R.; Matthews, B. W. *J. Mol. Biol.* **1977,** *114,* 119–132.
70. Mozingo, A. F.; Matthews, B. W. *Biochemistry* **1984,** *23,* 5724–5729.
71. Hangauer, D. G.; Monzingo, A. F.; Matthews, B. W. *Biochemistry* **1984,** *23,* 5730–5741.
72. Matthews, B. W. *Acc. Chem. Res.* **1988,** *21,* 333–340.
73. Brunning, P.; Kleemann, S. G.; Riordan, J. F. *Biochemistry* **1990,** *29,* 10488–10492.
74. Chen, Y-N. P.; Riordan, J. F. *Biochemistry* **1990,** *29,* 10493–1048.
75. Roark, W. H.; Tinney, F. J.; Cohen, D.; Essenberg, A. D.; Kaplan, H. R. *J. Med. Chem.* **1985,** *28,* 1291–1295.
76. Greenlee, W. J.; Allibone, P. L.; Perlow, D. S.; Patchett, A. A.; Ulm, E. H.; Vassil, T. C. *J. Med. Chem.* **1985,** *28,* 434–442.
77. Hangauer, D. G. In *Computer-Aided Drug Design, Methods and Applications*; Perun, T. J.; Propst, C. L., Eds.; Marcel Dekker: New York and Basel, 1989; pp 253–295.
78. Mumford, R. A.; Zimmerman, M.; tenBroeke, J.; Taub, D.; Joshua, H.; Rothrock, J. W.; Hushfield, J. M.; Springer, J. P.; Patchett, A. A. *Biochem. Biophys. Res. Commun.* **1982,** *109,* 1303–1309.
79. Northridge, D. B.; Alabaster, C. T.; Connell, J. M.; Dilly, S. G.; Lever, A. F.; Jardine, A. G.; Barclay, P. L.; Dargie, H. J.; Findlay, I. N.; Samuels, G. M. R. *Lancet* **1989,** *ii,* 591–593.
80. Haslanger, M. F.; Sybertz, E. J.; Neustadt, B. R.; Smith, E. M.; Neuchuta, T. L.; Berger, J. *J. Med. Chem.* **1989,** *32,* 737–739.
81. Roques, B.P.; Beaumont, A. *Trends Pharm. Sci.* **1990,** *11,* 245–249.
82. Rich, D. H. In *Comprehensive Medicinal Chemistry: The Rational Design, Mechanistic Study and Therapeutic Applications of Chemical*

Compounds; Hansch, C.; Sammes, P. G.; Taylor, J. B., Eds.; Pergamon Press: Elmsford, NY, 1990; Vol. 2, pp 391–441.

83. Lasdun, A.; Orlowski, M. *J. Pharmacol. Exp. Ther.* **1990,** *253,* 1265–1271.
84. Barrett, J.; Brown, M. A. *Biochem. J.* **1990,** *271,* 70170–70176.
85. Biollaz, J., Burnier, M., Turini, G. A., Brunner, D. B.; Porchet, M.; Gomez, H. J.; Jones, K. H.; Ferber, F.; Abrams, W. B.; Gavras, H.; Brunner, H. R. *Clin. Pharm. Ther.* **1981,** *29,* 665–670.
86. Ulm, E. H.; Hichens, M.; Gomez, H. J.; Till, A. E.; Hand, E.; Vassil, T. C.; Biollaz, J.; Brunner, H. R.; Schelling, J. L. *Br. J. Clin. Pharmacol.* **1982,** *14,* 357–362.
87. Engelhart, J.; Malkin, M.; Rhodes, R. *PM Network* **1989,** *3,* 13–28.
88. Noble, T. A.; Murray, K. M. *Clin. Pharm.* **1988,** *7,* 659–669.
89. Lancaster, S. G.; Todd, P. A. *Drugs* **1988, 35,** 646–669.
90. Thorsett, E. D.; Wyvratt, M. J. In *Neuropeptidase Inhibitors*; Turner, A. J., Ed.; Ellis Horwood: Chichester, England, 1987; pp 229–292.

Flecainide

Elden H. Banitt and Jack R. Schmid
3M Pharmaceuticals

In January 1986, a new, orally effective drug became available in the United States for the management of ventricular arrhythmias. Known as flecainide acetate (Tambocor, 3M Pharmaceuticals), this agent was among the first of a new generation of antiarrhythmics that profoundly affect both intensity and duration of sodium channel blockade in cardiac cell membranes.

Flecainide is a simple benzamide characterized by two trifluoroethoxy groups in the aromatic portion of the molecule and a piperidine ring in the amide side chain (Figure 1). The combination of these two features is essential to its biological activity.

Building the essential structural features of flecainide into a benzamide framework to produce a useful drug was a slow, deliberate process. Although luck and serendipity played a role during some phases of the process, flecainide evolved gradually in a definite stepwise fashion. Its development incorporated many distinctive elements of drug discovery, including a recognized therapeutic need, exploitation of a unique chemical reagent, useful lead compounds, satisfactory models for identification of biological activity, a logical progression of structure–activity relationships, a story that spans many years, and the cumulative contributions of many different individuals. This account describes the origin and evolution of flecainide.

2523–0/93/0163$07.75/0

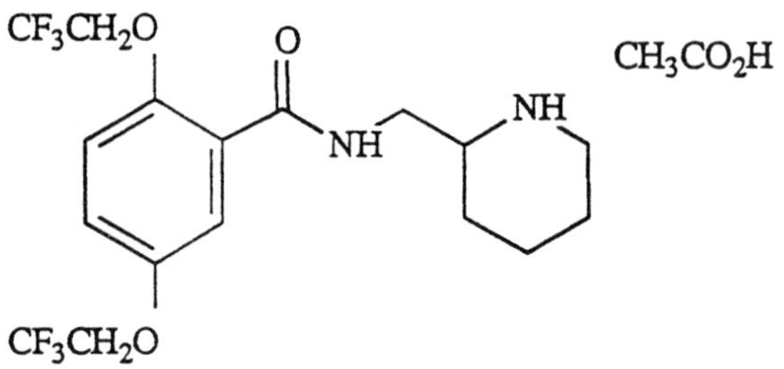

Figure 1. Flecainide acetate.

Need

The Nature of Arrhythmias

"Cardiac arrhythmia" is a very broad and imprecise term. All of the many different disorders of normal heart rhythm are included under this umbrella designation. Rhythm disorders may occur in the atria of the heart as well as in the ventricles. Disorders may involve gross changes in rate (either faster or slower), disturbances in timing, or failures in coordination. Some arrhythmias may even be characterized by combinations of these different irregularities. In a normal heart, cardiac impulses originate in the sinoatrial node, a small strip of tissue that serves as the primary cardiac pacemaker. An impulse originating in the sinoatrial node moves quickly through the atria, stimulates atrial activation, and then spreads rapidly throughout the ventricles by way of a specialized fibrous conduction system. The result is a smooth, coordinated contraction that is repeated 60 to 100 times every minute under normal conditions. The sequence is known as sinus rhythm because of its origin.

The smooth movement of impulses from the sinoatrial node to the ventricles depends on swift transmission of signals from cell to cell. On a cellular basis, the transmission is due to a very rapid movement of charged ions through cell membranes. Cardiac cells have an electrochemical potential across the cell membrane as a result of different relative concentrations of various ions on each side of the membrane. When stimulated, membranes of cardiac cells undergo a sudden change in permeability, allowing a massive surge of cations into the cell. The charge transfer amounts to a very weak electrical signal and is followed by a slower reverse movement to repolarize the membrane. In one way or another, most antiarrhythmic drugs act on the cells of this electrical conduction system by altering the normal functional mechanism or by making it more difficult for aberrant impulses to enter the pathway.

Although arrhythmias may be classified in many different ways, they are generally divided into disorders of impulse formation and disorders of impulse conduction. Formation of impulses normally occurs in the sinus node, but it is possible for abnormal impulses to originate spontaneously at other locations throughout the heart. Such impulses are known as ectopic. Ectopic beats may be superimposed on normal rhythms or may simply appear out of phase. Disorders of impulse conduction can be equally complex. Regular impulses originating in the sinus node spread out through the heart like waves formed by dropping a stone into water. But diseased tissue, congenital defects, or other factors can cause the signal to slow down as it passes through impaired areas. It may even change direction unilaterally or come to a complete stop at some junctions. The net effect of any of these changes in conduction is to interrupt the smooth wavelike progression of a normal heartbeat.

The variety of potential cardiac arrhythmias is clearly quite considerable. Although many of these arrhythmias are relatively harmless, others can be very dangerous. Probably the most common type of arrhythmia is premature ventricular contraction (extrasystoles). Nearly everyone experiences premature beats at times, and usually they are not troublesome. But when extrasystoles are frequent, when they occur in runs, or when they originate from more than one focus, they may presage serious complications. Premature ventricular contractions can progress to more serious ventricular arrhythmias, including tachycardia and fibrillation. Rhythm disturbances may also affect output of the heart and thereby alter blood pressure. Such abnormal changes may affect controlling mechanisms of the heart, which in turn can further complicate the normal cardiovascular function of supplying adequate oxygen and nutrients to all parts of the body. Overstimulation of the heart may lead to congestive heart failure. In addition, rhythm disturbances can easily be recognized by an individual and may lead to considerable emotional stress. Arrhythmias arising in the ventricles can be life-threatening and require prompt and aggressive treatment. Preventive treatment of less serious arrhythmias is usually desirable and may be important in avoiding progression to other, more serious rhythm problems.

Drug Therapy of Arrhythmias

No single medication has safely and effectively controlled arrhythmias in all cases, and solutions to the many problems presented by disorders in heart rhythm over the years have been less than satisfactory. If an ideal drug existed for the treatment of cardiac arrhythmias, it would have several important properties. Foremost among these should be high effectiveness and reliability in treating arrhythmias, because of the broad diversity of both atrial and ventricular rhythm disturbances. It should also be active intravenously (iv) for emergency use as well as orally for long-term prophylactic therapy. In this connection, it should have an extended duration of action to allow a convenient dosing schedule. Like other chronically used drugs, it should be well-tolerated, show a minimum of side effects, and be relatively free of interactions with other drugs. In addition, a lack of active or toxic metabolites is important in controlling dose and limiting toxic effects. None of the available antiarrhythmic drugs completely meets this list of requirements.

During the 1960s, the most commonly used antiarrhythmic drugs were the traditional agents quinidine, lidocaine, and procainamide. Digitalis was also used at this time for treating some types of atrial arrhythmias and is still currently used for certain specific disorders. Adrenergic β-blocking drugs such as propranolol were new, and investigations into their use as antiarrhythmics were just beginning. Although quinidine, lidocaine, and procainamide are effective drugs that are still widely used today, each has certain inherent properties that limit its usefulness. Quinidine, the oldest and one of the most effective antiarrhythmic agents, had been available for nearly 50 years. It was discovered coincidentally in 1914 by a Dutch physician who observed that a malaria patient being treated with cinchona alkaloids containing mainly quinine and quinidine showed improvement in severe rhythm disorders. Four years later it was established that quinidine was the best of the cinchona alkaloids for treating arrhythmias, although all have similar pharmacological effects. It has been in use ever since (*1*). While effective and widely used, quinidine is not well-tolerated for long-term applications because even at relatively small doses it can cause gastrointestinal problems as well as allergic rashes and reactions (*2*).

Lidocaine, a local anesthetic that has been in use since the early 1950s, has become the standard agent for iv applications. For years it has been the agent of choice in coronary care units for treating life-threatening ventricular arrhythmias associated with acute myocardial infarction or cardiac surgery. A rapid onset and short duration of action make lidocaine ideal in this setting, and its increased use has coincided with a reduction in arrhythmia-related mortality among patients hospitalized for cardiac emergencies (*3*). But lidocaine has not found wide use in the prolonged maintenance of patients who have chronic rhythm disorders because it is relatively ineffective in suppressing atrial arrhythmias and it must be given by iv infusion (*4*). In oral form, nearly 70% of lidocaine is metabolized by the liver in a first-pass effect (5).

Procainamide has been used since 1950 in treating ventricular arrhythmias. It was developed as a more stable and longer lasting alternative to the local anesthetic procaine. The cardiac effects of procainamide are quite similar to quinidine. Procainamide is effective orally, but patients must be dosed every 3–6 h. Like quinidine, it causes gastrointestinal and allergic reactions. In many patients procainamide also induces a syndrome resembling systemic lupus erythematosus; the reaction is a common reason for preferring quinidine, with its own set of side effects, to procainamide for long-term treatment (6).

At this time, then, a need clearly existed for improved antiarrhythmic drugs that could be used in oral dosage form over prolonged treatment periods, especially drugs that incorporated a better blend of ideal properties.

Prologue

Fluorochemical Approach

In the mid-1960s, when the story of flecainide properly begins, 3M was just setting into motion a vigorous effort to enter the pharmaceutical business. A small group of chemists and pharmacologists, working as part of Central Research Laboratories, was given the responsibility for new molecule discovery. One of the basic strategic goals of this group was to explore the possibility of new drugs containing fluorine.

The concept of fluorine-containing drugs was not particularly new in and of itself. It is well-known that substitution of fluorine for hydrogen in organic compounds imparts some very distinctive properties to the new compounds. Some of these properties are highly beneficial in terms of drug molecules. In particular, fluorinated compounds tend to have greater lipid solubility, giving rise to different absorption and transport rates in the body. Improved stability is also observed, owing to increased strength of the carbon–fluorine bond. As a result, one might reasonably expect a difference in metabolic stability or pathways of metabolic degradation. A third factor is the subtly altered electronic effects resulting from the strong electronegative properties of fluorine. Changes in the overall electronic pattern of a molecule can affect in a unique manner the way in which it interacts with neighboring molecules. All of these changes occur without grossly altering the overall bulk of the molecule. Because the van der Waals radius of fluorine is only slightly larger than that of hydrogen, the fluorine-for-hydrogen substitution does not change the size or steric shape of the molecule to a significant degree, yet it provides a molecule with considerably different properties (7).

If the idea of fluorine-containing drugs was not new, the basic premise still provided a valuable springboard. The real value in this approach rested with the considerable storehouse of technology and expertise that 3M had accumulated over many previous years of extensive work on nonmedicinal fluorochemical compounds. Much experience had been gained in both processes and materials. It was hoped that some of this information could be tapped and applied to the preparation of new and useful therapeutic drugs.

Hansen Alkylating Agent

Among the many fluorochemical resources available was the unique reagent 2,2,2-trifluoroethyl trifluoromethanesulfonate (**1**). R. L. Hansen, a 3M chemist, reported a convenient synthesis of this reagent in 1965 that made it available in large quantities for the first time (8). Hansen was primarily interested in the fundamental properties of fluorochemicals, and he studied the very powerful alkylating properties of the new reagent in a quantitative way. He found that it reacted with iodide anion in acetone more than 10^5 times as fast as CF_3CH_2Br. Although Hansen may not have anticipated its synthetic value, 2,2,2-trifluoroethyl trifluoromethanesulfonate proved to be an extremely useful tool for transferring the trifluoroethyl group and introducing it into a broad variety of compounds under very mild conditions.

With the advent of Hansen's alkylating agent, many new and unique substances that previously had been very difficult to obtain became readily available. The new and unique substances prepared with **1** eventually included compounds belonging to the flecainide series, and Hansen's reagent was a key factor in all of the synthetic work. Thus the discovery of flecainide, like the discovery of a number of other drugs, was made possible by new advances in basic chemical research.

$$CF_3SO_2OCH_2CF_3$$

1

Local Anesthetic Esters

One of the first to explore the synthetic promise of Hansen's reagent was Art Mendel, a chemist in the 3M Central Research group. Mendel was originally interested in local anesthetics related to procaine, and he prepared (*9*) a series of 2-dialkylaminoethyl esters of benzoic acid (**2**). The new compounds differed from procaine by the presence of one or more trifluoroethoxy groups on the aromatic ring. They were prepared by trifluoroethylation of simple methyl esters of the corresponding hydroxybenzoic acids (**3**) followed by saponification and reesterification. The initial report (*9*) on benzoate esters also included 2-dialkylaminoethyl esters of 1- and 2-naphthoic acids (**4**). From the large group of esters represented by general structures **2** and **4**, three compounds in particular emerged as promising leads (*9*). Each had excellent local anesthetic activity with an extended duration of action based on results from the corneal reflex test in rabbits (*10*). The best compounds were the disubstituted 2-diethylaminoethyl benzoate **5** and the two isomeric monosubstituted 2-diethylaminoethyl naphthoates **6** and **7**.

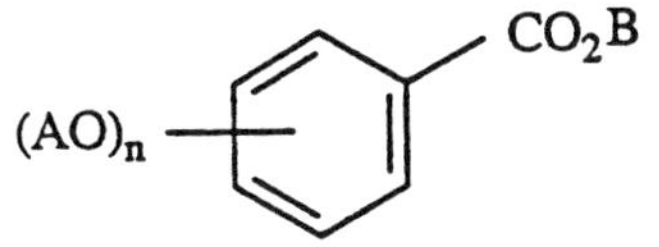

$(AO)_n$ — CO_2B

2: $A = CF_3CH_2$ $B = CH_2CH_2NR_2$

3: $A = H$ $B = CH_3$

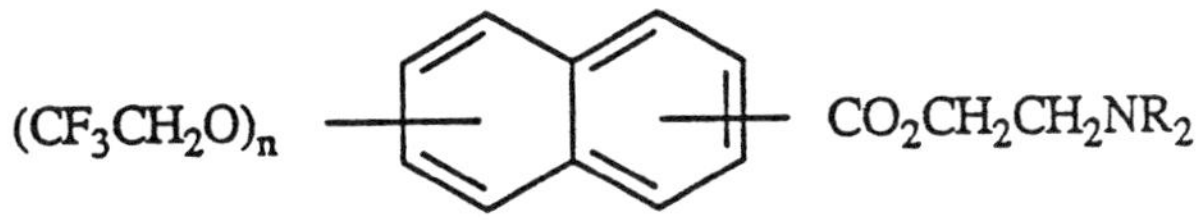

$(CF_3CH_2O)_n$ — $CO_2CH_2CH_2NR_2$

4

Discovery Phase

Antiarrhythmic Amides

A short time after the local anesthetic esters were prepared, Mendel and Coyne (*11*) extended this work and synthesized a series of *N*-(2-dialkylami-

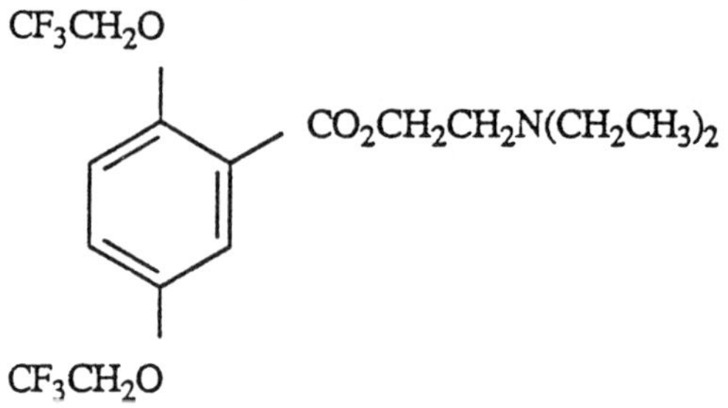

CF_3CH_2O

$CO_2CH_2CH_2N(CH_2CH_3)_2$

CF_3CH_2O

5

A

$CO_2CH_2CH_2N(CH_2CH_3)_2$

B

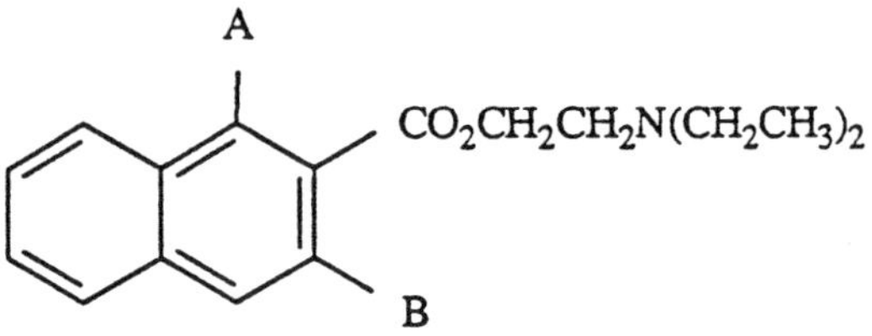

6: $A = CF_3CH_2O$, $B = H$

7: $A = H$, $B = CF_3CH_2O$

noethyl)benzamides and naphthamides corresponding to the original esters. Like the esters, many of these amides exhibited a high degree of local anesthetic activity and could be administered to test animals either by topical application or by injection. But of even greater interest than the local anesthetic effects was the potential application of compounds such as these with improved hydrolytic stability for other therapeutic purposes. J. R. Schmid, a cardiovascular pharmacologist working with Mendel and Coyne, recognized this added dimension and discovered that several of the compounds were active in animal models of arrhythmia at surprisingly low doses. All of the active compounds were benzamides; the *N*-(2-dialkylaminoethyl)naphthamides were significantly less potent (*12*). Among the best was compound **8**, the benzamide analogue of ester **5**.

Changing the link from ester to amide was a simple but fundamental change. The structural analogy is clear. Trifluoroethoxy-substituted benzamide (**8**) is related to ester (**5**) in the same way that the antiarrhythmic agent procainamide (**9**) is related to the local anesthetic procaine (**10**). At this point in time the thrust of our work shifted away from local anesthetics. A strong emphasis was placed on antiarrhythmic research, and the focus was narrowed to benzamides. The decade of the 1960s was just drawing to a close.

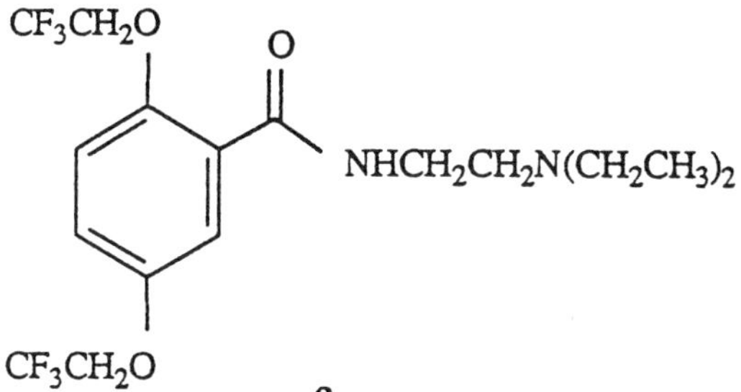

CF_3CH_2O O $NHCH_2CH_2N(CH_2CH_3)_2$ CF_3CH_2O

8

R-799

O $XCH_2CH_2N(CH_2CH_3)_2$ NH_2

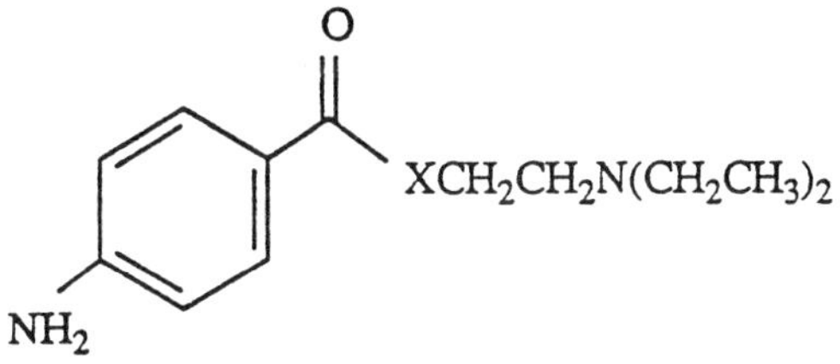

9: Procainamide (X = NH)

10: Procaine (X = O)

Animal Models of Arrhythmia

Rapid Screening. Before we could embark on an extensive search for improved antiarrhythmic agents, a good animal model was needed for the rapid screening of new compounds. Screening of chemicals for antiarrhythmic activity, as opposed to evaluation and final selection for potential use in the clinic, required a model that was easily and rapidly applied, inexpensive to operate, and capable of handling a large number of new compounds in an efficient manner. In addition, this model should detect both oral and parenteral antiarrhythmic effectiveness of a compound, accurately predict the potency of antiarrhythmic activity of experimental compounds relative to antiarrhythmic drugs currently available, and discriminate against inactive or toxic compounds.

Such a model became available in 1968. In this model, mice are exposed to chloroform vapors until cessation of respiration, at which time the chest is quickly opened and the heart is visually inspected for ventricular fibrillation. Prior treatment with effective antiarrhythmic agents will prevent this arrhythmia (*13*). Chloroform-induced ventricular fibrillation in mice probably has little relevance to sudden cardiac death in humans, but it meets the important criterion of rapid screening and identifies compounds with antifibrillatory activity. It was used throughout the antiarrhythmic program for initial evaluation of new compounds and for making basic structure–activity comparisons.

Secondary Dog Models. Compounds that performed well in the mouse protection screen were evaluated more definitively in a variety of canine models. These models had previously been used successfully to identify and evaluate the antiarrhythmic activities of both sotalol (MJ 1999) and verapamil (iproveratril) (*14*). It was reasoned that clinical value could be most reliably predicted by using a variety of antiarrhythmic models, each of which relied on a different mechanism for its production. Any test compound that showed a high degree of activity in all models and was orally active should be a promising candidate.

Induced arrhythmias in dogs are highly reproducible and have sharp end points of either blockade of the arrhythmia or conversion of the arrhythmia to sinus rhythm. They respond well to both orally and parenterally administered antiarrhythmic agents generally used in humans and simulate, at least to some extent, rhythm disturbances often seen in clinical settings. In addition, both atrial and ventricular antiarrhythmic activities can be evaluated with these models.

Four different procedures in dogs were used. Hydrocarbon-epinephrine ventricular arrhythmias were induced in pentobarbital-anesthetized dogs by intratracheally injected *n*-hexane followed shortly (within 15 s) by iv injection of epinephrine hydrochloride. This combination produced arrhythmias that were transient (3–5 min) and could be blocked by iv administered antiarrhythmic agents (*15*). Ouabain tachycardia was induced in pentobarbital-anesthetized dogs by the slow iv injection of ouabain until the arrhythmia began. After the arrhythmia stabilized, antiarrhythmic agents were administered until the arrhythmia was converted to sinus rhythm (*16*). Aconitine atrial arrhythmias were induced in open-chest pentobarbital-anesthetized dogs by direct application of aconitine nitrate to the right atrium. After stabilization of the arrhythmia, antiarrhythmic agents were infused iv at a constant rate until the arrhythmia was converted to sinus rhythm (*17*). Coronary ligation ventricular arrhythmias were induced by the Harris two-step procedure (*18*). Antiarrhythmic agents administered iv to conscious dogs approximately 24 h after coronary ligation converted the arrhythmias to sinus rhythm.

R-799

The lead compound (**8**), also known by the code name R-799, was many times more potent by oral administration in the mouse protection screen than any of the standard antiarrhythmic agents (Table 1). R-799 was also very effective in suppressing the variety of different arrhythmias in all four dog models. Although the results were very promising from a therapeutic point of view, troublesome signs of central nervous system (CNS) stimulation such as tremors, which were observed in some test animals, indicated that the safety margin of R-799 was probably too narrow for it to be considered as a candidate drug. Ninety-day toxicity studies in dogs later confirmed that stimulation of the CNS could indeed be a potential problem even at low doses. Although further development of R-799 was suspended at this time, interest remained strong in the unique family of compounds to which it belonged. As the first member of the trifluoroethoxybenzamide antiarrhythmic series, R-799 became a valuable and significant lead compound. As a prototype, it was the starting point for all the subsequent research aimed at finding a superior new candidate, and it served as a benchmark against which future compounds were measured.

From Lead to Optimum Compound

The allure of R-799 prompted several systematic structure–activity relationship studies designed to identify an improved benzamide candidate. At the outset, it was necessary to define the most favorable pattern of trifluoroethoxy substitution on the aromatic ring. A group of *N*-(2-diethylaminoethyl)benzamides that included the three monosubstituted isomers, all six disubstituted isomers, and several examples bearing three trifluoroethoxy substituents was therefore prepared (*12*). These compounds, together with test data from the mouse protection screen, are collected in Table 1. The reference antiarrhythmic agents quinidine, lidocaine, and procainamide are also included for comparison.

Among the monosubstituted compounds, a trifluoroethoxy group ortho (**11**) or meta (**12**) to the carboxamide function was better than substitution at

$$(CF_3CH_2O)_n-C_6H_{5-n}-C(=O)NH(CH_2)_2N(CH_2CH_3)_2$$

Table 1. Ring-Substituted N-(2-Diethylaminoethyl)-benzamides

Compound	*n*	*Position*	*ED₅₀ (μmol/kg, po)*[a]
8 (R-799)	2	2,5	62
11	1	2	186
12	1	3	137
13	1	4	>540
14	2	2,3	>540
15	2	2,4	>540
16	2	2,6	86
17	2	3,4	>440
18	2	3,5	>440
19	3	2,4,6	117
20	3	3,4,5	>400
Quinidine			217
Lidocaine			495
Procainamide			1,030

[a] Dose that prevented ventricular fibrillation in 50% of the mice.

the para position. Much more dramatic differences were observed among the disubstituted compounds. Only one other compound (**16**, 2,6-OCH_2CF_3) was in the same activity range as the lead compound **8** (2,5-OCH_2CF_3), but these two were very potent compared to the reference agents or other congeners in the series. A trifluoroethoxy group in the ortho position seemed to be a necessary but insufficient component for activity, and the importance of this feature in multisubstituted compounds was further illustrated by the activity of **19** compared to **20**. Although no distinction could be made between the two active compounds **8** and **16** on the basis of mouse data, a comparison of these two compounds in other arrhythmia models showed that **8** was clearly superior in terms of efficacy.

Because the most favorable aromatic substitution pattern in the initial set was 2,5-OCH_2CF_3, a second group of benzamides based on this ring nucleus was prepared to examine the effects of changing the amide side chain

(*12*). It was found that wide variations in the chain were possible without significant reductions in activity so long as the basicity of the amine nitrogen was not drastically altered. When this requirement was met, the amino group could be tertiary, secondary, or even primary. Whereas a two-carbon link between amide and amine nitrogen atoms is a characteristic often found in antiarrhythmic and local anesthetic compounds (*19*), extension of the link to three carbons had no adverse effect on activity.

As the number of compounds slowly increased, it became apparent that all of the best examples had one feature in common. Each was characterized by chain branching adjacent to the amine nitrogen. Most of these branched compounds showed substantially improved potency relative to the lead compound. Several examples are presented in Table 2.

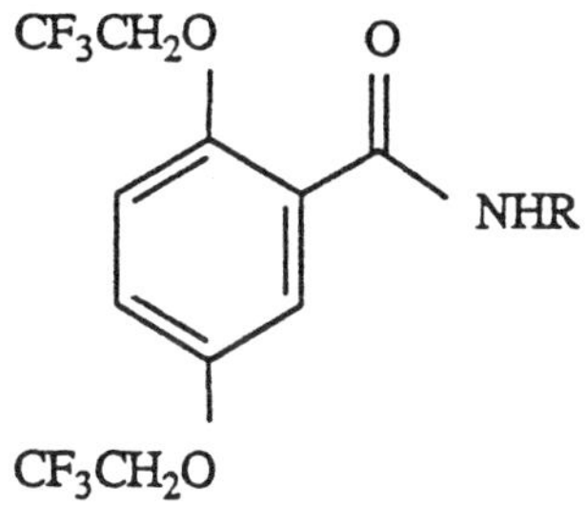

CF_3CH_2O O
NHR
CF_3CH_2O

Table 2. *N*-(Substituted aminoalkyl)-2,5-bis-(2,2,2-trifluoroethoxy)benzamides

Compound	*R*	*ED_{50} (μmol/kg, po)*
8 (R-799)	$CH_2CH_2N(CH_2CH_3)_2$	62
21	$CH_2CH(CH_3)N(CH_2CH_3)_2$	13
22	$CH_2C(CH_3)_2N(CH_2CH_3)_2$	33
23	$CH_2CH_2NHCH(CH_3)_2$	20
24	$CH_2C(CH_3)_2NHCH_2CH_3$	26

In a logical progression of structure modification, the next step involved joining branches of the chain together to produce cyclic structures. Our goal was to retain the advantages observed with branched-chain compounds while exploring the effects of restricted flexibility. It turned out to be a fortuitous combination. The result was a second broad set of benzamides from which flecainide eventually emerged (*20*).

The first example of the new cyclic series was derived in a formal sense from compound **21**. Joining together one ethyl group and the pendant methyl produces a five-membered ring. The actual compound, *N*-ethylpyrrolidine (**25**, Table 3), had an ED_{50} very similar to the original lead compound **8**. Expansion of the five-membered ring to the somewhat more flexible piperidine system resulted in lower activity (**26**). Earlier, however, it had been established that the amino function could be secondary as well as tertiary. This modification led directly to the secondary amine, **27**, which was considerably more active than **26**. Compound **27** eventually became flecainide. It was first synthesized in 1972.

Flecainide was conveniently prepared from 2,5-dihydroxybenzoic acid (**28**) in three steps (*20*). The route, illustrated in Figure 2, is typical of the synthetic methods used to prepare many other compounds belonging to the cyclic series. Trifluoroethylation of **28** produced the activated ester **29**, which could be transformed directly into a variety of benzamides under mild conditions.

Earlier work on ring substitution patterns had demonstrated that the most potent activity was observed with compounds bearing two trifluoroethoxy groups, provided that at least one group was ortho to the carboxamide function. A similar trend was found in the cyclic series, although the variation in potency with substituent changes was somewhat less pronounced. Among *N*-(2-piperidylmethyl)benzamides, the most favorable ring configuration was still 2,5-OCH_2CF_3. Replacement of the 5-OCH_2CF_3 with other substituents such as methyl or halogen resulted in a substantial reduction in activity (*20*).

Although the amine nitrogen in flecainide is part of a heterocyclic ring, a two-carbon link to the amide was found to be the optimum length, whether the amide was bonded to methylene (**27**) or directly to the ring (**31**). Extension of the chain, however, to three or four carbons resulted in marked reduction in activity; shortening of the chain to one carbon atom caused a complete loss of activity. As before, the need for an amino group of relatively strong basic strength was critical. A dramatic illustration of this factor is found in comparing **32** and **33** (Table 4). Compound **32**, with a weakly basic arylamino group, is inactive. But its saturated counterpart (**33**), which differs only in basic strength of the amine nitrogen, is highly active.

Flecainide was not as potent in the mouse protection screen as some of its congeners, or even several of the branched, open-chain compounds from which it evolved. But subsequent studies of flecainide and evaluation in other assays demonstrated that it was effective in controlling a broad spectrum of different arrhythmias and, equally important, was not plagued by the CNS liabilities of the original lead compound.

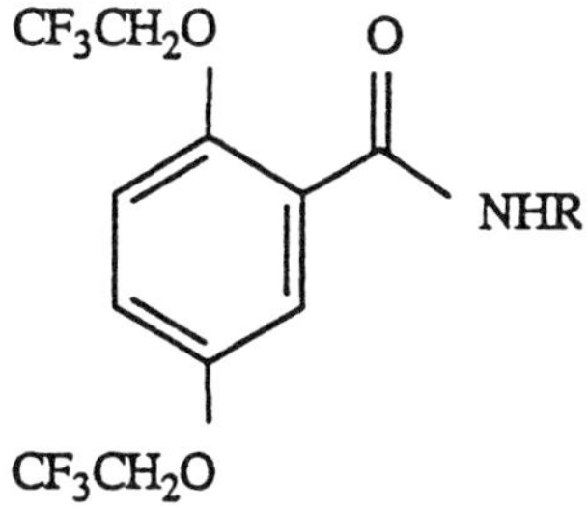

Table 3. *N*-(Heterocyclylmethyl)-2,5-bis-(2,2,2-trifluoroethoxy)benzamides

Compound	R	ED_{50} (μmol/kg, po)
25	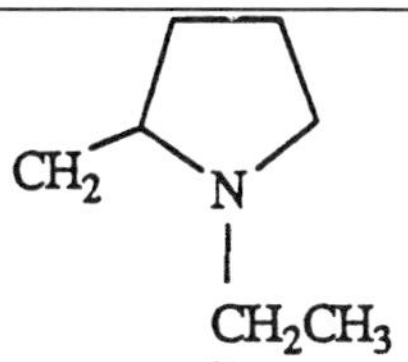 CH_2-(1-ethyl-2-pyrrolidinyl), N–CH_2CH_3	70
26	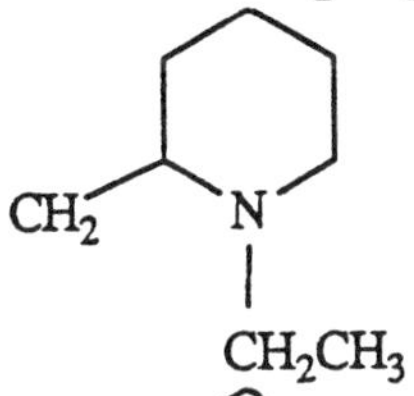 CH_2-(1-ethyl-2-piperidinyl), N–CH_2CH_3	102
27 (flecainide)	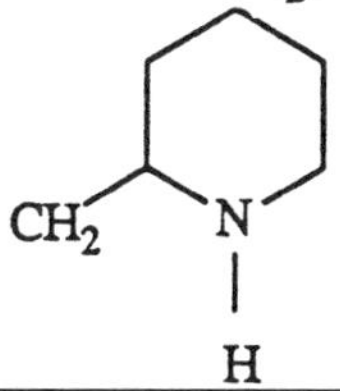 CH_2-(2-piperidinyl), N–H	48

Antiarrhythmic Properties of Flecainide

Efficacy in Dog Models

After satisfactory antiarrhythmic activity had been found in the mouse screen, the broad efficacy of flecainide was established in a battery of dog models and compared with the reference agents quinidine, lidocaine, and

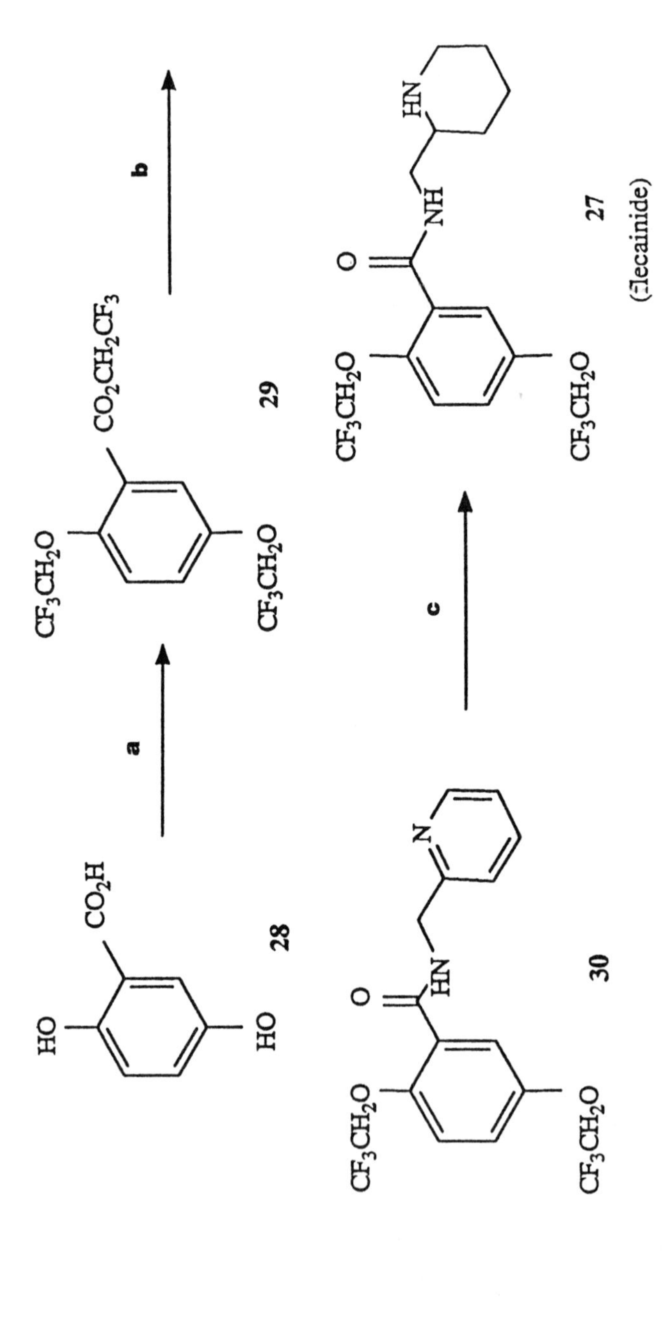

Figure 2. Synthesis of flecainide. **a,** *$CF_3SO_2OCH_2CF_3$, K_2CO_3/acetone;* **b,** *2-aminomethylpyridine/1,2-dimethoxyethane;* **c,** *H_2-PtO_2/acetic acid.*

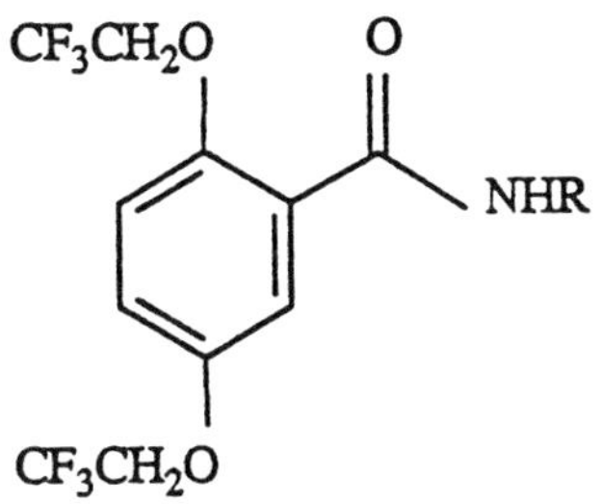

CF_3CH_2O O

NHR

CF_3CH_2O

Table 4. *N*-(Substituted)-2,5-bis(2,2,2-trifluoroethoxy)-benzamides

Compound	*R*	*ED_{50} (µmol/kg, po)*
27	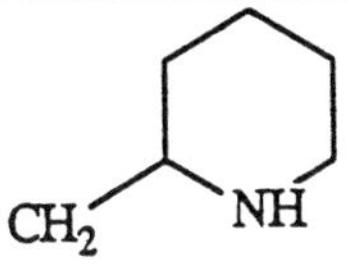 CH_2 NH	48
31	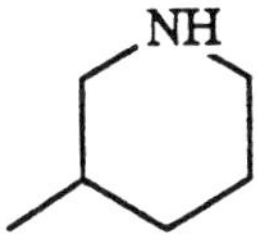 NH	57
32	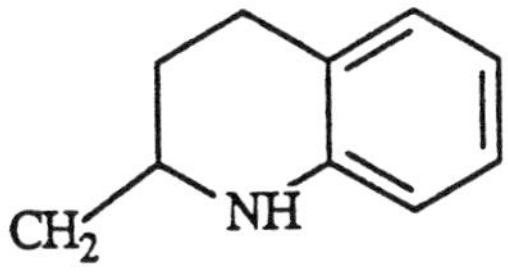 CH_2 NH	>450
33	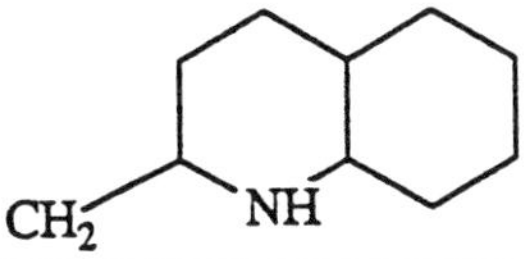 CH_2 NH	27

procainamide. Flecainide was more potent and had greater overall effectiveness in suppressing this spectrum of experimentally induced arrhythmias than any of the reference agents. The average iv dose of flecainide necessary to suppress all arrhythmias was approximately one-third that of the reference agents, and all arrhythmias treated with flecainide were completely suppressed, as shown in Table 5.

Table 5. Antiarrhythmic Activity of Flecainide and Reference Antiarrhythmic Drugs in Dog Models

	Hydrocarbon–Epinephrine		*Ouabain*		*Aconitine*		*Coronary Ligation*	
Drug	*Effectiveness*[a] *(%)*	*Dose*[b] *(mg/kg)*	*Effectiveness*[c] *(%)*	*Dose*[b] *(mg/kg)*	*Effectiveness*[c] *(%)*	*Dose*[b] *(mg/kg)*	*Effectvieness*[c] *(%)*	*Dose*[b] *(mg/kg)*
Flecainide	100	3.4	100	1.0	100	7.2	100	3.2
Quinidine	100	5.3	83	15.8	80	14.1	100	9.2
Lidocaine	60	10.0	100	4.0	57	13.9	60	9.0
Procainamide	100	6.4	57	12.8	20	12.0	40	13.5

NOTE: At least five dogs were tested for each drug in each dog model.
[a] Percentage of dogs whose arrhythmia was fully blocked.
[b] Average total iv dose that produced indicated percent effectiveness.
[c] Percentage of dogs with conversion to sinus rhythm.
SOURCE: Reproduced with permission from ref. 21. Copyright 1984 Cahners Publishing Company.

In addition, the antiarrhythmic activity of flecainide was compared with several newer class I antiarrhythmic agents in controlling ouabain-induced ectopic ventricular tachycardia. Flecainide was the most effective compound in this comparative evaluation (*21*). The oral efficacy of flecainide was established by intraduodenal administration of the drug to conscious coronary-ligated dogs and conscious and anesthetized ouabain-treated dogs.

Electrophysiological Effects

Isolated Tissues. The electrophysiological effects of flecainide have been studied in a variety of isolated tissues and reviewed (*21*). The results of all the studies reviewed show remarkable similarities in electrophysiological effects. Flecainide causes concentration-dependent decreases in both amplitude and maximum rate of rise (V_{max}) of the action potential with very little or no effect on the resting potential. The overshoot of the action potential was decreased. The duration of the action potential was increased in atrial and ventricular muscle, but was either decreased or showed no change in Purkinje fibers. The effective refractory periods of atria, Purkinje system, and ventricles were prolonged. Flecainide had only a slight effect on repolarization. Automaticity in the sinoatrial node and Purkinje fibers was decreased. Decreases in maximum rate of depolarization were marked and were present even at low concentrations of flecainide. All of these effects of flecainide in various isolated cardiac tissues are consistent with inhibition of the fast sodium current and place flecainide in the group of antiarrhythmic agents labeled as class I in the scheme originally defined by Vaughan-Williams (*22*).

The effects of flecainide on the slow inward current system, represented by the force of ventricular contraction and slow channel potentials, showed only minor depression at rather high concentrations of flecainide (*23, 24*). This indicates that flecainide is a weak calcium antagonist that causes a concentration-dependent depression of contractile force in isolated healthy cardiac muscle.

Intact Animals. The electrophysiological effects of flecainide in intact animals were first studied by Hodess et al. (*25*). They found that conduction was slowed in all tissues of the heart, with the greatest slowing observed in the His–Purkinje system and ventricles. The effective refractory period of the atrium and the effective and functional refractory periods of the atrioventricular node were slightly prolonged. These effects were plasma-concentration-related over a plasma level range of 0.4–0.7 μg/mL. Subsequently, clinical studies demonstrated that this plasma concentration range of flecainide was effective in suppressing ventricular arrhythmias.

Flecainide generally increased ventricular fibrillation threshold during supraventricular pacing at plasma concentrations above 1.0 μg/mL. The ventricular fibrillation threshold when measured during premature ventricular beats was increased at plasma concentrations of flecainide below 1.0 μg/mL, indicating that flecainide may be more effective in increasing the ventricular fibrillation threshold in the latter setting. Flecainide had very little effect on sinus node or junctional pacemaker function even at high concentrations, but slowed ectopic atrial and ectopic ventricular pacemaker rates. This slowing was related to plasma concentrations of flecainide. The electrophysiological effects observed in isolated tissues and in intact animals suggested that flecainide should be highly effective in suppressing arrhythmias in the clinic.

Class Assignment of Flecainide. The most widely used system of classification of antiarrhythmic agents is a modification of the system originally proposed by Vaughan-Williams (*26*). This system separates antiarrhythmic agents into four classes based on mechanism of action. Class I agents, such as flecainide, exert their electrophysiological and antiarrhythmic effects by blocking membrane sodium channels. The subtleties of classifying newer antiarrhythmic agents subsequently led Vaughan-Williams (*27*) to further subdivide class I drugs into three separate groups based on intensity and duration of sodium channel blockade. The intensity of sodium channel blockade is reflected in the degree of depression of the rate of rise and amplitude of phase O of the cardiac action potential, in conduction of the cardiac impulse, and in decreased excitability of the heart (*28*). Agents in class IA (e.g., quinidine, procainamide, and disopyramide) have an intermediate intensity and duration of sodium channel blockade. Agents in class IB (e.g., lidocaine, mexiletine, and tocainide) have the lowest intensity and duration. Agents in class IC (e.g., flecainide, encainide, and lorcainide) have the highest intensity and duration. Flecainide is the first of the class IC antiarrhythmic agents.

The electrophysiological and hemodynamic profiles of flecainide in humans were reported by Anderson et al. (*29*) and show clear similarities to results observed in preclinical experiments. Flecainide depresses all parts of the heart, with the strongest effects occurring on the His–Purkinje system. The effects on refractory periods are less pronounced and are primarily limited to the ventricles. Sinus node recovery times are usually unaffected except in sick sinus syndrome, where recovery time is increased. Prolongation of PR and QRS intervals is relatively greater with flecainide than with other antiarrhythmic drugs, but effects on the QT interval are small. Flecainide is well-tolerated hemodynamically. Blood pressure and heart rate are generally unchanged. Flecainide has an intermediate negative inotropic potential compared with other antiarrhythmic agents. In stable patients, the ejection fraction does not change during maintenance therapy. Acute administration to patients with compromised left ventricular function decreases the ejection fraction approximately 10% and is not associated with clinical failure; maintenance therapy with flecainide in this setting can worsen existing congestive heart failure.

Summary Description of Flecainide Studies

Perhaps the most important property of flecainide is its ability to antagonize completely all arrhythmias induced in all four dog models. This favorable profile alone suggested that flecainide would perform well in clinical studies, especially because none of the reference agents demonstrated a similar level of performance. Another significant finding was the outstanding oral effectiveness of flecainide. Small increases in dose above the effective dose generally increased the duration of antiarrhythmic action in all dog models.

The results of electrophysiological studies in both isolated and intact animals showed marked similarities. The most prominent finding was a decrease in conduction in all cardiac tissues and especially in the His–Purkinje system and ventricles. Refractory periods in these cardiac tissues were slightly prolonged. Flecainide had only a slight effect on repolarization and little effect on normal pacemakers, but slowed ectopic pacemakers. An effective iv antiarrhythmic dose of flecainide in anesthetized dogs produced no evidence of significant myocardial depression, no adverse effects on hemodynamics, no alteration of coronary arteriovenous blood gas or plasma potassium concentration, and had no apparent effect on regional blood flow in several vascular beds.

Flecainide is classified as a class IC antiarrhythmic agent on the basis of its marked depressant effect on rate of rise of the action potential and a minor increasing effect on the action potential duration. An increase in QRS duration is considered by some to be a hallmark of the IC class of antiarrhythmic drugs.

Metabolism

The favorable antiarrhythmic effects of flecainide described in the previous section were highly encouraging, but its development into a useful drug probably would not have occurred without equally satisfactory metabolic properties. Comprehensive metabolic studies were performed in humans as well as in laboratory animals, and from these studies a fortunate combination of favorable properties emerged, particularly in the areas of oral absorption, pharmacokinetics, and biotransformation. In many ways the metabolic properties of flecainide turned out to be very compatible with the needs and demands of antiarrhythmic therapy (*30*).

Flecainide appears promptly in plasma when a single oral dose, in either capsule or tablet form, is administered to healthy human subjects. On average, more than 90% is absorbed and delivered to the systemic circulation as unchanged flecainide (*31*). No first-pass effect or presystemic biotransformation of any consequence occurs. Absorption of flecainide is fairly rapid, with a peak plasma level reached within 3–4 h (*31*), and plasma levels of flecainide are proportional to dose within the therapeutic range. No significant changes in these properties were observed in patients with cardiac arrhythmias, renal disease, or congestive heart failure (*30*).

Once absorbed and delivered to the systemic circulation, flecainide is eliminated rather slowly. In healthy subjects, the plasma half-life is about 13 h after a single dose (*31*) and 16 h following multiple doses. For patients with premature ventricular contractions, the half-life increases to approximately 20 h (*30*). These values suggested that flecainide would be suitable for a convenient, twice-daily dosage schedule.

Elimination of flecainide and its metabolites in humans occurs primarily by urinary excretion. A substantial amount of an oral dose is excreted as unchanged drug, although biotransformation produces two major and several minor urinary metabolites. The two major metabolites were isolated from urine and identified as meta-*O*-dealkylated flecainide (**34**) and the meta-*O*-dealkylated lactam of flecainide (**35**) by comparison with samples synthesized independently (*32*). Each metabolite was found in urine in free and conjugated forms. Dealkylation of flecainide in humans takes place exclusively at the meta position, as none of the isomeric ortho compound could be detected. Ring oxidation and phenolic conjugation follow. The two metabolites **34** and **35** were evaluated for antiarrhythmic activity in the ouabain and coronary-ligated dog models. Metabolite **34** had some measurable activity but was only about one-fifth as effective as flecainide. The lactam metabolite **35** had no detectable antiarrhythmic activity even at high doses (*33*). A separate study that compared the electrophysiological properties of the metabolites gave similar results (*34*). Thus the metabolites present in any long-term therapy program would not be expected to potentiate the antiarrhythmic activity of flecainide or contribute to any other pharmacological problems.

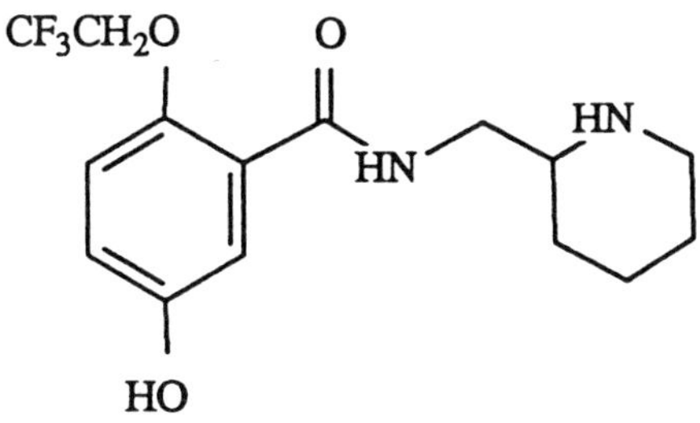

34

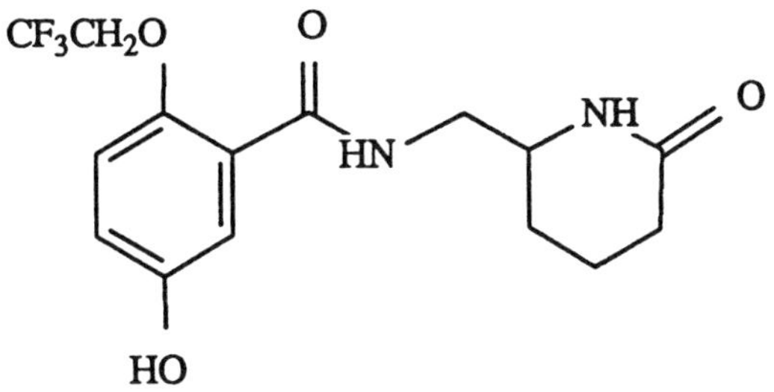

35

Optical Isomers

Flecainide has a chiral center in the piperidine ring. Normally flecainide is administered as the racemate, but optical isomers of a chiral drug may well behave as different pharmacological entities (*35*). In order to study the properties of its individual enantiomers, racemic flecainide acetate was converted to its free base and resolved by fractional crystallization of diastereomeric salts (*36, 37*). Absolute configuration was determined by optical rotatory dispersion techniques; the R and S configurations were assigned to (−) and (+)-flecainide, respectively (*37*).

Animal studies in the mouse protection screen and ouabain dog model revealed no significant differences between the two enantiomers (*36*). Both are highly effective in suppressing ouabain-induced ventricular tachycardias at equivalent doses. In addition, the dog studies did not reveal any differences between enantiomers in their effects on blood pressure or ventricular rate. A comparison of the myocardial effects of the two enantiomers with racemic flecainide in guinea pig isolated driven left atrium showed similar negative inotropism and no evidence of stereoselectivity.

The similarity of the two enantiomers was further confirmed by in vitro electrophysiological studies using isolated canine cardiac Purkinje fibers (*38*). No differences could be detected. The only reported differences in enantiomers occurred in long-term studies with human subjects. In these studies, a selective disposition of enantiomers was observed, leading to a modest but statistically significant predominance of R-(−)-flecainide in plasma. The reason for this enantioselective disposition is not known (*38*). Because both enantiomers are active, the accumulated data suggest no advantage in administering a single enantiomer of flecainide rather than the racemic mixture.

Epilogue

The first clinical studies with flecainide were carried out in Germany during the mid- to late 1970s. Favorable results in terms of both efficacy and tolerance became the basis for a subsequent clinical plan and much broader studies. In extensive clinical studies worldwide, flecainide was found to be effective in the prompt control and long-term prevention of ventricular tachycardias, premature ventricular contractions, and ventricular arrhythmias resistant to other treatments. It was particularly effective and useful in the control of premature ventricular contractions. Among patients with this arrhythmia, more than 90% responded favorably to flecainide, and, in the majority of responders, the arrhythmia was almost entirely suppressed (*29*). Clinical experience also showed that flecainide was relatively free of noncardiac side effects and other complications that limit the long-term use of many known antiarrhythmics. More recent clinical studies (*39*) indicate that flecainide may also find an important role in the treatment of supraventricular arrhythmias, but its use for this indication will necessarily have to wait for more experience with the drug.

The discovery and development of flecainide were made possible by a timely convergence of several key components. The roots of the project were firmly embedded in basic fluorocarbon technology, which made available for the first time many new and unique compounds. But it was the early identification of useful therapeutic activity among these compounds that permitted efforts to focus on the antiarrhythmic area, where a clear need for improved agents existed. A rapid screening method, new at the time, made possible the systematic progression from lead to optimum compound. The discovery of potent oral activity in a variety of arrhythmia models led to broader studies, and in a fortuitous way, the favorable metabolic properties of flecainide strengthened its therapeutic profile. All of these factors, working together, were required to make flecainide available for the treatment and prevention of human arrhythmias.

Acknowledgments

The list of people who played an active role in the successful development and introduction of flecainide is very long. Among those who were heavily involved in the discovery and early development phases we remember in particular chemist W. R. Bronn and pharmacologists B. D. Seebeck, C. W. Henrie, and A. N. Welter. It is a pleasure to acknowledge their important and valuable technical contributions.

Literature Cited

1. Sneader, W. *Drug Discovery: The Evolution of Modern Medicines*; John Wiley & Sons: New York, 1985; p 144.
2. Helfant, R. H. *Bellet's Essentials of Cardiac Arrhythmias*, 2nd ed.; W. B. Saunders: Philadelphia, PA, 1980; p 312.
3. Kniffer, F. I.; Lomas, T. E.; Nobel-Allen, N. L.; Lucchesi, B. R. *Circulation* **1974,** *49,* 264–271.
4. Mason, D. T.; DeMaria, A. N.; Amsterdam, E. A.; Zelis, R.; Massumi, R. A. *Drugs* **1973,** *5,* 261–291.
5. Helfant, R. H. *Bellet's Essentials of Cardiac Arrhythmias*, 2nd ed.; W. B. Saunders: Philadelphia, PA, 1980; p 318.
6. Helfant, R. H. *Bellet's Essentials of Cardiac Arrhythmias*, 2nd ed.; W. B. Saunders: Philadelphia, PA, 1980; p 315.
7. Sheppard, W. A.; Sharts, C. M. *Organic Fluorine Compounds*; W. A. Benjamin: New York, 1969; p 455.
8. Hansen, R. L. *J. Org. Chem.* **1965,** *30,* 4322–4324.
9. Mendel, A. U.S. Patent 3,655,728, 1972.
10. Luduena, F. P.; Hoppe, J. O. *J. Pharmacol. Exp. Ther.* **1952,** *104,* 40–53.
11. Mendel, A.; Coyne, W. E. U.S. Patent 3,719,687, 1973.
12. Banitt, E. H.; Coyne, W. E.; Schmid, J. R.; Mendel, A. *J. Med. Chem.* **1975,** *18,* 1130–1134.
13. Lawson, J. W. *J. Pharmacol. Exp. Ther.* **1968,** *160,* 22–31.
14. Schmid, J. R.; Hanna, C. *J. Pharmacol. Exp. Ther.* **1967,** *156,* 331–338.
15. Garb, S.; Chenoweth, M. B. *J. Pharmacol. Exp. Ther.* **1948,** *94,* 12–18.

16. Lucchesi, B. R.; Hardman, H. F. *J. Pharmacol. Exp. Ther.* **1961,** *132,* 372–381.
17. Scherf, D. *Proc. Soc. Exp. Biol. Med.* **1947,** *64,* 233–239.
18. Harris, A. S. *Circulation* **1950,** *1,* 1318–1328.
19. Buchi, J.; Perlia, X. In *Drug Design*; Ariens, E. J., Ed.; Academic: New York, 1972; Vol. 3, p 243.
20. Banitt, E. H.; Bronn, W. R.; Coyne, W. E.; Schmid, J. R. *J. Med. Chem.* **1977,** *20,* 821–826.
21. Kvam, D. C.; Banitt, E. H.; Schmid, J. R. *Am. J. Cardiol.* **1984,** *53,* 22B–25B.
22. Vaughan-Williams, E. M. In *Symposium on Cardiac Arrhythmias*; Sandoe, E.; Flensted-Jensen, E.; Olsen, K. H., Eds.; AB Astra: Sodertalje, Denmark, 1970; p 449.
23. Schulze, J. J.; Knops, J. *Arzneimittelforschung* **1982,** *32,* 1025–1029.
24. Josephson, M. A.; Ikeda, N.; Singh, B. N. *Am. J. Cardiol.* **1984,** *53,* 95B–100B.
25. Hodess, A. B.; Follansbee, W. P.; Spear, J. F.; Moore, E. N. *J. Cardiovasc. Pharmacol.* **1979,** *1,* 427–439.
26. Vaughan-Williams, E. M. *J. Clin. Pharmacol.* **1984,** *24,* 129–147.
27. Vaughan-Williams, E. M. In *Progress in Pharmacology*; Gustav Fischer: Stuttgart, 1979; Vol. 2/4, pp 13–23.
28. Anderson, J. L.; Bigger, J. T.; Laidlaw, J.; Morganroth, J. *Pract. Cardiol.* **1987,** *13,* 108–116.
29. Anderson, J. L.; Stewart, J. R.; Crevey, B. J. *Am. J. Cardiol.* **1984,** *53,* 112B–119B.
30. Conard, G. J.; Ober, R. E. *Am. J. Cardiol.* **1984,** *53,* 41B–51B.
31. Conard, G. J.; Carlson, G. L.; Frost, J. W.; Ober, R. E. *Clin. Pharmacol. Ther.* **1979,** *25,* 218.
32. McQuinn, R. L.; Quarfoth, G. J.; Johnson, J. D.; Banitt, E. H.; Pathre, S.V.; Chang, S. F.; Ober, R. E.; Conard, G. J. *Drug Metab. Dispos.* **1984,** *12,* 414–420.
33. Investigator's Brochure, R-818 (flecainide acetate, Tambocor), Oral and Parenteral Antiarrhythmic, July 1983, on file at Riker Laboratories, Inc., St. Paul, MN.
34. Guehler, J.; Gornick, C. C.; Tobler, H. G.; Almquist, A.; Schmid, J.R.; Benson, W.; Benditt, D. G. *Am. J. Cardiol.* **1985,** *55,* 807–812.
35. Simone, M. *Med. Res. Rev.* **1984,** *4,* 359–413.
36. Banitt, E. H.; Schmid, J. R.; Newmark, R. A. *J. Med. Chem.* **1986,** *25,* 299–302.
37. Blaschke, G.; Scheidemantel, U.; Walther, B. *Chem. Ber.* **1985,** *118,* 4616–4619.

38. Kroemer, H. K.; Turgeon, J.; Parker, R. A.; Roden, D. M. *Clin. Pharmacol. Ther.* **1989,** *46,* 584–590.
39. Anderson, J. L.; Jolivette, D. M.; Fredell, P. A. *Am. J. Cardiol.* **1988,** *62,* 62D–66D.

Esmolol

Paul W. Erhardt
Berlex Laboratories

The esmolol story began in a two-man laboratory that in 1976 constituted the entire chemical R&D effort of a rather humble Arnar-Stone Laboratories. Nearby, Abbott and Searle loomed as legitimate pharmaceutical players on plush campuses. Whatever we lacked in size, however, we more than made up for with an entrepreneurial spirit, a spirit that dated back to our founders. For Arnar and Stone, it was said, had managed to discover a local anesthetic concoction while conducting formulation research in the corner of a garage.

Besides, Arnar-Stone Laboratories had just been adopted by American Hospital Supply Corporation, thereby gaining a strong financial parent as well as a specialized sales force tailored for the hospital audience. Eventually our name was changed to American Critical Care, reflecting this affiliation and our unique mission to provide intravenous (iv), critical care therapeutics. Thus, a legitimate stage had been set for our small cast of spirited researchers to soon play roles in the discovery of esmolol, a clinically useful, ultra-short-acting β-adrenergic-receptor-blocking agent.

2523–0/93/0191$06.00/0

Rationale and Early Designs

The concept of an ultra-short-acting β-blocker was first brought to our attention by John Zaroslinski, then Vice President of Research, who speculated that a drug of this type could be given iv to attenuate the high adrenergic tone that can exacerbate an acute myocardial infarction (MI). Its function would be analogous to shielding the heart during an adrenergic storm. As the storm subsided, the iv flow could be halted, and the drug's built-in lability would quickly terminate its action. At that time, no studies substantiating that β-blockers could reduce cardiac damage and mortality subsequent to an MI had been reported. In fact, the use of β-blockers in this setting was contemplated only with considerable hesitancy because of the potential for inducing cardiac failure and death (*1*). Reservations were further heightened by the long durations of action that many pharmaceutical houses had striven for in their antihypertensive β-blockers.

An ultra-short-acting β-blocker therefore seemed an ideal research target. Such a drug would enhance the position of American Critical Care in both the cardiovascular and critical care arenas because, at the time, every large company wanted to have a β-blocker in its product portfolio and because a short-duration version would be ideally suited for critical care use in hospitals. The therapeutic notion seemed almost too good to be true, so the Medicinal Chemistry department quickly took up the task of converting this concept into an actual drug molecule.

Bob Borgman, the Director of Medicinal Chemistry, suggested that we could take two approaches toward the design of candidate molecules: polar compounds that might be rapidly excreted as such, and ester-containing structures in which the ester is placed between the aryl and amino pharmacophores that are requisite in all prototypical β-blockers (e.g. propranolol, **1**). When the ester linkage is hydrolyzed, the two pharmacophores separate and become inactive. Because esterases are ubiquitous in the body, the second possibility was more intriguing, although I had some reservations pertaining to the key distance between the two pharmacophores that this approach seemed to violate. Our initial approaches toward the design of ultra-short-acting β-blocking compounds are illustrated in **2**, **3**.

Borgman decided to set up an extramural collaboration to have some initial structures made quickly. He also gave Dave Stout, my fellow chemist in our two-man laboratory, the task of exploring ways to put the ester-type β-blockers together. At the time I was heavily engaged in synthetic work with a pair of *E*- and *Z*-cyclopropyl dopamine analogues.

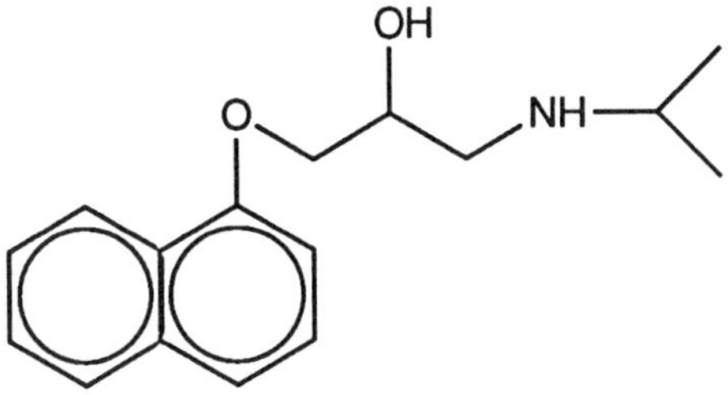

OH

O NH

1 (Propranolol, a prototype β-blocker with a long duration of action)

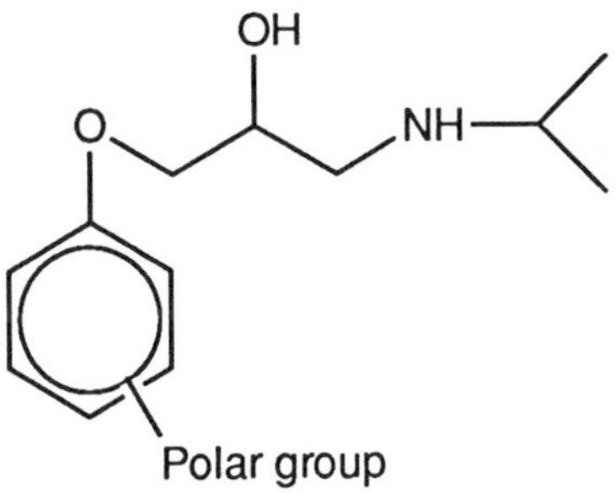

OH

O NH

Polar group

2 (Active polar analogue that might be rapidly excreted)

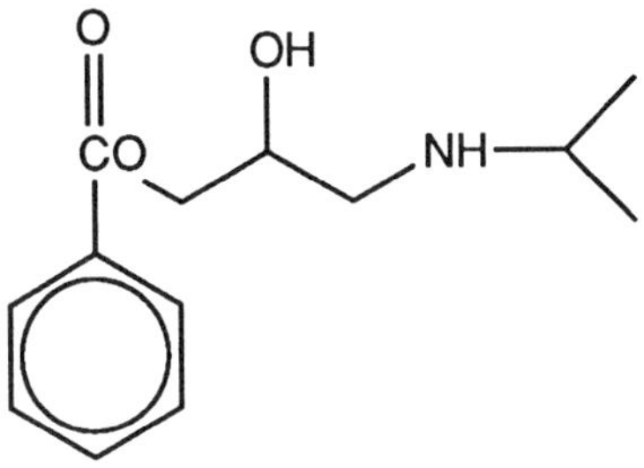

O

OH

CO NH

3 (Active internal ester analogue that would be inactivated upon hydrolysis)

Impasse

After approximately 6 months, several polar β-blocking compounds had been made by our external partner, whereas the ester approach had bogged down in a variety of unwieldy syntheses attempted both in-house and externally. The polar compounds produced weak β-blockers with only somewhat shortened durations of action, particularly when determined after 3-h infusions (unpublished data). Therefore, this approach was abandoned. Later, after American Critical Care had established that other chemical approaches did validate the concept, and after a patent strategy had been formulated, the initial polar β-blocker effort was cleared for publication (2).

Stout, along with Larry Black, our new support staff in chemistry, had taken a few shots at the esters, but to no avail. Our efforts had been sporadic, however, because Stout and Black had also been busy synthesizing compounds for our dopamine program as well as some new antiarrhythmic agents to replace bretylium. I was beginning the final synthetic work with the cyclopropyl dopamine targets and had not been able to devote any time to the ultra-short-acting β-blockers. Thus, for one reason or another, everyone in the chemistry group was beginning to experience some degree of frustration.

External versus Internal Esters

Another 6 months passed, and still there were no short-acting esters. Frustration had now grown to the point of discomfort. However, an alternative solution to the ester approach had begun forming in my mind. I finally mentioned it to Rick Gorczynski, a pharmacology colleague, while we were enjoying a couple of beers at my house after work on a hot summer's evening. I suggested that there may be a different way to construct short-acting β-blockers. I thought of the compounds as "external esters," as opposed to the "internal esters" that we had been trying to make thus far. I explained that one shouldn't have to make internal esters to get inactive fragments after hydrolysis. If esters were simply placed close to either the amine or aryl pharmacophores, then, when the compounds were hydrolyzed, the resulting carboxylic acids would be ionized at physiological pH. This full-blown anionic charge would be too foreign to be recognized by the β-adrenergic receptor. In other words, the resulting anions could be used to mask, rather than separate, the desired amino group or the lipophilic aromatic system. Even if the full-blown anionic charge were relieved somewhat by the aqueous environment, the β-receptor, when presented with a solvated cluster, should still reject these species.

External Esters Advance

Despite this new insight, the esmolol story nearly came to an end at our next research meeting. The meeting was attended by seven investigators—the entire senior staff of Biology and Medicinal Chemistry—and chaired by Bob Lee, the new Vice President of Research. Because Bob Borgman had moved into Process Chemistry and the director of Biology had moved into Toxicology, neither Medicinal Chemistry nor Biology had managers at the time.

Lee began by reminding us that a clock was ticking. We had been working on ultra-short-acting β-blockers for over a year, with essentially no definitive results. Medicinal Chemistry had delivered only a few compounds, none of short duration. He then suggested there might be a logical inconsistency in attempting to synthesize a compound that is supposed to fall apart. Were we trying to put a square peg in a round hole? As we discussed the situation, we realized that most short-duration therapeutics, such as our dopamine product, were derived from natural sources, and that there was little precedent for the design and synthesis of such agents. On the other hand, we knew that short-duration compounds were undesirable to most companies, and consequently there had been little industry effort in that direction. Therefore, we would not be able to assess the potential for achieving our goal based on an analysis of the prior art.

But Lee posed another negative question: Could we afford to spend more time in an unprecedented area that drew on speculative chemical considerations? This seemed to be the final strike for the ester approach. In fact, the entire ultra-short-acting β-blocker program appeared to have its head on the block, with the guillotine poised to drop.

Although I had only begun to explore the external esters, I decided to place the idea before the group. I indicated that a couple of other structural approaches should be tried in order to fully assess the chemical feasibility of our program. These new approaches involved what I called external esters, and they offered considerably more flexibility than the internal esters in terms of both designing compounds and being able to try different synthetic methods. Drawing on the analogy of square pegs and round holes, I suggested that instead of trying to modify peg or hole, we could use this relationship to advantage. If we were to place ester groups near the pharmacophores of the β-blockers, once hydrolyzed the ester groups might become the square pegs and prevent the pharmacophores from additional interaction with the round, β-receptor holes. Lee agreed to extend the program an additional 3 months. My time and that of Sheung-Tsam Kam, a new senior chemist in our group, were to be devoted entirely to preparing several external and internal ester compounds, respectively.

Synthesis of External Ester Compounds

The next experiments probably represented the last chance for developing ultra-short-acting β-blockers, and the clock was ticking loudly. Within 2 months I had synthesized several representative compounds having ester functions on the nitrogen alkyl substituent and on the aromatic nucleus of the typical β-blocker framework. In most cases the corresponding acids were prepared for comparison. The chemistry associated with much of this work "followed well-trodden paths" (3). The three types of structures that I prepared initially are shown in **4–6**.

For the first time, ester structure-activity relationships were possible. Front-line testing was conducted by Bill Anderson, who employed two in vitro screens using guinea pig right atria and trachea, respectively, to assess β_1- and β_2-blocking potencies. Reasonably potent β-blockers were then studied by Gorczynski in an in vivo dog model to determine their durations of action after 40-min and 3-h infusions. Based on the behavior of ultra-short-acting standards such as dopamine ($t_{1/2}$ approximately 10 min), Gorczynski had devised the 3-h protocol as a reasonable pseudo-equilibrium or steady state that helped to rule out time-dependent distribution and compartmentalization short-duration artifacts.

For the N-external esters, β-blocking potency appeared to be highest when the ester was located beta to the amine—for example, when n equaled 2. This finding was similar to that reported for several structurally related amides (4). Important to the concept, the N-external acids were found to be completely inactive in the front-line screen. As I had hoped, the round holes were not accepting the square pegs. For n equaling 3 or 4, the free-amine esters were found to undergo spontaneous cyclization to their lactams. Therefore, a corresponding δ-lactam was prepared and, not unexpectedly, was found to be inactive as a β-blocker. Thus, these esters potentially provided a second, nonenzymatic pathway for their inactivation, which might be beneficial in extreme disease states if esterase function was impaired.

An Ortho-Methyl Framework

Although our progress was exciting, problems had also appeared as we tested many of our candidate compounds. Anderson complained about the intrinsic

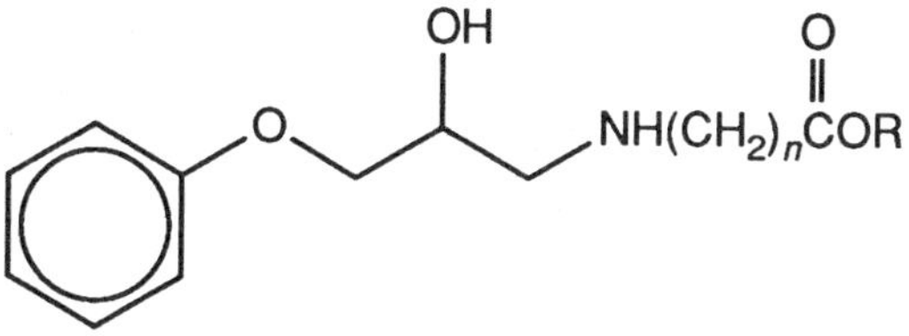

4 (N-external esters)

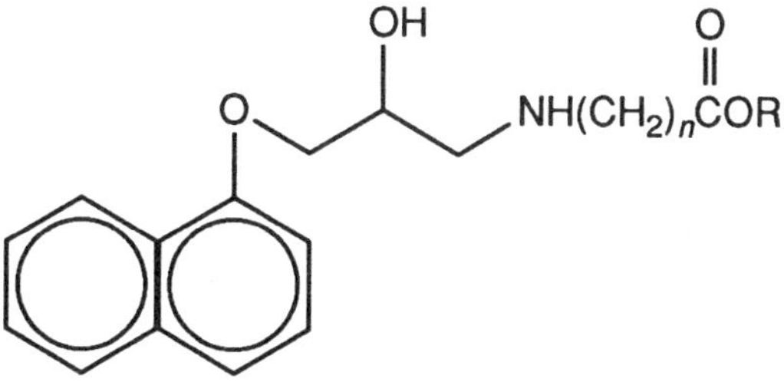

5 (Propranolol-like N-external esters)

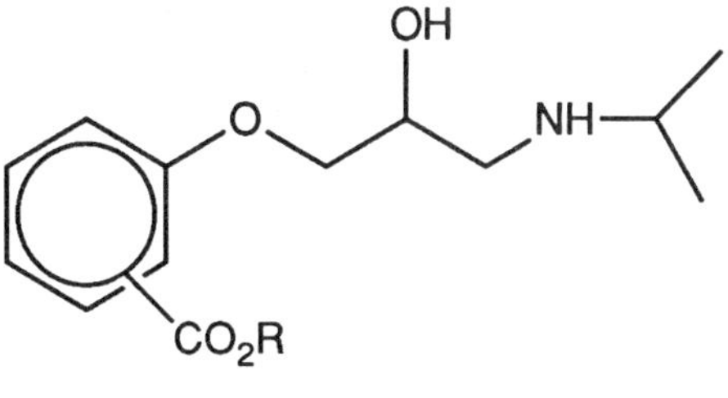

6 (Aryl external esters)

activity that plagued his data analysis for the entire phenoxypropanolamine N-external ester series and about the poor solubility exhibited by the propranolol-like compounds. Therefore, I decided to optimize a β-blocker nucleus for the in vitro models so that we could attach our esters on that framework and proceed without these concerns. Close examination of the extensive literature on β-agonists and β-blockers uncovered two citations in which intrinsic activity had been observed for other aryl-unsubstituted phenoxypropanolamine systems (*4, 5*). Interestingly, substitution anywhere on the aromatic ring with just about anything seems capable of removing the intrinsic activity problem.

At this time Chi Woo joined my group as an assistant chemist. The nonester (nonlabile), aryl-substituted standards seemed an ideal set of compounds for him to prepare as his first assignment. Within 2 weeks he had completed synthesis of this group of about ten standards. In addition, I had made enough intermediate epoxide in many of the cases to allow him to attempt several reactions with ethyl β-alanate. Woo delivered several examples of the $n = 2$, N-external ester targets as well.

None of our standards exhibited intrinsic activity. Picking the best blend of potency and aqueous solubility led to the ortho-methyl system shown in **7**. This system was then utilized as a constant frame from which Woo and I explored variables associated with the N-external esters such as connecting moiety (CM) length and type, ester leaving groups (LG), and multiple (M) external ester arrangements. Although much of our chemistry continued to follow well-trodden paths (3), some extremely practical methods that exploited the ability of "migrant" benzyl groups to control the extent of amine alkalytions were delineated and reported (6) as part of this overall effort (7).

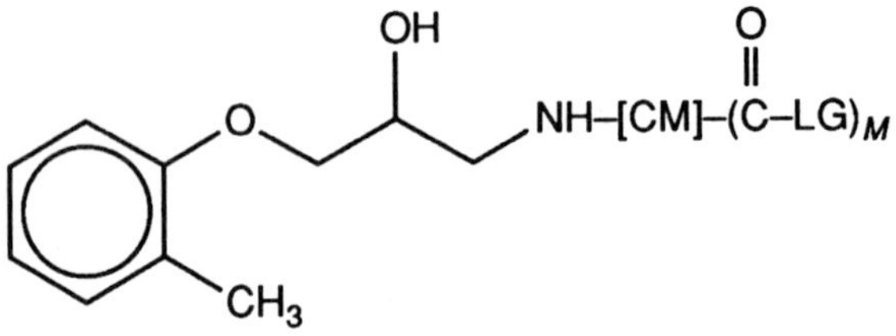

OH
O
NH–[CM]–(C(=O)–LG)$_M$
CH_3

7

In vitro studies confirmed that the acids resulting from hydrolysis were inactive, and both in vitro and in vivo studies confirmed that the parent ester molecules could be designed to retain potent β-blocking properties. In addition, short-duration compounds had finally been obtained. For example, after a 40-min infusion, propranolol exhibited an approximately 45-min duration of activity to 80% recovery (t 80) from 50% blockade of an isoproterenol-induced tachycardia, whereas a typical N-external ester (e.g., CM = $-CH_2CH_2-$, LG = $-CH_2CH_3$, and M = 1) exhibited an approximately 15-min t 80 duration. However, with longer dosing, such as a 3-h infusion, none of the N-external esters exhibited the desired 10-min ultra-short duration of action.

Aryl External Esters

Concurrent with preparation of the initial unsubstituted phenoxypropanolamine N-external esters, I had prepared several external aryl esters and acids as well. At this point, these compounds seemed to be at about the same level of success as the N-external esters. The acids were inactive as β-blockers in vitro, whereas their ester parents were able to retain activity both in vitro and in vivo. In addition, the structure–activity relationships for the esters paralleled what would be expected from the β-blocker literature. For example, substitution in the ortho position provided very potent and nonselective compounds (8), whereas esters in the para position yielded compounds that were less potent but tended to exhibit selectivity for β_1- over β_2-receptors (9). Durations for an ortho-methyl ester, a para-methyl ester, and a para-ethyl ester after 40-min infusions were about 15, 30, and 60 min, respectively.

Molecular Architectures

Although these results were encouraging, the 3-month extension of the ultra-short-acting β-blocker program had just elapsed, and we had not produced a usable compound. We needed a compound that was closer to our goal, and we needed it immediately. At issue was the most efficient way to proceed toward optimizing the initial findings.

Aesthetic Symmetry

From the outset, we had recognized the virtues of a cardioselective β-blocker in the various cardiac indications envisioned for our drug. However, because N-substituents can also bestow cardioselectivity, it was more the aesthetic appeal associated with the symmetry of the para-substituted aromatic ring, rather than its ability to endow cardioselectivity, that caused me to focus on this system. I was also intrigued by the shorter duration of the methyl ester in comparison with its ethyl ester, and I wondered if simple steric effects could explain the difference of 30 min. But how should one analyze this complex in vivo metabolic situation? How should the key structure–activity relationships be derived?

A Black Box

Because we did not have time to study several candidate enzymes separately, and there was no way of telling which system, in the end, might be the best one to exploit, we decided to take an alternative approach that also seemed to make sense. We simply lumped together all of the in vivo hydrolytic systems that might be attacking our compounds, because this condition is what a drug would eventually see anyway. Thus, by regarding the situation as a single black box of esterases, we were able to produce a net duration data point for each compound. This approach, in turn, allowed us to explore structure–activity relationships within a specific series of modifications. While we had used an in vivo model to validate the concept, now we would use it to refine our notions and fine-tune our compounds. The testing would have to be done quickly.

Extended Methyl Esters

My formal training in medicinal chemistry had taught me how to design and synthesize drugs to cause changes in the body, but it was from Robert Smith, my postdoctoral mentor in the area of drug metabolism, that I gained practical experience on how the body can cause changes in drugs. Fortuitously, Smith was in town on business, and he joined me for dinner at my home. To facilitate open conversation, he signed a confidentiality agreement. When I asked him what he thought about extending the para-carboxy group away from the bulky aromatic ring and using a simple methyl ester to obtain an ultra-short-duration compound, his response was favorable, and he even suggested that I extend the ester by more than just one methylene unit to ensure that I fully test my idea. I had already ordered these starting materials, and with his additional encouragement I was hoping to quickly produce both the methylene- and ethylene-extended methyl esters. Because Lee had called another research meeting, now only 2 weeks away, these compounds would represent our last shots in the ultra-short-acting β-blocker program.

Synthesis of ASL-8052

Once the starting materials arrived, I immediately synthesized the key steric probes shown in **8**. In vitro and in vivo testing were undertaken in an attempt to provide data for the next research meeting. The biological studies indicated that the durations after 40-min infusions were reduced in linear fashion with the insertion of each methylene unit: *t* 80 values were 31, 20, and 12 min, respectively, for the $n = 0$ (parent system that I had prepared earlier as part of the first set of *o*, *m*, and *p*-aryl external esters), $n = 1$, and $n = 2$ cases. Most remarkable, however, was that the 12-min duration observed for the $n = 2$ case (ASL-8052) was retained in the 3-h infusion protocol. Because a thorough search of the β-blocker patent literature had not revealed structures of this type, we were excited by the possibility that this simple steric probe might be the compound we were looking for. When the results of the biological studies with ASL-8052 became available, the research meeting was no longer needed.

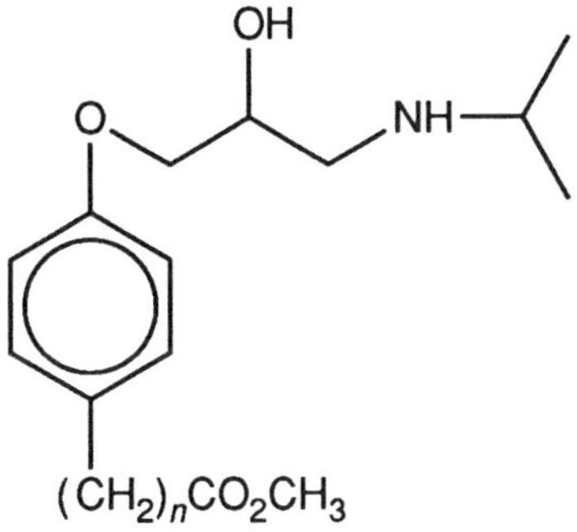

$(CH_2)_nCO_2CH_3$

8 $n = 1, 2$ (ASL-8052)

As we had for the N-external esters, Chi and I spent the next several months exploring variables associated with the aryl external esters. None of this additional work (*10, 11*), however, produced a compound superior to ASL-8052. The next higher homologue ($n = 3$) exhibited a duration of 40 min, an abrupt departure from the straight line observed for the lower level homologues. Thus, ASL-8052 seemed to represent the best compound in that neither of its immediately larger or smaller homologues was as short-acting. Likewise, various multiple esters were found to offer little, if any, improvements in duration. Finally, providing esters that contained good leaving groups failed to benefit either in both the N- and aryl-external ester series. The synthetic scheme that we employed to prepare ASL-8052 is shown in Scheme 1.

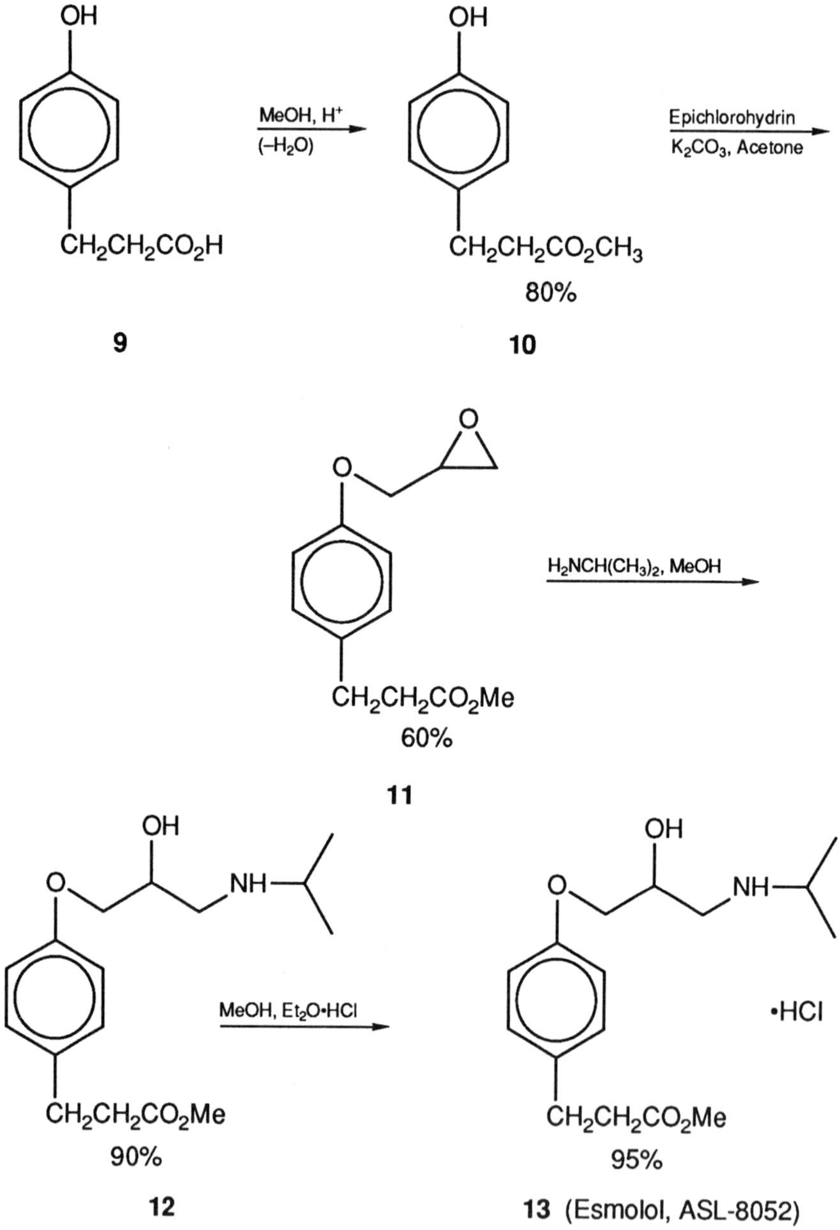

OH

$CH_2CH_2CO_2H$

9

MeOH, H^+
$(-H_2O)$

OH

$CH_2CH_2CO_2CH_3$
80%

10

Epichlorohydrin
K_2CO_3, Acetone

O

O

$CH_2CH_2CO_2Me$
60%

11

$H_2NCH(CH_3)_2$, MeOH

OH
O
NH

$CH_2CH_2CO_2Me$
90%

12

MeOH, Et_2O•HCl

OH
O
NH

•HCl

$CH_2CH_2CO_2Me$
95%

13 (Esmolol, ASL-8052)

Scheme 1

Return to Internal Esters

Kam had maintained his effort on the internal esters with diligence and, through sheer persistence, had finally provided the key internal ester system that had escaped our group for so long. A representative compound from his initial series is shown in **14**. With an example of this class of esters in hand, we soon determined that one of the stability problems was due to an intramolecular cyclization between its secondary amine and ester carbonyl moiety. By this time I had become extremely conscientious about the advantages and disadvantages of steric factors during chemical synthesis (*6*) and during catabolism by enzymes (*10*), respectively. I suggested to Kam that he might try employing a bulky N-alkyl substituent to hinder the problematic attack by the amine in his system. In addition, such functionality actually tends to increase blocking potency (e.g., α-gem-dimethyl substitution with tertiary-butyl versus isopropyl). Even though this suggestion required another run through a tedious synthesis, Kam immediately embarked on this

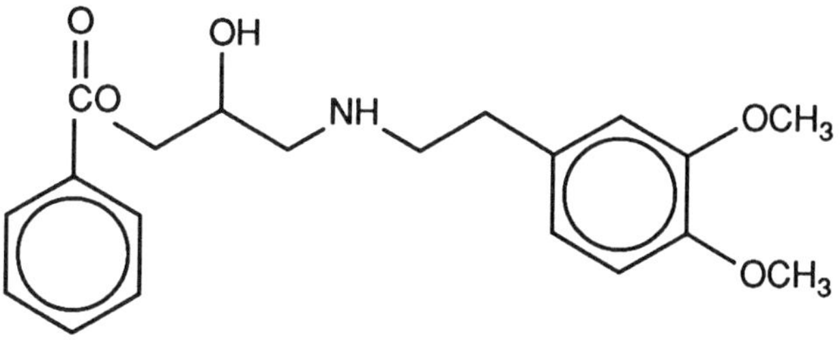

14

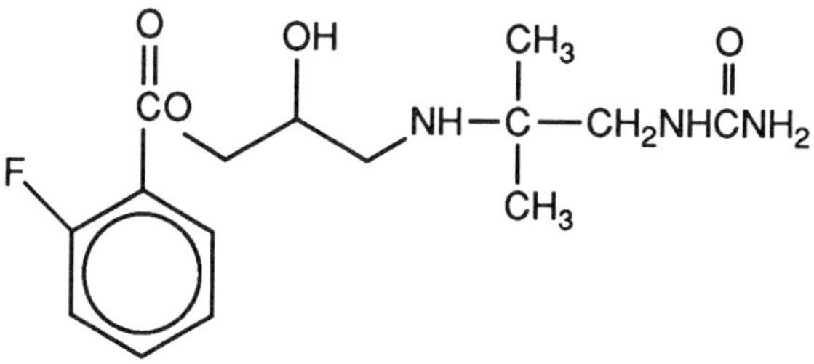

15

effort and completed it in short order. The second-generation, internal ester **15**, eventually became the backup compound for ASL-8052 (*12, 13*).

Patenting and Clinical Development

Despite the small size of the chemistry section, we were provided with excellent resources in the patenting area. The chemists largely wrote the patents, but worked closely with an attorney who was assigned full-time to assist R&D. Several patent applications were submitted to protect all three types of ester placements (*14–23*).

After I synthesized the kilogram quantities of raw material needed for toxicology and phase 1 studies in space rented in Canada, clinical formulation work was accomplished in France. Studies on the first clinical batch of ASL-8052 were completed in Germany. The ultra-short duration initially observed in dogs was duplicated exactly in humans. Pharmacological responses were appropriate, and the pharmacokinetic data were superb. The distribution half-life, elimination half-life, volume of distribution, and total clearance averaged 2 min, 9 min, 3.4 L/kg, and 285 mL/kg/min, respectively. Our 45-pound investigatory new drug application was submitted in March 1982 and was quickly granted; ASL-8052 soon became esmolol (USAN). A new drug application was approved early in 1987, and the drug was finally introduced to the marketplace as Brevibloc. It is used "to moderate abnormal heart activity in critical care situations and during surgery" (*24*). Brevibloc as the racemic mixture continues to be useful clinically (*25, 26*), with applications in both young (*27*) and adult (*28*) patients.

During the final scenes of the esmolol story, Du Pont Pharmaceuticals purchased American Critical Care, and as reported in an industry journal, "$190 million of the $425 million purchase price was paid in installments, based on steps toward commercialization of Brevibloc" (*24*).

Acknowledgments

The most important acknowledgments belong to all of the researchers who participated at various times on the esmolol discovery team. An acknowledgment of practical relevance to this writing is also extended to Dot Gula, who typed and retyped the manuscript's many initial drafts.

Literature Cited

1. Zaroslinski, J.; Borgman, R. J.; O'Donnell, J. P.; Anderson, W. G.; Erhardt, P. W.; Kam, S. T.; Reynolds, R. D.; Lee, R. J.; Gorczynski, R. J. *Life Sci.* **1982,** *31,* 899–907.
2. O'Donnell, J. P.; Parekh, S.; Borgman, R. J.; Gorczynski, R. J. *J. Pharm. Sci.* **1979,** *68,* 1236–1238.
3. Woodward, R. B.; Bader, F. E.; Bickel, H.; Frey, A. J.; Kierstead, R. W. *J. Am. Chem. Soc.* **1956,** *78,* 2023–2025.
4. Smith, L. H. *J. Appl. Chem. Biotechnol.* **1978,** *28,* 201–212.
5. Augstein, J.; Cox, D. A.; Ham, A. L.; Leeming, P. R.; Snarey, M. *J. Med.Chem.* **1973,** *16,* 1245–1251.
6. Erhardt, P. W. *Synth. Commun.* **1983,** *13,* 103–113.
7. Erhardt, P. W.; Woo, C. M.; Gorczynski, R. J.; Anderson, W. G. *J. Med. Chem.* **1982,** *25,* 1402–1407.
8. Barrett, A. M. *Drug Design* **1972,** *3,* 205–228.
9. Ablad, B.; Brogard, M.; Carlsson, E.; Ek, L. *Eur. J. Pharmacol.* **1970,** *13,* 59–64.
10. Erhardt, P. W.; Woo, C. M.; Anderson, W. G.; Gorczynski, R. J. *J. Med. Chem.* **1982,** *25,* 1408–1412.
11. Erhardt, P. W.; Woo, C. M.; Matier, W. L.; Gorczynski, R. J.; Anderson, W. G. *J. Med. Chem.* **1983,** *26,* 1109–1112.
12. Kam, S. T.; Matier, W. L.; Mai, K. X.; Barcelon-Yang, C.; Borgman, R. J.; O'Donnell, J. P.; Stampfli, H. F.; Sum, C. Y.; Anderson, W. G.; Gorczynski, R. J.; Lee, R. J. *J. Med. Chem.* **1984,** *27,* 1007–1116.
13. Gorczynski, R. J.; Vuong, A. *J. Cardiovasc. Pharmacol.* **1984,** *6,* 555–564.
14. Erhardt, P. W.; Borgman, R. J.; O'Donnell, J. P. U.S. Patent 4,387,103, 1983.
15. Kam, S. T.; Erhardt, P. W.; Borgman, R. J.; O'Donnell, J. P. U.S. Patent 4,405,642, 1983.
16. Erhardt, P. W.; Borgman, R. J.; U.S. Patent 4,450,173, 1984.
17. Erhardt, P. W.; Woo, C. W. U.S. Patent 4,556,668, 1985.
18. Erhardt, P. W.; Borgman, R. J.; O'Donnell, J. P. U.S. Patent 4,593,119, 1986.
19. Borgman, R. J.; Erhardt, P. W.; Kam, S. T.; O'Donnell, J. P. U.S. Patent 4,604,481, 1986.
20. Matier, W. L.; Erhardt, P. W.; Patil, G. U.S. Patent 4,508,725, 1985.
21. Erhardt, P. W.; Matier, W. L. U.S. Patent 4,623,652, 1986.
22. Erhardt, P. W.; Matier, W. L. U.S. Patent 4,692,446, 1987.

23. Erhardt, P. W.; Matier, W. L. U.S. Patent 4,804,677, 1989.
24. *Chem. Eng. News,* Jan. 12, 1987; p 13.
25. Turlapaty, P.; Laddu, A.; Murthy, V. S.; Singh, B.; Lee, R. *Am. Heart J.* **1987,** *114,* 866–885.
26. Benfield, P.; Sorkin, E. M. *Drugs* **1987,** *33,* 392–412.
27. Wiest, D. B.; Trippel, D. L.; Gillette, P. C.; Garner, S. S. *Clin. Pharmacol. Ther.* **1991,** *49,* 618–623.
28. Gray, R. J. *Chest* **1988,** *93,* 3984–1103.

Diltiazem

Hirozumi Inoue and Taku Nagao
Tanabe Seiyaku Company, Ltd.

Diltiazem, a derivative of 1,5-benzothiazepine, was developed by Tanabe Seiyaku Co. Ltd. as an antianginal, antihypertensive, and antiarrhythmic agent with a new mechanism of action, calcium antagonism, different from that of the conventional coronary vasodilators. It was launched in Japan in 1974 and then in more than 100 countries worldwide, including North America and all European countries. Today it is widely used in the treatment and study of cardiovascular diseases.

Background

Around the middle of the 1960s, central nervous system (CNS)-active drugs such as tranquilizers and antidepressant agents appeared in large number on the Japanese drug market and commanded huge sales. Because Tanabe Seiyaku lacked drugs of that category in its product line, there was strong interest in the marketing department in the development of an excellent CNS-active drug.

At about the same time, it was reported that thiazesim, a 1,5-benzothiazepine derivative, had antidepressant activity and that its 2,3-dehydro derivative had a tranquilizing effect (*1–3*). H. Kugita of the Organic Chemistry Research Laboratory of Tanabe Seiyaku turned his attention to

2523–0/93/0207$09.00/0

the unique chemical structure of thiazesim and its clinically confirmed antidepressant activity. In addition, there had been only few reports on the analogous compounds. Consequently, the researchers tried to develop a new CNS-active drug by synthesizing novel 1,5-benzothiazepine derivatives. At that time, 1,4-benzodiazepines bearing the seven-membered heterocyclic ring were highly valued as tranquilizers. It was reported that this series of compounds (e.g., oxazepam) showed an intensified CNS effect when a hydroxyl group was introduced into the α-position of the lactam carbonyl group (Chart 1) (4). In the expectation of increased CNS activity, therefore, the synthesis of new 1,5-benzothiazepines bearing a hydroxyl group at the 3-position was begun in the Organic Chemistry Research Laboratory in September 1966.

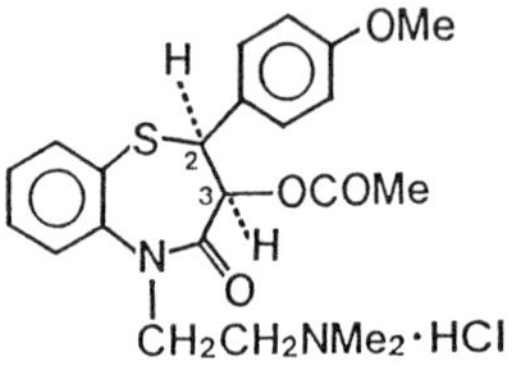

diltiazem hydrochloride

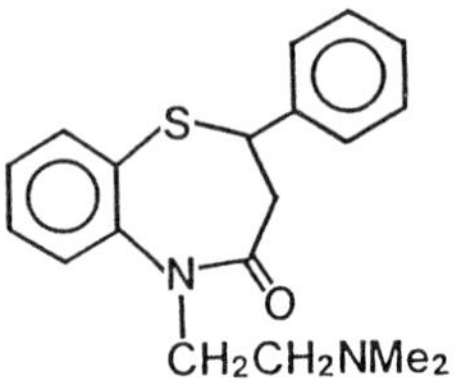

thiazesim

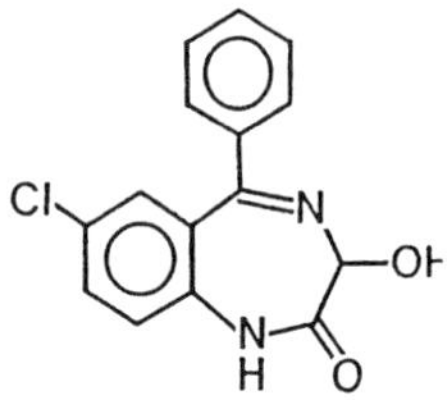

oxazepam

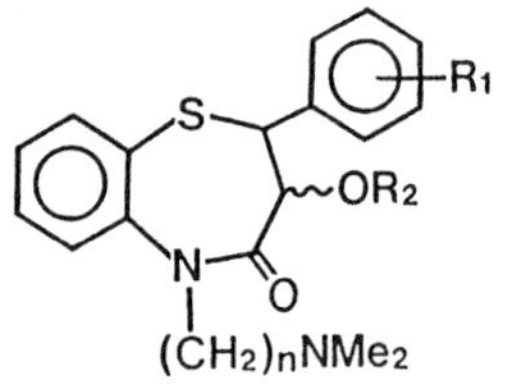

1

Chart 1

The CNS activities of the newly synthesized 1,5-benzothiazepines were examined at the Clinical Pharmacological Department of the Central Inspection Laboratory of Tanabe Seiyaku. Early animal tests demonstrated that some of the compounds had CNS effects comparable to those of thiazesim. However, the compounds lacked sufficient efficacy as CNS drugs to warrant clinical evaluation.

A new compound does not always have the biological activity for which it was designed, and occasionally one may exhibit unexpected properties. Therefore, when the efficacy of a new compound is investigated, it is necessary to screen for activities other than the primary one sought. A systematic means of evaluating new compounds that addresses this concern is called a random screening test. Although it is said by some to have a low probability of finding biologically active compounds, this method is indeed capable of screening substances that have unique chemical structures.

The Biological Research Laboratory made an effort to establish a screening system for broad biological activities of compounds within the limited quantity of substances. The laboratory was already using an original biological activity screening system based on the concept of random screening. The evaluation of a wide variety of pharmacological effects of the new 1,5-benzothiazepines by this screening system was conducted from the beginning in parallel with the evaluation of their effect on the CNS. As a result, the 3-*O*-acyl derivatives were noted to have a strong coronary vasodilating effect along with minimal effects on the CNS. In the secondary screening test, in which the drugs were administered intravenously to anesthetized dogs, these compounds increased not only coronary blood flow but also the partial pressure of oxygen in the coronary sinus blood, as well as renal blood flow.

With these findings, we turned our efforts to exploring the synthesis of these derivatives as new coronary vasodilators. Finally, *cis*-3-acetoxy-5-[2-(dimethylamino)ethyl]-2,3-dihydro-2-(4-methoxyphenyl)-1,5-benzothiazepin-4(5H)-one (CRT-401; the code name derived from coronary, renal, Tanabe) was chosen as the most promising compound because of its high efficacy, low acute toxicity, good absorption from the digestive tract, and other features. However, because CRT-401 was a racemate, its optically active isomers were synthesized and the efficacy of each isomer was further investigated. Only the dextro isomer was found to have a coronary vasodilating effect, and the acute toxicity of the dextro isomer was almost equal to that of the racemate. Based on these data, the dextro isomer of CRT-401, CRD-401, was selected as a final candidate for development. This compound later became diltiazem.

Synthetic Study of Diltiazem

The synthesis of the 3-oxygenated 1,5-benzothiazepine derivative represented by general formula **1** required the seven-membered lactam (**6**) bearing a phenyl group at the C_2-position and a hydroxyl group at the C_3-position as a key intermediate. The 2-phenyl-1,5-benzothiazepine and the corresponding 2,3-dihydro derivative had been synthesized by the reaction of 2-aminothiophenol with phenylpropiolic acid or cinnamic acid (*1–3, 5*). However, there had been no report on the synthesis of the 3-hydroxyl derivative. Therefore, the reaction of the 3-phenylglycidic ester (**4**) with 2-aminothiophenol (**2**) or 2-nitrothiophenol (**3**) was examined as a potential route to **6**.

The first problem to be solved was the regioselectivity in this reaction. When the thiol group of the thiophenol attacks C_3 of the glycidic ester (**4**), the reaction should proceed via course *a* in Chart 2 to give the desired seven-membered 1,5-benzothiazepine (**6**). Its attack on C_2 of the glycidic ester, however, would follow course *b*, leading to the formation of the six-membered 1,4-benzothiazine (**8**) via the β-hydroxyester. In addition, because two diastereoisomers are possible for the compound **6** and its precursor (**5**), a method for the stereoselective synthesis of each diastereoisomer had to be established. Finally, two diastereoisomers, trans and cis, are also possible for the glycidic ester (**4**), the starting material of this reaction. It was presumed that differences in the geometric structure of the two isomers might affect the course of the reaction with **2** and **3**. Therefore, this study was interesting both pharmacologically and chemically.

Challenge to Create a New Skeleton: Synthesis of Seven-Membered Lactam by Reaction with Aminothiophenol or Nitrothiophenol

At the time, there was only one example of the reaction of the *trans*-glycidic ester with 2-aminothiophenol. Heating of both compounds in ethanol in the presence of potassium hydroxide yielded the three products **10**, **11**, and **12** (*6*). All these compounds resulted from attack of the thiol group of 2-aminothiophenol on C_2 of the glycidic ester (Chart 3).

Accordingly, various conditions in the reaction of *trans*-3-phenylglycidic ester (**9**) with 2-aminothiophenol (**2**) were examined. When a mixture of **9** and **2** was heated at 160 °C without solvent, compounds **13** and **14** were obtained (*7*) that were different from any of the compounds **10**, **11**, or **12**. Elemental analysis and spectroscopic data for these products, as well as identification of the desulfurized product (**15**) with an authentic sample, proved that **14** was the desired 1,5-benzothiazepine derivative with a seven-membered lactam moiety. Subsequently the 2-phenyl group and 3-hydroxy group of **14** proved to be cis (*8*). It was later found that the cis isomer has coronary vasodilating effects whereas the corresponding trans isomer is almost inactive (*9*). Therefore, it was fortunate that the pharmacologically

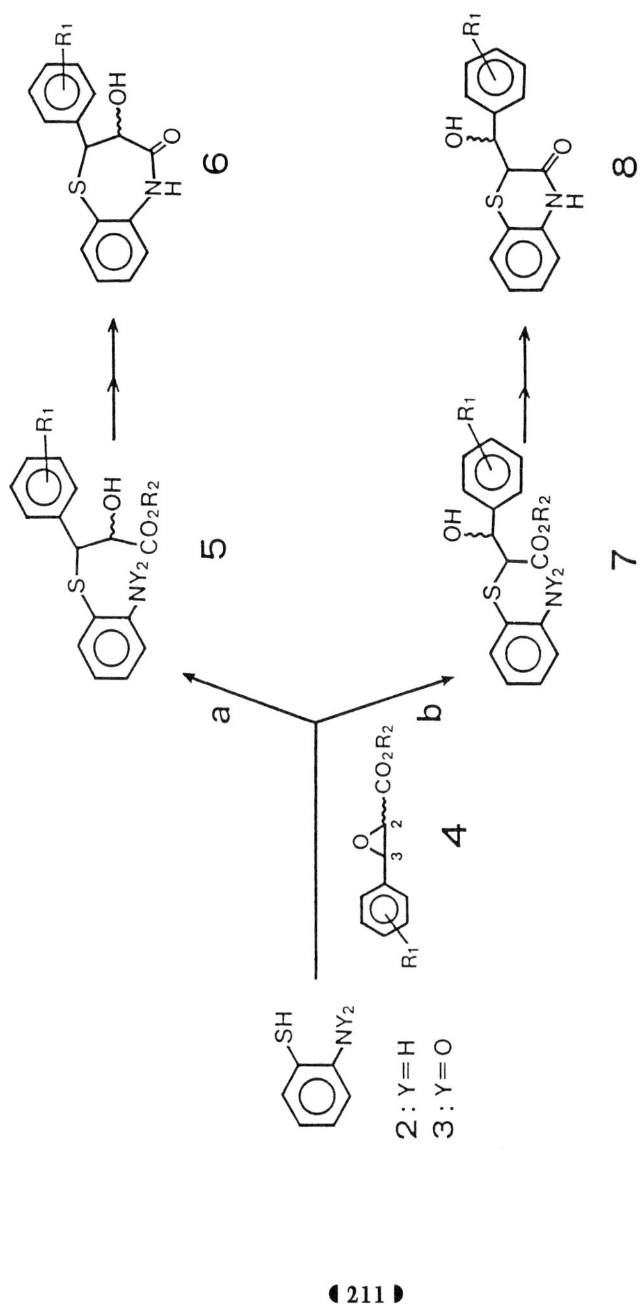

Chart 2

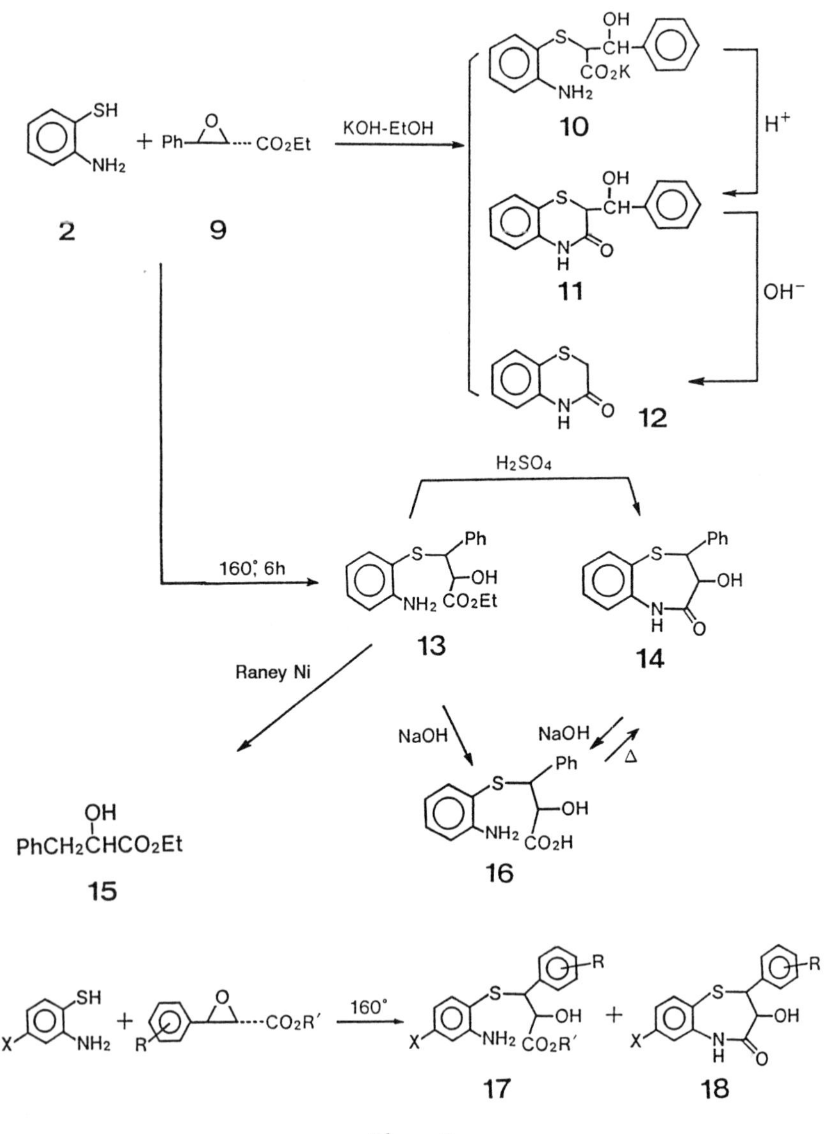

SH
NH2
+ Ph
O
CO2Et
2
9
KOH-EtOH
OH
S
CH
CO2K
NH2
10
H+
OH
S
CH
N
H
O
11
OH−
S
N
H
O
12
H2SO4
160°, 6h
S
Ph
OH
NH2
CO2Et
13
S
Ph
OH
N
H
O
14
Raney Ni
NaOH
NaOH
Δ
OH
PhCH2CHCO2Et
15
S
Ph
OH
NH2
CO2H
16
SH
X
NH2
+
R
O
CO2R′
160°
S
R
OH
X
NH2
CO2R′
17
+
S
R
OH
X
N
H
O
18

Chart 3

active cis isomer was selectively synthesized by this method in an early stage of the study.

Although this method enabled the synthesis of the 2,3-*cis*-1,5-benzothiazepine skeleton, it was unsatisfactory in terms of yield. However, the method involved only a short sequence of reactions, and the desired product could be obtained even when less reactive phenylglycidic esters were used. Therefore, in an early stage of the study, this method was utilized to produce a variety of *cis*-lactams (**18**) using the *trans*-3-phenylglycidic esters with various substituents. The reaction conditions of this method were then studied in detail to improve yield. When the reaction was carried out in a solvent under mild conditions, a high yield of intermediate aminoester (**17**) was obtained. Alkaline hydrolysis of **17** followed by thermal cyclization in xylene gave a high yield of cis lactam (**18**) (Chart 3).

Second, to obtain better yield of the lactam, *trans*-3-phenylglycidic esters with 2-nitrothiophenol were examined (7, 8). Heating of 2-nitrothiophenol (**3**) and methyl *trans*-3-(4-methoxyphenyl)glycidate (**19**) in acetonitrile at 50 °C for 2 days gave a 56% yield of the nitroester (**20a**). Reduction of **20a** gave the aminoester (**21a**) identical to the sample obtained by the reaction of **19** with 2-aminothiophenol (**2**) (Chart 4). On the other hand, the reaction of **3** with **19** in ethanol at room temperature in the presence of a catalytic amount of sodium bicarbonate readily gave an 80% yield of the nitroester (**20b**), which was different from **20a**. Elemental analysis and spectroscopic data, as well as the results of desulfurization of its reduction product (**21**), revealed that this compound (**21b**) was a diastereoisomer of **21a**.

Reduction and hydrolysis of the isomeric nitroesters followed by cyclization produced a high yield of the isomeric seven-membered lactams (**23a, b**). Their stereochemistry was examined by comparing the NMR spectra of the respective *N*-methyl derivatives (**24a, b**). The coupling constants (*J* values) between C_2–H and C_3–H of **24a** and **24b** were 7 Hz and 11 Hz, respectively. Inspection of a Dreiding model revealed that two conformations, A and A′, were the most probable for the 2,3-*cis*-lactams, and the conformations B and B′ were the most probable for the trans isomer (Chart 5).

Of these four conformers, the large *J* value between C_2–H and C_3–H is only possible with B. Therefore, the *N*-methyl derivative **24b** with a *J* value of 11 Hz was assigned for the trans isomer and the compound **24a** with a *J* value of 7 Hz was assigned for the cis isomer. This assignment was later unequivocally confirmed by X-ray crystallographic analysis (*10*).

Establishing the stereochemistry of these lactams allowed the stereochemical consideration of the reaction of the *trans*-3-phenylglycidic ester (**19**) with 2-nitrothiophenol. Because the 2-phenyl and 3-hydroxyl groups of

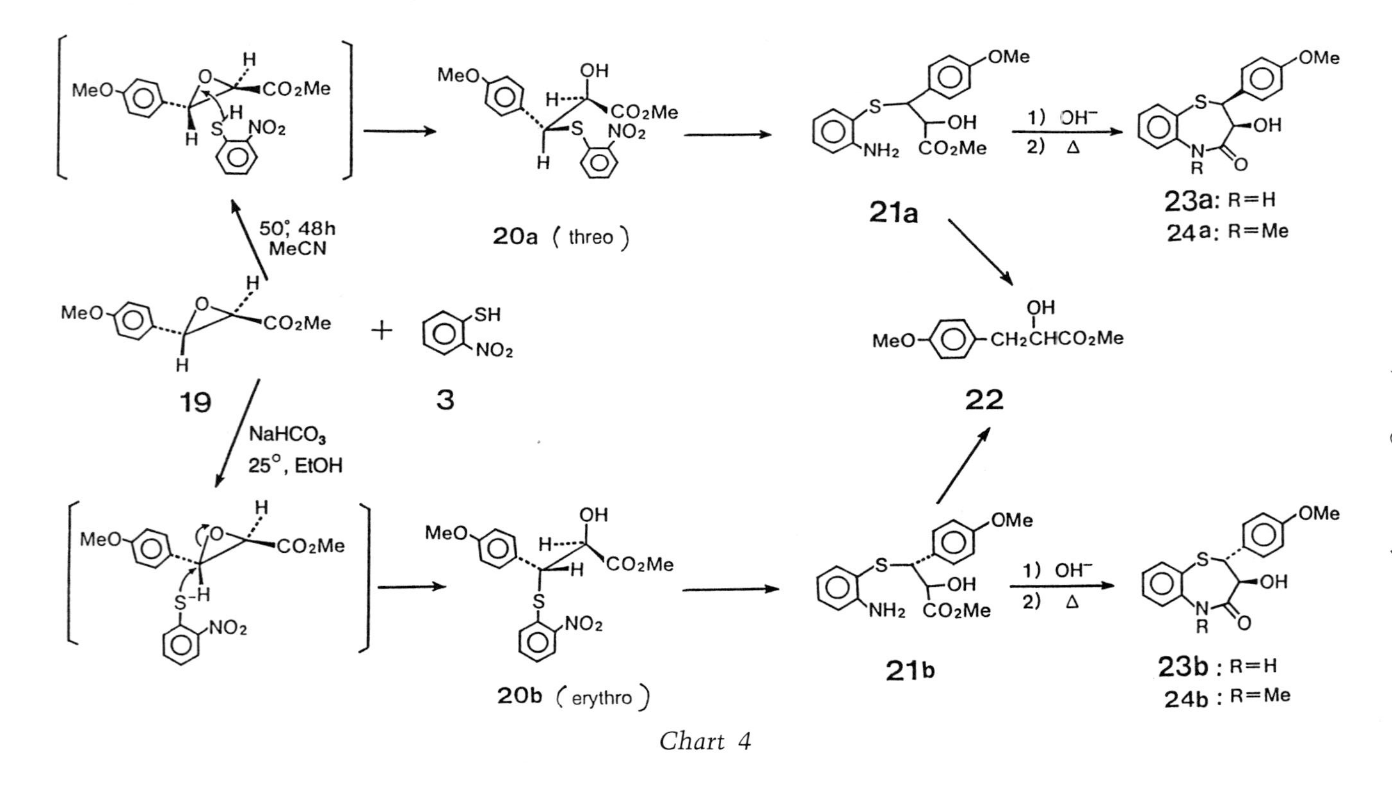
19
3
50°, 48h
MeCN
NaHCO3
25°, EtOH
20a (threo)
20b (erythro)
21a
21b
22
1) OH⁻
2) Δ
23a: R=H
24a: R=Me
23b : R=H
24b : R=Me

Chart 4

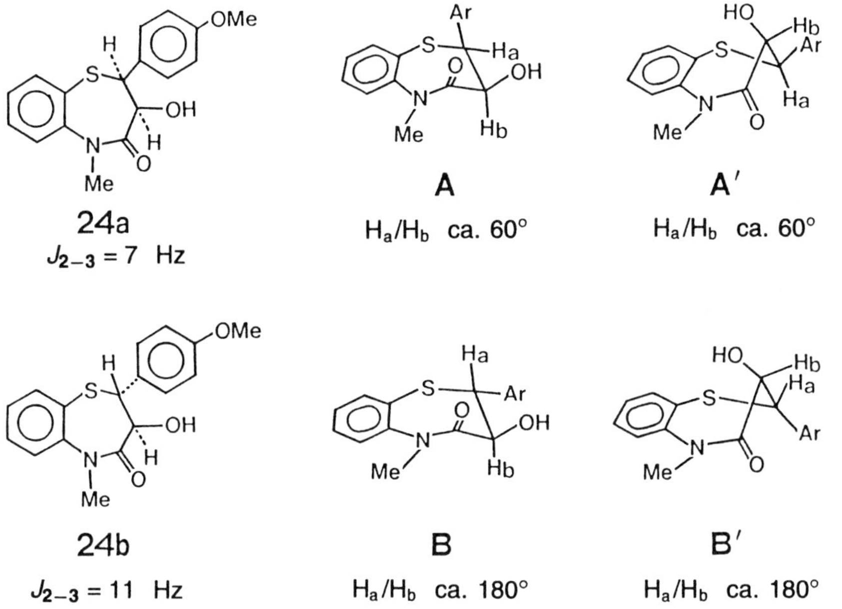

24a J_{2-3} = 7 Hz

A H_a/H_b ca. 60°

A′ H_a/H_b ca. 60°

24b J_{2-3} = 11 Hz

B H_a/H_b ca. 180°

B′ H_a/H_b ca. 180°

Chart 5

the seven-membered lactam (**23a**) proved to be cis, the nitroester (**20a**), the precursor of **23a**, should be the threo isomer. Therefore, the thiol group of 2-nitrothiophenol attacked the benzylic carbon of the glycidic ester from the side of the oxirane ring. Accordingly, upon heating in acetonitrile, the cis opening of the oxirane ring of the glycidic ester occurred, with retention of configuration (Chart 4). On the other hand, in the presence of sodium bicarbonate, since the lactam (**23b**) proved to be trans, attack of thiolate anion occurred from the opposite side of the oxirane ring (trans opening), leading to the formation of the erythro isomer (**20b**). The method of producing the *threo*-nitroester (**20a**) by heating **3** with **19** in acetonitrile was later applied to the industrial production of diltiazem.

Using *trans*-3-phenylglycidic esters with various substituents on the phenyl ring, we next examined the effects of those substituents of the reaction (*11*). In the case of glycidates with electron-donating substituents, a high yield of *threo*-nitroesters was generally obtained. In the aforementioned reaction of the 4-methoxy derivative (**19**), precise examination of the reaction products revealed that the unwanted erythro isomer (**20b**) occurred at a rate approximately one-third that of the threo isomer (**20a**). The reaction of the 4-methyl analogue gave a mixture of threo and erythro isomers in nearly equal amounts. Upon introduction of electron-withdrawing groups such as nitro and chlorine, the reaction hardly proceeded. It was also found that, even in the case of compounds having electron-donating groups, the threo isomer was preferentially produced as the reaction temperature increased (*11*). The use of a dipolar aprotic solvent such as hexamethylphosphamide (HMPA) or dimethylformamide (DMF) was found to increase the formation of the erythro isomer and of the regioisomer that resulted from the attack of the thiol group on C_2 of the glycidic ester (*11*).

On the other hand, the reaction in the presence of a catalytic amount of sodium bicarbonate was not significantly affected by the nature of the substituents on the phenyl group of glycidic esters. In every case, erythro isomers were formed stereoselectively (*12*).

Studies of the reaction of the *trans*-3-phenylglycidic esters with 2-amino or 2-nitrothiophenol were extended to the reaction of the corresponding *cis*-glycidic esters (*13*). The mode of opening of the *cis*-glycidic ester and the steric structure of the product are just opposite to those of the *trans*-glycidic ester. Thus, cis and trans openings of the *cis*-glycidic ester are expected to yield the erythro and threo isomers, respectively (Chart 6). The study with *cis*-3-(4-methoxyphenyl)glycidic ester (**25**), however, revealed that this compound is less reactive and has lower stereoselectivity than the corresponding trans isomer (**19**).

Improvements in the Synthesis of Diltiazem: Effect of Acidic Catalyst in the Reaction of Methyl 3-(4-Methoxyphenyl)glycidic Ester with 2-Nitrothiophenol

The most important step in the synthesis of diltiazem is the reaction of the *trans*-glycidic ester (**19**) with 2-nitrothiophenol to form the *threo*-ester (**20a**). However, as described in the previous section, insufficient stereoselectivity of this reaction caused concomitant formation of the unwanted erythro isomer (**20b**) as a minor product. Moreover, the reaction required an extremely long time. Thus, this step was a major obstacle in the synthesis of diltiazem. The catalytic effects of various acids on this reaction were examined, and the reaction was found to proceed very rapidly at a low temperature in the presence of Lewis acids or Brönstead acids (*12*). However, the stereoselectivity of the reaction in the presence of sulfuric acid, perchloric acid $BF_3{\bullet}Et_2O$, or zinc chloride still remained unsatisfactory. Surprisingly, adding magnesium or calcium chloride led to the trans opening of the oxirane ring of the glycidic ester. Additional screening tests of the catalytic

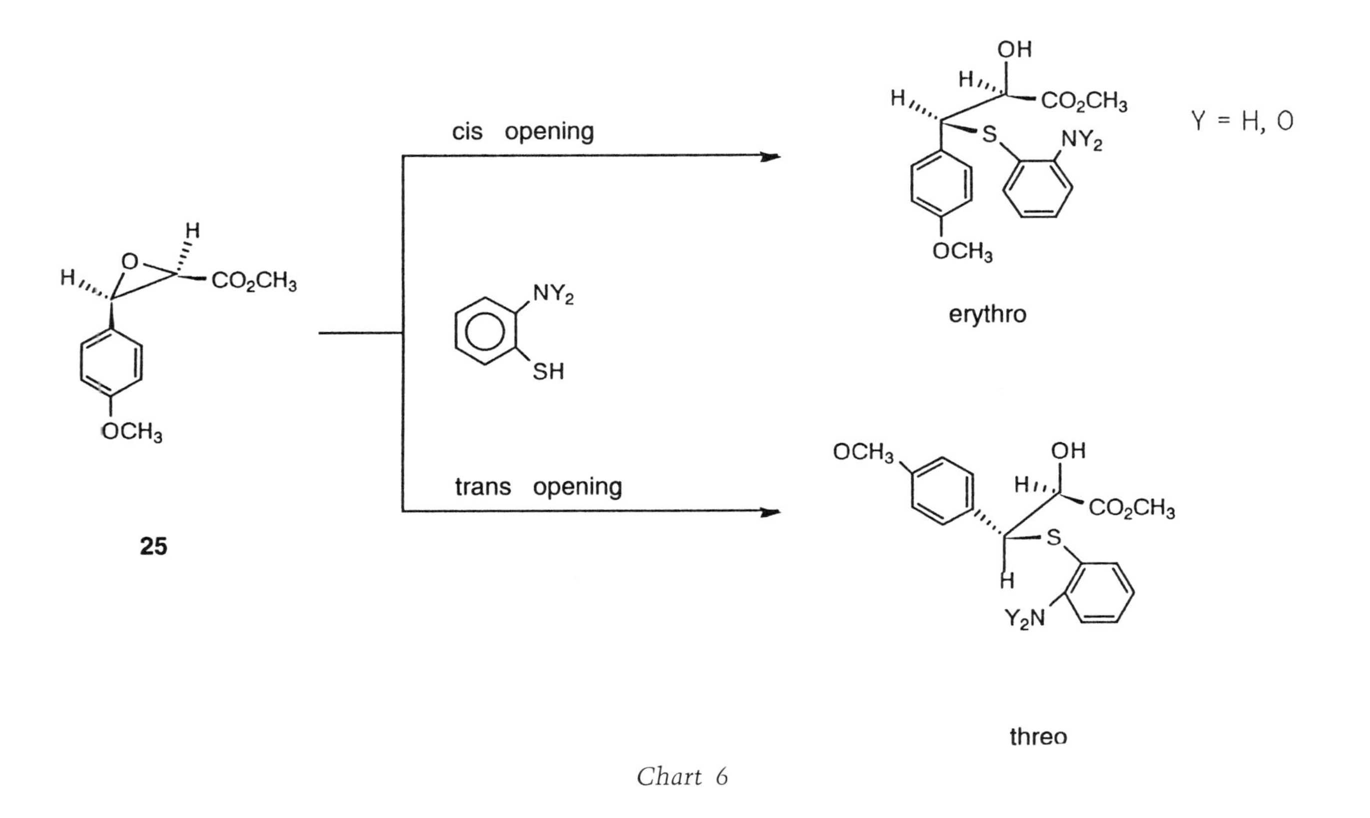

Chart 6

effects of various Lewis acids revealed that the Lewis acids containing tin are excellent catalysts. They greatly accelerated the reaction, with extremely high stereoselectivity. For instance, the reaction in the presence of stannous 2-ethyl hexanoate at room temperature readily gave the necessary threo isomer (**20a**) in more than 80% yield (Table 1).

The mechanism of the excellent catalytic action of tin compounds was studied in detail. The reaction of stannic chloride with 2-nitrothiophenol was found to form a crystalline adduct. Examination of the structure and reactivity of this adduct led to the mechanistic explanation shown in Chart 7. Through this plausible transition state, the reaction easily yielded the necessary threo isomer (**20a**) with extremely high stereoselectivity. Further study of the Lewis acid catalyzed reaction was then continued in the Technical Department of Onoda Factory, where zine acetate was found to be a more appropriate catalyst.

The Search for Better Derivatives: Introduction of Substituents at 5- and 3-Positions, and Optical Resolution

In the structure of the new 3-oxygenated 1,5-benzothiazepines, structural parameters possibly related to pharmacological activity were the substituents at the 2-,3-, and 5-positions, those on the fused benzene ring, and the stereochemistry of the 2- and 3-substituents. A large number of derivatives were synthesized by varying these parameters (*14*). Reaction of the *cis-* (**28a**) and *trans*-lactam (**28b**) with dimethylaminoalkyl halides in the presence of methylsulfinyl carbanion in dimethyl sulfoxide (DMSO), or in the presence of potassium hydroxide in DMSO, produced 5-dimethylaminoalkyl derivatives (**29a, b**) in high yield. Later, the use of K_2CO_3 in acetone or ethyl acetate was found to be a more practical method (*15*). Then, these amino derivatives were allowed to react with alkyl halides to give quaternary salts.

Acylation of the hydroxyl group of **29** with acid halides, acid anhydrides, or isocyanates gave the corresponding O-acyl derivatives (**30a, b**) (Chart 8). Because these O-acyl derivatives exhibited increased coronary vasodilating activity, a number of O-acyl derivatives of various carbon chain lengths and 3-alkoxy derivatives were synthesized.

Among a series of compound, *cis*-3-acetoxy-5-[2-(dimethylamino)ethyl]-2,3-dihydro-2-(4-methoxyphenyl)-1,5-benzothiazepin-4(5H)-one was found to have the highest coronary vasodilating activity. The pharmacological activity of this compound was further studied as a possible candidate for clinical trial under the code name of CRT-401.

Table 1. Effect of Catalyst in the Reaction of Methyl 3-(4-Methoxyphenyl) glycidate (19) with 2-Nitrothiophenol (3)

			Condition		Product (Isolated Yield)		
Entry	Catalyst	Solvent	Temp.	Time	*threo* (%)	*erythro* (%)	*threo*/*erythro* (Whole Product)
1	—	MeCN	50 °C	2 days	56		3
2	Conc. H_2SO_4	Dioxane	10 °C	10 min	43	15	
3	$HClO_4$	Dioxane	10 °C	10 min	48	18	
4	$BF_3{\bullet}Et_2O$	ET_2O	15 °C	20 min	51	12	
5	$ZnCl_2$	Dioxane	15 °C	1 h	50	25	
6	$MgCl_2$	Dioxane	25 °C	40 h		65	
7	$CaCl_2$	Dioxane	25 °C	40 h		63	
8	$SnCl_4$	Dioxane	25 °C	18 h	69		6.7
9	$SnBr_4$	Dioxane	25 °C	20 h	63		
10	SnI_4	Dioxane	25 °C	22 h	63		
11	$SnCl_2$	Dioxane	25 °C	21 h	69		12.5
12	SnI_2	Dioxane	25 °C	17 h	68		
13	SnF_2	Dioxane	25 °C	22 h	72		
14	$Sn(OCOC_7H_{15})_2$	Dioxane	25 °C	19 h	74 (82)[a]		9.3
15	$Sn(OCOC_7H_{15})_4$	Dioxane	25 °C	19 h	50		
16	$Sn(OCOC_{17}H_{35})_2$	Dioxane	25 °C (18 h) + 50 °C (2 h)		73		

[a] Corrected yield based on the purity of 2-nitrothiophenol, which was determined by titration.

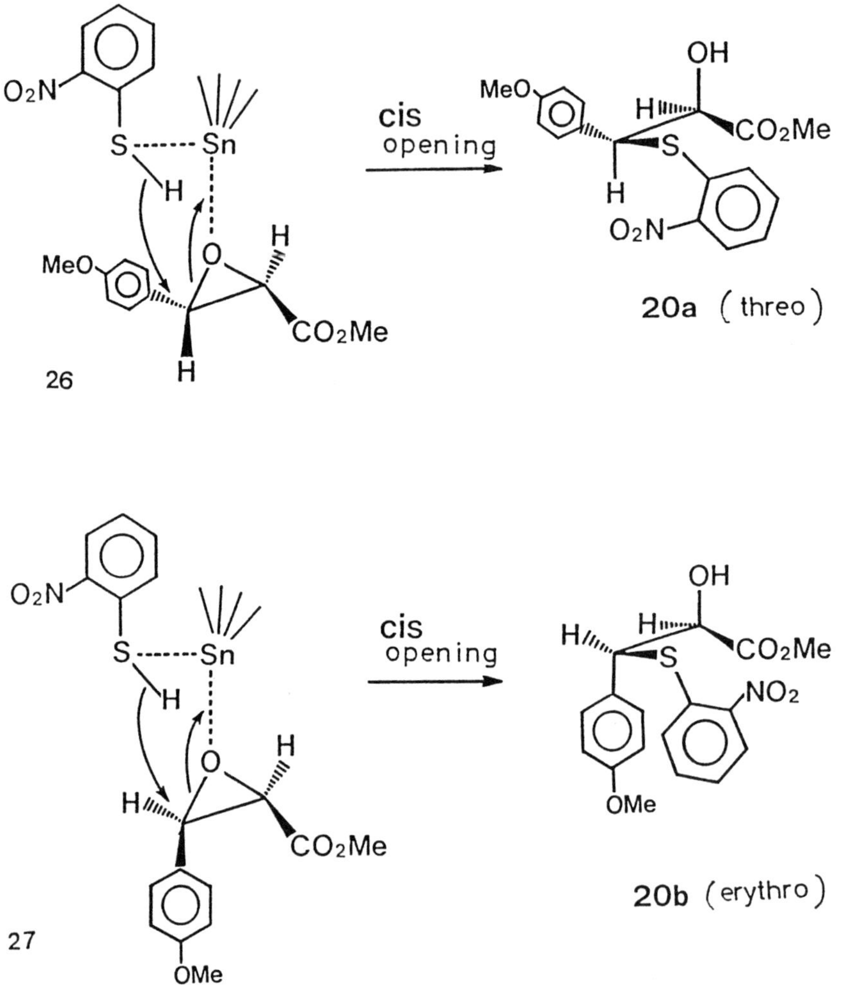

26 → cis opening → 20a (threo)

27 → cis opening → 20b (erythro)

Chart 7

The optical resolution of CRT-401 was then examined. Optical resolution of the intermediate *threo*-nitrocarboxylic acid (**31**) was achieved by using cinchonidine (*10*). The levo and dextro isomers (*l*- and *d*-**31**) were converted to the desired levo and dextro isomers of CRT-401, respectively, without any racemization (Chart 9). Of the two optical isomers, only the dextro isomer (*d*-**32**) exhibited coronary vasodilating activity. The absolute

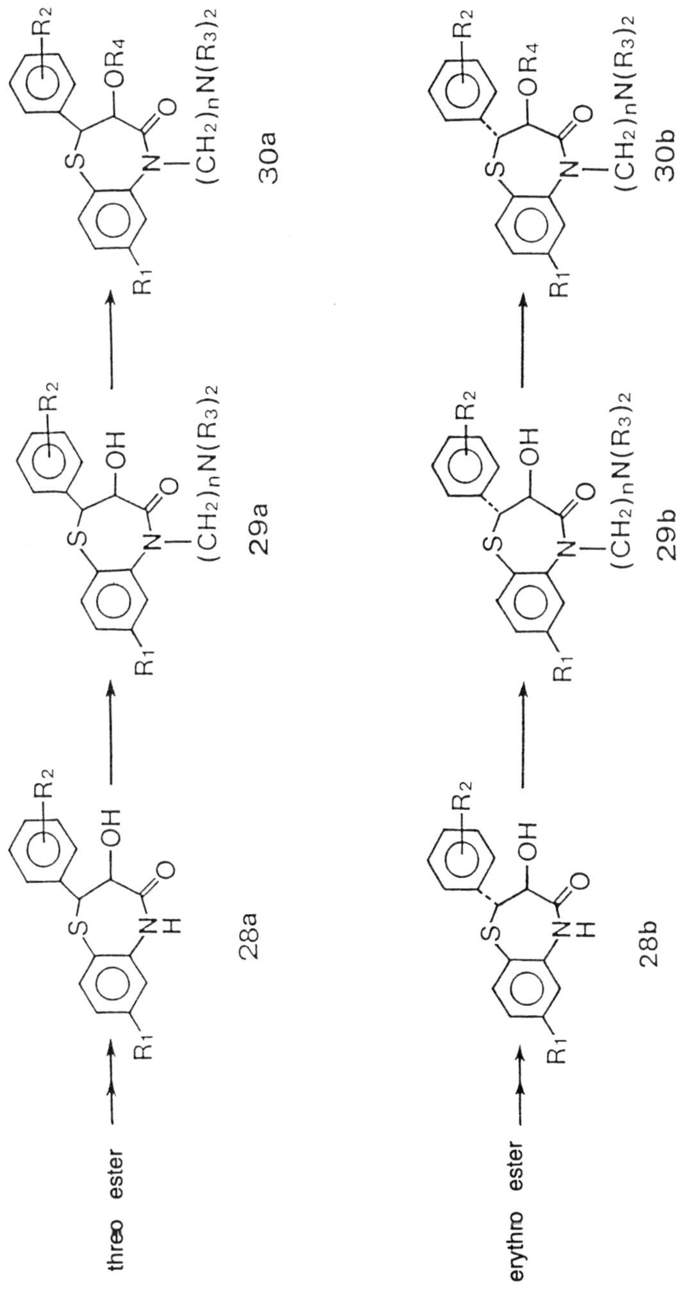

Chart 8

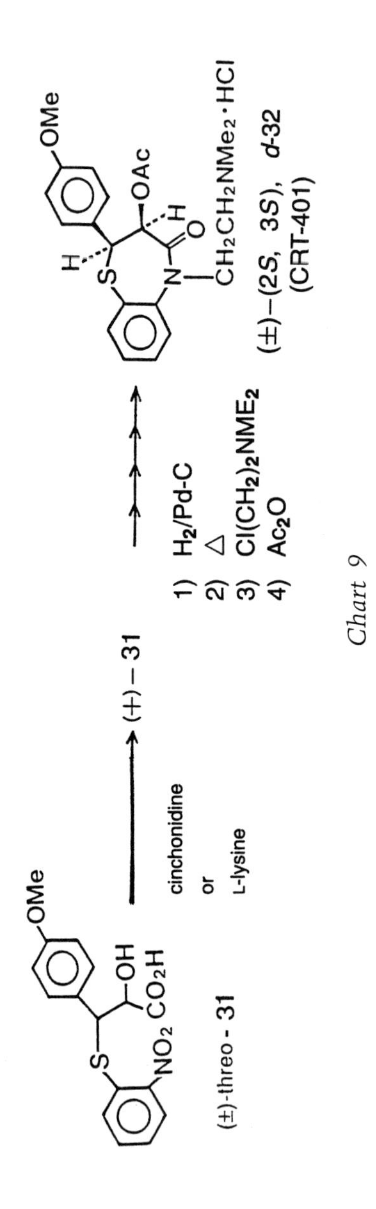
OMe
S
NO2
OH
CO2H
(±)-threo - 31
cinchonidine
or
L-lysine
(+) – 31
1) H2/Pd-C
2) △
3) Cl(CH2)2NME2
4) Ac2O
H
OAc
O
CH2CH2NMe2·HCl
(±)–(2S, 3S), d-32
(CRT-401)

Chart 9

stereochemistry of *d*-**32** was determined to be 2*S*, 3*S* by X-ray crystallographic analysis.

Subsequently, L-lysine was found to be an excellent resolving agent for **31** (*16*). With the use of 0.5 equivalent of L-lysine, the L-lysine salt of the desired dextro isomer of **31** was obtained with good optical purity and in nearly quantitative yield. This finding enabled the practical resolution of **31**.

Structure–Activity Relationships

The structure–activity relationships determined by examining the coronary vasodilating effect of a large number of derivatives are summarized in the following list (*9*) and in Figure 1:

1. With respect to the effect of the substituents on the 2-phenyl ring, the 4-methoxy and 4-methyl derivatives exhibited potent activity. Removing the substituents, introducing chlorine, and increasing the number of a methoxy group all resulted in a marked decrease in activity.
2. With regard to the effect of the substituents on the oxygen atom at the 3-position, acylation or alkylation caused an increase in activity in many of the compounds. In particular, the acetyl or ethoxycarbonyl derivative showed the highest activity. In a series of alkoxy derivatives, the activity decreased with increasing length of the carbon chains.
3. With regard to the effect of the substituents on the nitrogen atom at the 5-position, the presence of a dialkylaminoalkyl group was essential for activity. Dealkylation or quaternization of the tertiary amino group in the side chain caused a marked decrease in activity. The optimal length of aminoalkylene chain was two.
4. The coronary vasodilating activity of this series of compounds was greatly influenced by the stereochemistry of 2- and 3-substituents. The cis isomer exhibited high activity, whereas the corresponding trans isomer was almost inactive.
5. With regard to optical resolution, the activity of **32** (CRT-401) was found to reside almost entirely in the dextro isomer (CRD-401, diltiazem). The levo isomer showed only marginal activity.

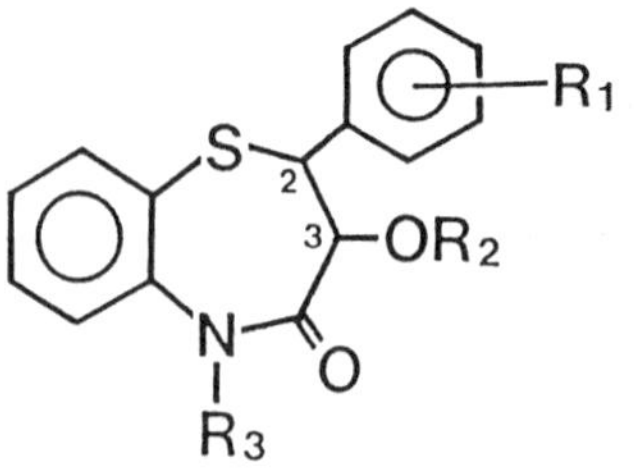

1) 2,3-cis ≫ 2,3-trans
2) *d*-cis ≫ *l*-cis
3) R_1: MeO(p) ≧ Me(p) > diMeO > triMeO = OH(p)
4) R_2: COMe = COOEt > Me > Et > H
5) R_3: $CH_2CH_2NMe_2$ > $CH_2CH_2CH_2NMe_2$ > $CH_2CH_2N^+Me_3$ = CH_2CH_2NHMe > H

Figure 1. Structure–activity relationships in derivatives.

These results made it clear that of the four possible stereoisomers for the planar structure, the *d*-cis isomer (2*S*, 3*S*) selectively exhibits potent activity.

The CNS activity of thiazesim has been reported not to differ between its enantiomers (*3*). The strict enantiomeric stereoselectivity shown in the coronary vasodilating effect of diltiazem is therefore of significant interest.

The duration of action of the *d*-cis (2*S*, 3*S*) and *dl*-cis isomers (2*RS*, 2*RS*) was apparently longer than that of the *l*-cis (2*R*, 3*R*) and *dl*-trans isomers (2*RS*, 2*SR*) (Figure 2).

The acute toxicity and the local anesthetic activity of the *d*-cis isomer (diltiazem) were almost equal to those of the racemate (*17, 18*). The smooth-muscle-relaxing effect of diltiazem was almost twice that of its racemate (CRT-401). In addition, high doses of the racemate or of the *l*-cis isomer caused severe clonic convulsion in mice, but the *d*-cis isomer did not.

Optical resolution of the trans isomer of CRT-401 did not show any remarkable increase in coronary vasodilating activity, compared with the corresponding racemic compound (*19*).

6. The compounds without oxygenated function at the 3-position had almost no activity. Although introduction of a hydroxyl group at the 3-position of 1,5-benzothiazepine skeleton was initiated in an attempt to obtain a more potent

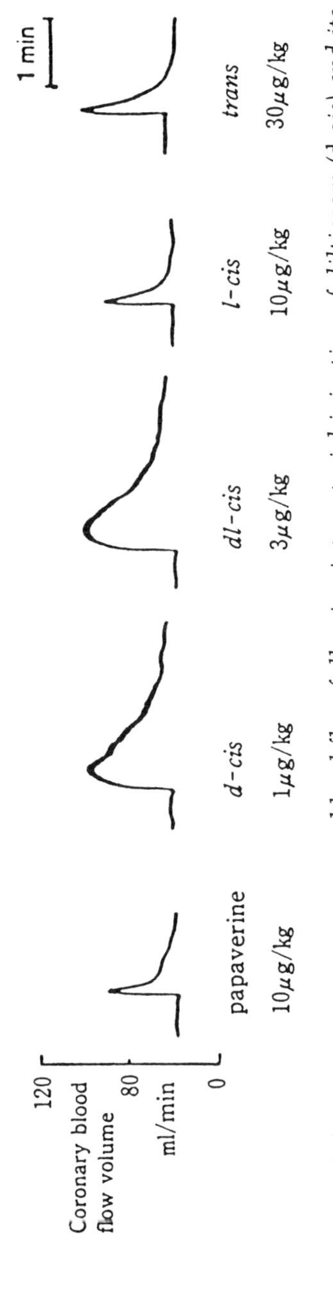

Figure 2. Increase in coronary artery blood flow following intraarterial injection of diltiazem (d-cis) and its stereoisomers in anesthetized dogs. (Reproduced with permission from ref. 18 Copyright 1972 Japanese Pharmacological Society.)

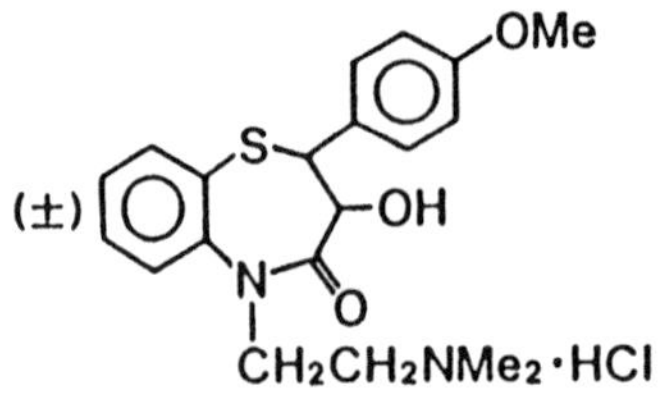

Figure 3. Substitution of the skeleton.

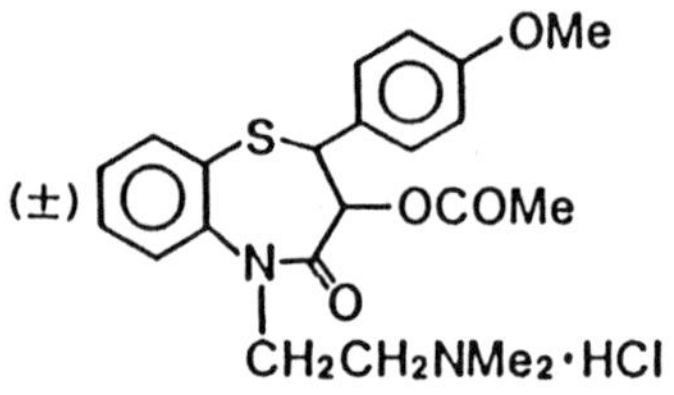

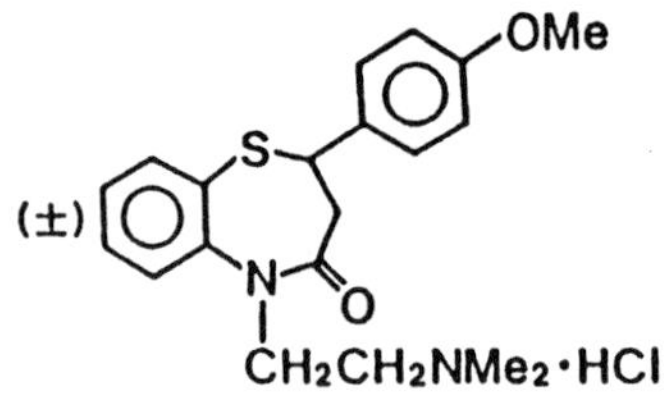

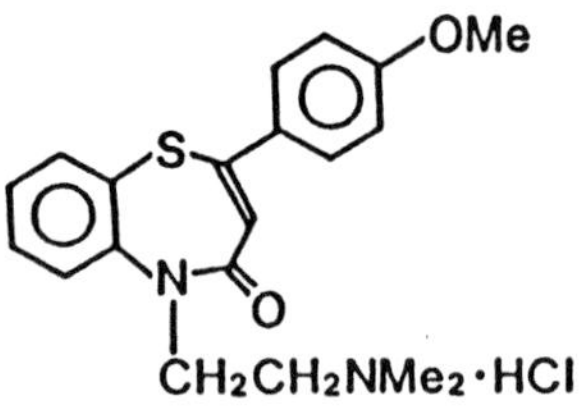

CNS agent, it unexpectedly resulted in the emergence of potent coronary vasodilating activity (Figure 3).

7. The oxa (*20*) and aza (*21*) analogues, in which the sulfur atom of 1,5-benzothiazepine was replaced with an oxygen or nitrogen atom, were much less active in coronary vasodilation.

Pharmacological Studies of Diltiazem

Background of Discovery

Angina pectoris occurs when there is an imbalance between cardiac oxygen supply and demand and the myocardium becomes transiently ischemic. Nitroglycerin had been reported to increase total coronary blood flow in healthy people but not in patients with angina pectoris (*22*), and β-blocking drugs were reported to be efficacious in the treatment of this disease.

Therefore, it was believed insufficient for antianginal agents to have only vasodilating effects, and that their action to diminish myocardial oxygen consumption was also important.

In the 1950s to 1960s, many coronary vasodilators were developed. Many of them did not increase myocardial oxygen consumption but dilated coronary arteries selectively, and their use was aimed at the treatment of patients with angina pectoris. However, the clinical efficacy of these compounds was not satisfactory, although they increased coronary blood flow remarkably in animal experiments. It is claimed that some coronary vasodilators decrease myocardial oxygen consumption and thus possess both β-blocking and coronary vasodilating effects.

Improving the known coronary vasodilators did not appear to be a promising approach. New chemical entities had to be found that possessed distinct properties from the known coronary vasodilators. Therefore, pharmacologists in the Biological Research Laboratory joined the in-house random screening test for seeking novel compounds.

Screening Method

In an effort to improve the test method, the pharmacologists devised methods of perfusing the isolated heart of a guinea pig with a solution that contained a small amount of rabbit blood, and also devised a heart preparation with ventricular fibrillation to accurately measure coronary blood flow. These techniques, which could also determine cardiac function and the volume of coronary blood flow over a long period of time, allowed the effects of coronary vasodilators to be determined. The Biological Research Laboratory immediately incorporated this system into the screening test and evaluated newly synthesized derivatives with oxygenated function at the 3-position of 1,5-benzothiazepines. In March 1968, the coronary vasodilating activity of these compounds was successfully found, and the results of the first screening test were immediately confirmed in a canine model (Figure 4) (23) that had been originally designed for studying the side effects of a drug on the heart.

Progress in Pharmacological Study

During a random screening test, compounds with a 1,5-benzothiazepine skeleton were found to have a strong coronary vasodilating effect, and a candidate compound was selected after the structure–activity studies were completed. The main target of research was coronary dilation and related cardiovascular actions. In the early days of the investigation we found that,

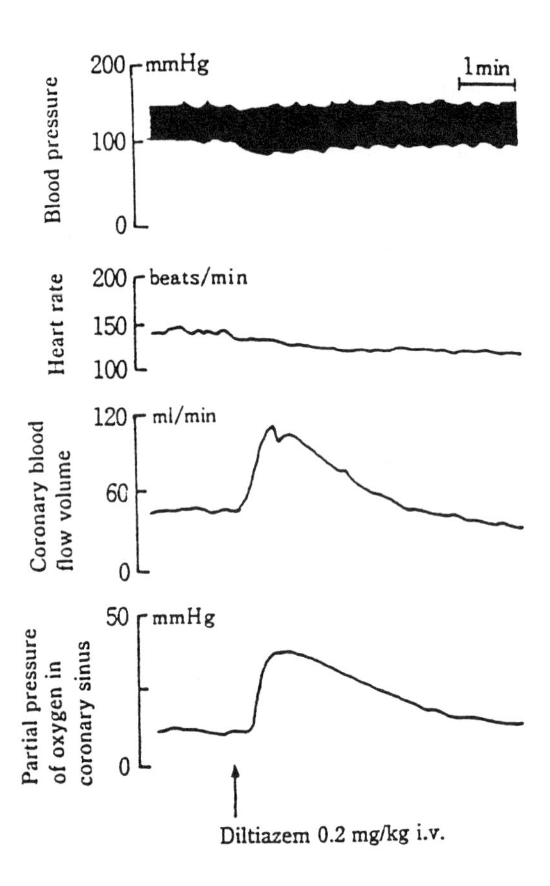

Figure 4. Newly synthesized 1,5-benzothiazepine derivatives, as exemplified by diltiazem, selectively increased coronary blood flow and increased the pressure of oxygen in the coronary sinus in anesthetized dogs. (Reproduced from ref. 23. Copyright 1971 Editio Cantor KG.)

in contrast to dipyridamole, the candidate compound could increase renal blood flow substantially. As the identifying abbreviation, CRD-401 (i.e., diltiazem), implies, the candidate compound was investigated as a drug to dilate both coronary and renal blood vessels.

Particularly at the begining of the research, pharmacological studies were conducted to test possible uses of diltiazem as an antianginal agent. In various experimental models the drug was found to increase coronary collateral blood flow and reduce myocardial oxygen consumption.

The vasodilating action of diltiazem is caused by a direct action of the drug on vascular smooth muscles. In the process of analyzing the mechanism of action of the drug, we found a calcium antagonistic property. Subsequent electrophysiological studies showed that diltiazem blocks calcium channels in cardiac and smooth muscles. In addition to these properties, antianginal/antiischemic, antihypertensive, and antiarrhythmic actions of diltiazem were proved. Diuretic, cerebrovasodilating, anti-platelet-aggregating, and antiarteriosclerotic effects were also found. Thus, the usefulness of diltiazem was elucidated step by step.

Basic Experiments Predicting Antianginal Effect

The most important aspect of the pharmacological studies of diltiazem was predicting clinical efficacy from animal models during drug development. At that time, nitroglycerin had been generally approved as an effective drug for the treatment of angina pectoris. Therefore, we considered that diltiazem should have a similar effect as nitroglycerin. Several mechanisms of action for nitroglycerin had been proposed, such as a reduction in venous return and a dilating effect on the coronary collateral vessels. Pharmacologists in the Biological Research Laboratory focused on the latter effect, because it suggested the effective distribution of blood flow to ischemic myocardium (*24*). To test this effect, it was necessary to develop coronary collateral vessels without causing myocardial infarction. For this purpose, ameroid constrictors were obtained from the United States and the experiment shown in Figure 5 was conducted. When the effect of diltiazem on the developed coronary collateral circulation was studied, the drug caused a longer vasodilating effect on coronary collateral vessels than on total coronary blood flow (*25*). Diltiazem apparently has a relatively high selectivity for the collateral arteries, and in this respect it differs from the coronary vasodilators that increase total coronary blood flow. Diltiazem was therefore thought to increase blood flow to ischemic areas without causing the "steal phenomenon"—the stealing of blood from ischemic areas.

The various findings and hypotheses on the manner of diltiazem's actions convinced pharmacologists that diltiazem would have a positive effect in double-blind clinical studies. The experimental work also suggested that diltiazem acts directly on large coronary arteries that are independent of autoregulation. In this connection, diltiazem was later found to improve intramyocardial distribution of blood flow and to be effective for vasospasm

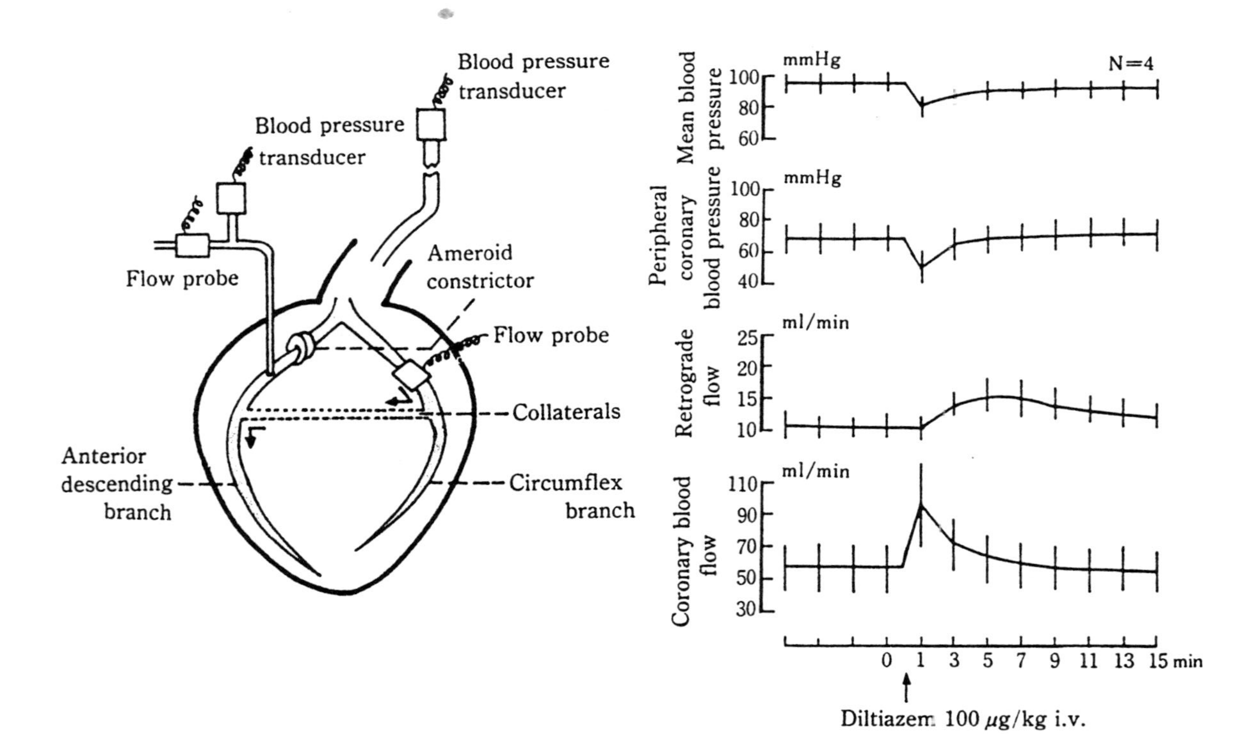

Figure 5. Diltiazem dilates the collaterals of coronary arteries in dogs.

of the coronary artery. Investigation of the collateral circulation, therefore, was the crucial study in the development of diltiazem.

The effect of diltiazem on coronary collateral circulation was reexamined at several institutes at home and abroad, and by several methods. The selectivity of diltiazem for the coronary collateral circulation was of central interest in the international development of diltiazem. Although redistribution of myocardial blood flow is also possibly affected by a decrease in heart rate, few calcium antagonists have so far shown the improvement in redistribution of myocardial blood flow. The effects of diltiazem may contribute to its efficacy for non-Q-wave myocardial infarction (*26*).

Calcium Antagonistic Effect

Discovery. A major problem in the development of diltiazem was to elucidate its mechanism of action. The coronary vasodilating effect of diltiazem was considered to be a direct effect on vascular smooth muscle, because the effect was not due to an increase in myocardial oxygen consumption and was not antagonized by a variety of receptor antagonists (*18*). The coronary vasodilating effect was observed in the isolated perfused heart as well as in the canine heart lung preparation, in which the coronary vasodilating effect was much more pronounced than diltiazem's negative inotropic effect (*23*). In the pharmacological study, diltiazem showed noncompetitive inhibition of various receptor agonists in isolated smooth-muscle preparations (*17*). This finding indicated that diltiazem is a papaverinelike drug.

Following Ebashi's demonstration in the 1960s that the factor that controls the contraction and relaxation of skeletal muscles is the concentration of intracellular free calcium ions, the role of calcium ions in cellular function attracted much attention, especially in Japan. The researchers in the Biological Research Laboratory were educated by H. Kumagai, Ebashi's teacher, and some of them had studied with Ebashi at the University of Tokyo. Thus, the pharmacologists in the Biological Research Laboratory were in a good position to study the effects of calcium ions.

Diltiazem strongly depressed the potassium-induced contraction in isolated intestines and decreased the maximum response, whereas the *l*-cis isomer showed a much weaker effect (*17*). Of interest, these non-receptor-mediated actions are stereoselective. The pharmacologists in the Biological Research Laboratory thought that diltiazem probably inhibited transmembrane calcium influx. After discussion of this data with K. Tagagi of the University of Tokyo, who had studied papaverinelike action in smooth muscle, an experiment was designed to study the effect of diltiazem under conditions of a changing external calcium concentration.

The calcium antagonistic action of diltiazem was thus first discovered during experiments on isolated rabbit ears (*27*). To evaluate the effect of diltiazem on calcium-induced vascular contractions, rabbit ear arteries were perfused with high-potassium Locke solutions containing various concentrations of calcium ions. In this preparation, diltiazem shifted the concentration-response curves for calcium ions to the right; the *l*-cis isomer showed a similar effect but to a lesser extent (Figure 6). These observations demonstrated that calcium and diltiazem are antagonistic, and moreover pharmacologists thought that some kind of drug receptor should exist (*28*). The antagonistic relationship was then studied in detail in canine coronary arteries, and it was concluded that diltiazem and calcium ions have a competitive antagonistic relationship (*29*). Thus, the calcium antagonistic effect of diltiazem was found in the course of studies on the mechanism of its vasodilating action. Then diltiazem was the first drug to be called a "calcium antagonist" on the market.

Electrophysiological Studies. The calcium antagonistic effect of diltiazem was also confirmed in cardiac muscles. Diltiazem decreased the contraction of isolated cardiac muscles of guinea pigs at concentrations higher than those needed to relax the vascular smooth muscle of coronary arteries. By analyzing the relationship between the concentration of diltiazem and the amount of calcium ions needed to cause the original level of contraction, we identified a constant 1:100 molar ratio between the two (*30*). Thus, a competitive antagonism was observed between diltiazem and calcium ions.

To study the mechanism of the apparent calcium antagonistic effect of diltiazem, a partially depolarized cardiac muscle preparation was used. In a high-potassium solution, the sodium channels were blocked, and the action potentials depended on the calcium channels. Papillary muscle contraction was inhibited by the addition of diltiazem and recovered after the addition of calcium ions. Likewise, both overshoot and maximum rate of rise in the action potential were reduced by diltiazem and then recovered with the addition of calcium ions (*31*). This was the first evidence that diltiazem has a blocking effect on transmembrane calcium influx. Then, from electrophysiological studies using taenia coli (*32*), portal vein (*33*), and cardiac muscle (*34, 35*), it was found that the calcium antagonistic action of diltiazem was due to the blocking of voltage-dependent calcium channels.

In the meantime, Fleckenstein had posited calcium antagonists in a study of the cardiac depressant mechanism of some coronary vasodilators (*36*). In 1972, his group reported that D-600, a methoxy derivative of

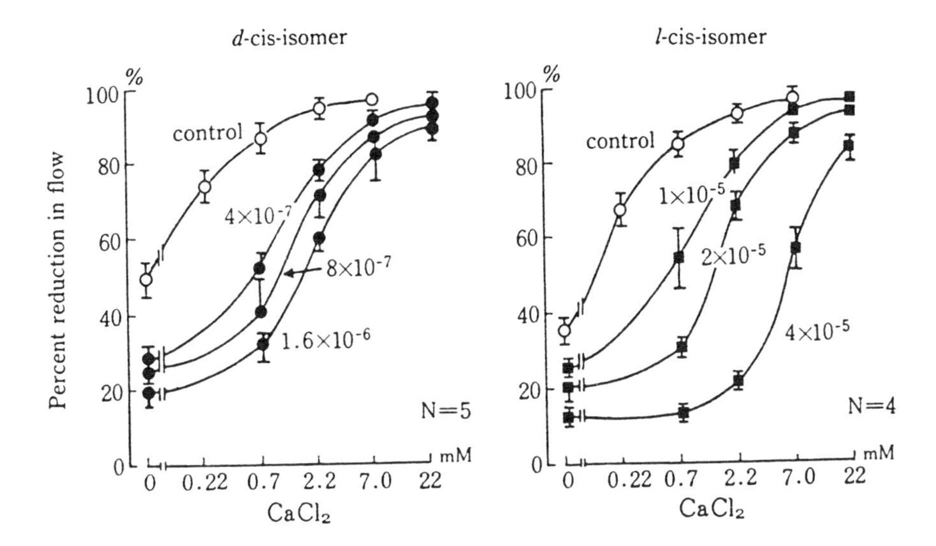

*Figure 6. Diltiazem (*d-cis *isomer) competitively inhibits the Ca^{2+}-induced contractions of the rabbit's ear artery (perfused with potassium chloride-Locke solution). (Reproduced from ref. 27. Copyright 1972.)*

verapamil, selectively blocked the calcium current (*37*). In 1975 he recognized, based on Tanabe's electrophysiological studies, that diltiazem is one of the calcium antagonists (*38*). Thus, diltiazem contributed to the establishment of a new drug group, the calcium antagonists.

Using the fluorescence of aequorin, diltiazem was found to reduce the calcium ion content in cardiac muscle (*39*). Electrophysiological studies finally determined that diltiazem acts on L-type calcium channels. This is the basis for the antiarrhythmic effect of diltiazem, which is also a typical class IV antiarrhythmic drug.

The differences in the effects of diltiazem, verapamil, and 1,4-dihydropyridine calcium antagonists can be analyzed by using new electrophysiological methods. Diltiazem is more closely related to verapamil; however, it can also exhibit some of the intermediate effects of both 1,4-dihydropyridine and verapamil, depending on the experimental conditions (*40*).

Radioreceptor Binding to L-Type Calcium Channels. Radioligands of selective calcium antagonists, including the 1,4-dihydropyridine derivatives, have been used to determine the exact binding sites of the three representative calcium antagonists, the phenylalkylamines, 1,4-dihydropyridines, and 1,5-benzothiazepines (Figure 7). The binding sites as well as the interactions between sites are different for each drug class (*41, 42*). For instance, for the specific binding of ^{3}H-nitrendipine, nifedipine competitively inhibits binding, verapamil and *l-cis*-diltiazem partially inhibit binding, and diltiazem potentiates binding. This unexpected result is regarded as one of the reasons why diltiazem was classified into an independent group in the World Health Organization's classification of calcium channel blockers (*43*). The structure of calcium channels has been determined, and specific binding sites for representative calcium antagonists have been studied. In an experiment with azidobutyryl diltiazem, the binding site of diltiazem proved to be the α_1 subunit of the T-tubule calcium channel of rabbit skeletal muscle, different from the binding sites of other calcium channel blockers (*44*). The stereoselectivity of diltiazem and a receptor-binding site for 1,5-benzothiazepine, presumed in early studies, have now been proved.

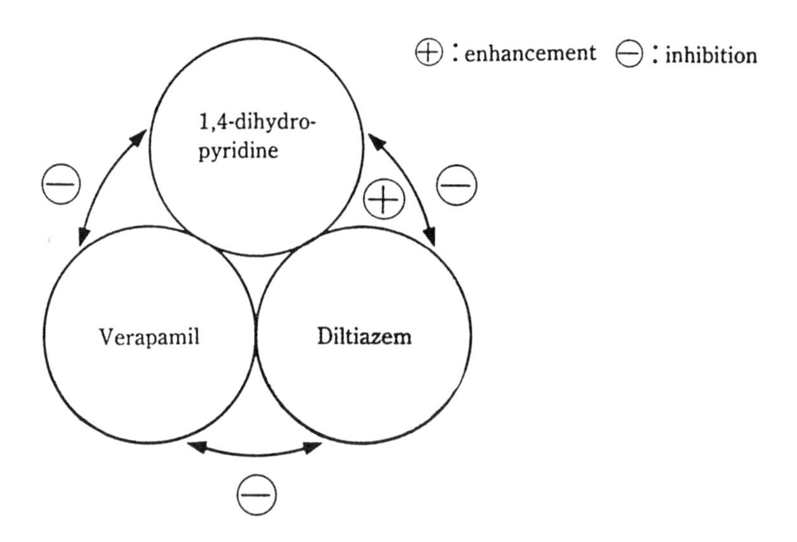

Figure 7. Different Ca^{2+} antagonists have different binding sites in Ca^{2+} channels. Interactions between antagonists are also reported to be different.

Antihypertensive Effect

Diltiazem is an effective antihypertensive agent. It reduces total peripheral resistance at a dose that dilates the coronary arteries. In addition to its antihypertensive action, diltiazem also dilates the renal arteries (*23*). Both its renal vasodilating effect and its diuretic effect were enhanced under angiotensin II-loaded conditions (*45*). The hypotensive effect of diltiazem was also noted in spontaneously hypertensive rats, even in low doses, but the duration of action was rather short. (This was later found to be a species difference.)

The arterial blood pressure in spontaneously hypertensive rats increases during the 4th to 12th weeks of age. When 30 mg/kg of diltiazem was administered at the begining of the 5th week and continued for 30 days, the increase in blood pressure was significantly reduced after the 21st day of administration. Even after termination of drug administration, the blood pressure did not immediately rise to the level seen in control rats (*46*).

Although the antihypertensive effect of diltiazem in renal hypertension and renal vascular hypertension had been confirmed in the clinical stage of drug development, its usefulness in essential hypertension had not yet been studied. After the introduction of diltiazem as an antianginal agent, it was observed that blood pressure fell slowly but definitely when ischemic heart disease accompanied by hypertension was treated with diltiazem. At the time, there was no plan to develop diltiazem as an antihypertensive agent. In 1977, however, a double-blind clinical trial of diltiazem against a diuretic agent was started. At the same time, pharmacological studies of diltiazem as an antihypertensive agent resumed and several effects were confirmed.

Treatment with vasodilators has not been the first choice in many cases, owing to their side effects, such as increases in heart rate, plasma renin activity, and the retention of sodium and water. When diltiazem was administered orally to conscious, spontaneously hypertensive rats, the blood pressure decreased in a dose-dependent manner and the heart rate decreased slightly (*47*). Thus, reflex tachycardia was not readily seen with diltiazem, compared to the 1,4-dihydropyridine calcium antagonists.

Diltiazem was more effective in rats with renal hypertension and DOCA-salt hypertension than in normal rats (*48*). Unlike hydralazine, diltiazem showed a more potent hypotensive effect in spontaneously hypertensive rats than in a normotensive Wistar–Kyoto strain (*47*). These results indicated that the hypotensive effect of diltiazem is enhanced in the hypertensive state. This is one of the reasons why the antihypertensive effect of diltiazem is easily observed in patients with hypertension.

Summary

Diltiazem was derived from a lead compound found in an original random screening test and developed as an excellent drug by Tanabe researchers (*49*). Tanabe researchers synthesized it and found its pharmacological effects, including its calcium antagonistic action. Its stereochemistry and the stereoselectivity of its biological activities attracted the interest of researchers. Finally, we realized that diltiazem had a specific binding site in the L-type calcium channel.

Diltiazem is one of a small number of original, important drugs that Japan has contributed to the world. Japanese clinical researchers played a central role in uncovering its clinical efficacy. U.S. and European scientists and clinical researchers, together with pharmaceutical companies, conducted additional basic research and clinical studies on diltiazem, a representative calcium antagonist.

Acknowledgments

The work described in this chapter represents the efforts of many Tanabe scientists, all of whom we acknowledge and thank. We also thank Ichiro Chibata, President of Tanabe Seiyaku Co., Ltd., Japan, for his encouragement during the discovery and development of diltiazem.

Literature Cited

1. Krapcho, J.; Spitzmiller, E. R.; Turk, C. F. *J. Med. Chem.* **1963,** *6*, 544–546.
2. Krapcho, J.; Turk, C. F. *J. Med. Chem.* **1966,** *9*, 191–195.
3. Krapcho, J.; Turk, C. F.; Piala, J. J. *J. Med. Chem.* **1968,** *11*, 361–364.
4. Bell, S. C.; Childress, S. *J. Org. Chem.* **1962,** *27*, 1691–1695.
5. Mills, W. H.; Whitworth, J. B. *J. Chem. Soc.* **1927,** 2738–2753.
6. Culvenor, C. C. J.; Davies, W.; Heath, N. S. *J. Chem. Soc.* **1949,** 278–282.
7. Kugita, H.; Inoue, H.; Ikezaki, M.; Takeo, S. *Chem. Pharm. Bull.* **1970,** *18*, 2028–2037.
8. Kugita, H.; Inoue, H.; Ikezaki, M.; Konda, M.; Takeo, S. *Chem. Pharm. Bull.* **1970,** *18*, 2284–2289.

9. Nagao, T.; Sato, M.; Nakajima, H.; Kiyomoto, A. *Chem. Pharm. Bull.* **1973,** *21,* 92–97.
10. Inoue, H.; Takeo, S.; Kawazu, M.; Kugita, H. *Yakugaku Zasshi* **1973,** *93,* 729–732.
11. Hashiyama, T.; Inoue, H.; Aoe, K.; Kotera, K.; Takeda, M. *J. Chem. Soc., Perkin Trans. 1* **1985,** 421–427.
12. Hashiyama, T.; Inoue, H.; Konda, M.; Takeda, M. *J. Chem. Soc., Perkin Trans. 1* **1984,** 1725–1732.
13. Hashiyama, T.; Inoue, H.; Konda, M.; Takeda, M. *Chem. Pharm. Bull.* **1985,** 33, 1256–1259.
14. Kugita, H.; Inoue, H.; Ikezaki, M.; Konda, M.; Takeo, S. *Chem. Pharm. Bull.* **1971,** *19,* 595–602.
15. Gaino, M.; Iijima, I.; Nishimoto, S.; Ikeda, K.; Fujii, T. *Jpn. Kokai* **1983,** 58-99471; *Chem. Abstr.* **1983,** *99,* 175819v.
16. Senuma, M.; Shibazaki, M.; Nishimoto, S.; Shibata, K.; Okamura, K.; Date, T. *Chem. Pharm. Bull.* **1989,** *37,* 3204–3208.
17. Nagao, T.; Sato, M.; Iwasawa, Y.; Takada, T.; Ishida, R.; Nakajima, H.; Kiyomoto, A. *Jpn. J. Pharmacol.* **1972,** *22,* 467–478.
18. Nagao, T.; Sato, M.; Iwasawa, Y.; Takada, T.; Ishida, R.; Nakajima, H.; Kiyomoto, A. *Jpn. J. Pharmacol.* **1972,** *22,* 1–10.
19. Tanaka, T.; Inoue, H.; Date, T.; Okamura, K.; Aoe, K.; Takeda, M.; Kugita, H.; Murata, S.; Yamaguchi, T.; Kikkawa, K.; Nakajima, S.; Nagao, T. *Chem. Pharm. Bull* **1992,** *40,* 1476–1480.
20. Hashiyama, T.; Watanabe, A.; Inoue, H.; Konda, M.; Takeda, M.; Murata, S.; Nagao, T. *Chem. Pharm. Bull.* **1955,** *33,* 634–641.
21. Hashiyama, T.; Inoue, H.; Takeda, M.; Murata, S.; Nagao, T. *Chem. Pharm. Bull.* **1985,** *33,* 2348–2358.
22. Gorlin, R.; Brachfeld, N.; MacLeod, C.; Bopp, P. *Circulation* **1959,** *19,* 705–718.
23. Sato, M.; Nagao, T.; Yamaguchi, I.; Nakajima, H.; Kiyomoto, A. *Arzneimittelforschung* **1971,** *21,* 1338–1343.
24. Fam, W. M.; McGregor, M. *Circ. Res.* **1964,** *15,* 355–365.
25. Nagao, T.; Murata, S.; Sato, M. *Jpn. J. Pharmacol.* **1975,** *25,* 281–288.
26. Gibson, R. S.; Boden, W. E.; Theroux, P.; Strauss, H. D.; Pratt, C. M.; Gheorghiade, M.; Capone, R. J.; Crawford, M. H.; Schechtman, R. C.; Perryman, M. B.; Roberts, R.; The Diltiazem Reinfarction Study Group. *N. Engl. J. Med.* **1986,** *315,* 423–429.
27. Nagao, T.; Sato, M.; Nakajima, H., Kiyomoto, A. *Abstracts of Papers,* The First Symposium in Pharmacological Activities, Tokushima, Japan, 1972, pp 97–100.

28. Nagao, T.; Sato, M.; Nakajima, H.; Kiyomoto, A. In *Structure–Activity Relationships in Drug Actions*; Ishida, U.; Muraoka, S., Eds.; Hirokawa: Tokyo, 1974; pp 157–173.
29. Nagao, T.; Ikeo, I.; Sato, M. *Jpn. J. Pharmacol.* **1977,** *27,* 330–332.
30. Nakajima, H.; Hoshiyama, M.; Yamashita, K.; Kiyomoto, A. *Jpn. J. Pharmacol.* **1975,** *25,* 383–392.
31. Nakajima, H.; Hoshiyama, M.; Yamashita, K.; Kiyomoto, A. *Jpn. J. Pharmacol.* **1976,** *26,* 571–580.
32. Magaribuchi, T.; Nakajima, H.; Kiyomoto, A. *Jpn. J. Pharmacol.* **1977,** *27,* 361–369.
33. Takenaga, H.; Magaribuchi, T.; Nakajima, H. *Jpn. J. Pharmacol.* **1978,** *28,* 457–464.
34. Yatani, A. *Rinsho Kenkyu* (*Jpn. J. Clin. Exp. Med.*) **1976,** *53,* 562–569.
35. Morad, M.; Tung, L.; Greenspan, A. M. *Am. J. Cardiol.* **1982,** *49,* 595–601.
36. Fleckenstein, A. *Circ. Res.* **1983,** *52* (Suppl. 1), 3–16.
37. Kohlhardt, M.; Bauer, B.; Krause, H.; Fleckenstein, A. *Pflugers Arch.* **1972,** *335,* 309–322.
38. Fleckenstein, A. *Annu. Rev. Pharmacol. Toxicol.* **1977,** *17,* 149–166.
39. Morgan, J. P.; Weir, W. G.; Hess, P.; Blinks, J. R. *Circ. Res.* **1983,** *52* (Suppl. 1), 47–52.
40. Lee, K. S.; Tsien, R. W. *Nature (London)* **1983,** *302,* 790–794.
41. Glossman, H.; Ferry. D. R.; Goll, A.; Rombush, M. J. *J. Cardiovasc. Pharmacol.* **1984,** *6,* s608–s621.
42. Depover, A.; Matlib, M. A.; Lee, S. W.; Dube, G. P.; Grupp, I. L.; Grupp, G.; Schwartz, A. *Biochem. Biophys. Res. Commun.* **1982,** *108,* 110–117.
43. Vanhoutte, P. M.; Paoletti, R. *Trends Pharmacol. Sci.* **1987,** *8,* 4–5.
44. Naito, K.; McKenna, E.; Schwartz, A.; Vaghy, P. L. *J. Biol. Chem.* **1989,** *264,* 21211–21214.
45. Yamaguchi, I.; Ikezawa, K.; Takada, T.; Kiyomoto, A. *Jpn. J. Pharmacol.* **1974,** *24,* 511–522.
46. Sato, M.; Murata, S.; Narita, H.; Tomita, M.; Yamashita, K.; Yamaguchi, I. *Folia Pharmacol. Jpn.* **1979,** *75,* 99–106.
47. Narita, H.; Nagao, T.; Yabana, H.; Yamaguchi, I. *J. Pharmacol. Exp. Ther.* **1983,** *227,* 472–477.
48. Yamaguchi, I.; Ikezawa, K.; Murata, S.; Narita, H.; Ikeo, T.; Sato, M. *Folia Pharmacol. Jpn.* **1979,** *75,* 191–199.
49. *Diltiazem from Birth to Today*; Tanabe Seiyaku Co., Ltd.: Osaka, Japan, 1987.

Aztreonam

C. M. Cimarusti
Bristol Myers Squibb
Pharmaceutical Research Institute

Historical Setting

The microbial world provided the initial knowledge that led to successful drug discovery and development programs in each of the first three decades of existence of the Squibb Institute for Medical Research. The *Penicillium notatum* culture that Paul Florey brought to New Brunswick in 1941 led to the development and marketing of penicillin G several years later. Microbiologists working at a New York state laboratory provided a streptomycete culture that involved Institute scientists in the early 1950s in a second successful development program: nystatin (NY-STATe-IN). This program complemented the internal effort that led to the discovery and development of amphotericin B from another streptomycete. During the 1960s, the Institute's synthetic antimicrobial group considered the synthesis of analogues of bacilysin (**1**). In the course of retrosynthetic analysis, attention initially focused on the potential utility of analogue **2** as a chemical precursor of **1**. The same analysis that suggested **3** as a precursor of **2** also prompted speculation on the structure–activity relationship between aromatic and dihydroaromatic moieties in compounds with biological activity. This insight eventually led to the synthesis of cephradine (**5**) (*1*) as an analogue of cephalexin (**4**) and, somewhat later, to "dihydro"-propranolol (**7**) as an analogue of propranolol (**6**). A structure–activity analysis of many oxygenat-

2523–0/93/0239$14.85/0

ed analogues of **7** finally resulted in the synthesis of nadolol (**8**) (2). As a new member of the antimicrobial group in 1973, I found a general expectation that the microbial world would, once again, inform the discovery process. This time, however, the decade would be almost gone before the event.

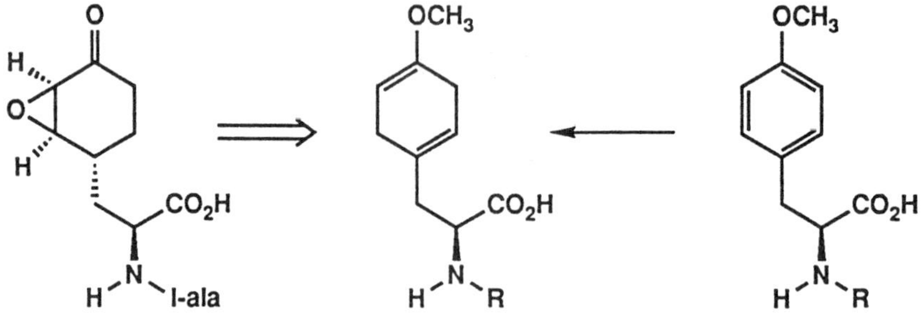

1 (Bacilysin) **2** **3**

Worldwide, natural products programs provided a variety of new structures during the 1970s: cephamycins were described in 1971 (3), nocardicin and clavulanic acid in 1976 (4, 5), and thienamycin in 1978 (6). Similarly, the description of penems (7) and cefotaxime (8) provided additional impetus for synthetic work. Several of these lines of research terminated, in our laboratories, in the cephamycin SQ 14,359 (**9**) and the aminothiazole-containing cephem SQ 25,668 (**10**). Work on the former compound was abandoned when the superior anti-Gram-negative activity of cefotaxime was fully appreciated. The latter compound was investigated in an effort to exploit this activity; it was not developed because of its limited spectrum of activity (anti-Gram-negative only) and the perceived difficulty of establishing a clear-cut compound- and process-patent position. As will be apparent later, the experience and synthetic intermediates generated during these programs played a key role in the subsequent discovery and development of aztreonam. In addition, having thought through the issues that prevented development of **10**, we were in a better position to address the same issues as they arose in the development of aztreonam.

Also in the 1970s, a parallel effort in the antimicrobial area was established at Squibb with the aim of providing basic information on transpeptidases and β-lactamases, the key enzymes involved in the struggle between humans and bacteria. This program, directed by Miguel Ondetti, was established after his basic work on the angiotensin-converting enzyme led to the discovery and development of captopril (9). Ondetti and his

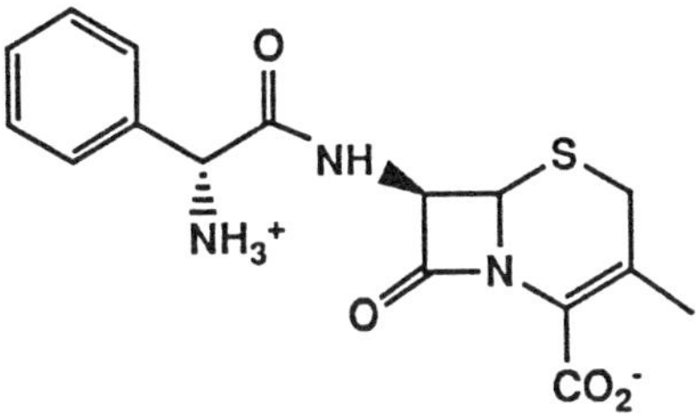

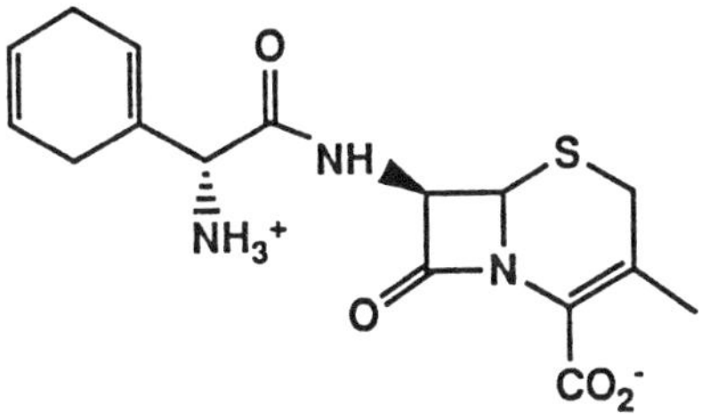

4 (Cephalexin)

5 (Cephradine)

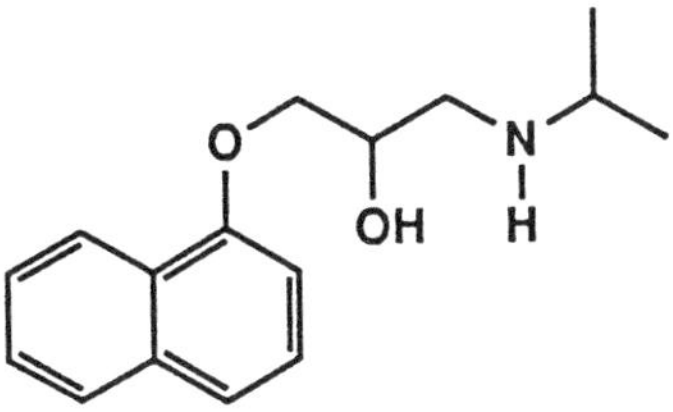

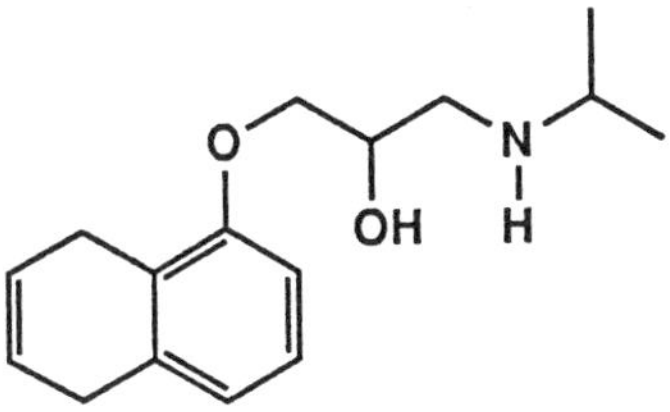

6 (Propranolol)

7

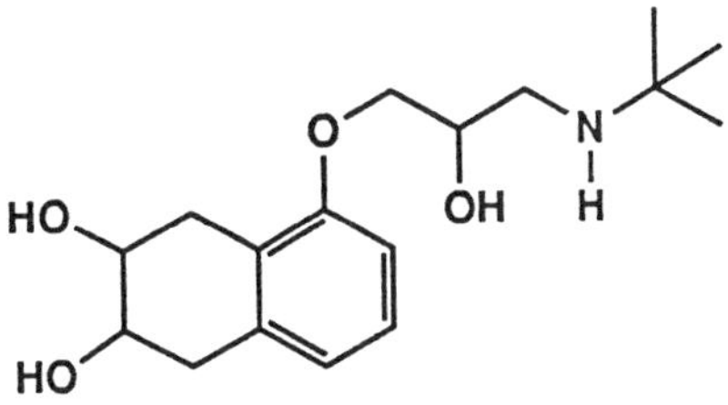

8 (Nadolol)

colleagues, many of whom played a role in the subsequent development of monobactams, envisioned a designed inhibitor of one or both of these enzyme groups. The systems they put into place to evaluate synthetic inhibitors informed the early investigations of the mechanism of action of the monobactams.

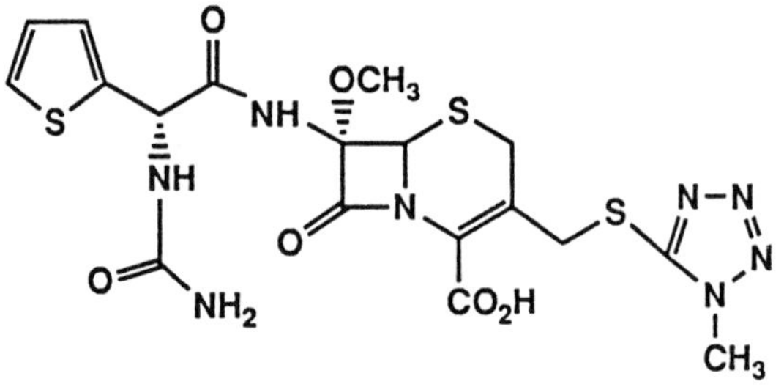

9 (SQ 14,359)

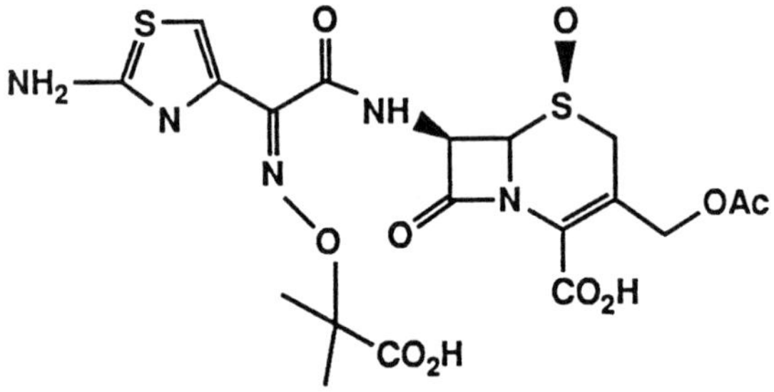

10 (SQ 25,668)

Another enzyme selected for examination was alanine racemase. Synthetic inhibitors of this enzyme included compounds with various leaving groups at the β-carbon atom of D-alanine (*10*). The intermediate **11**, first prepared as a precursor of potential inhibitor **12**, was eventually converted to the inactive enantiomer of nonmethoxylated monobactams. The absence of activity of these materials provided additional assurance that monobactams shared a penicillinlike mechanism of action. A homologue of **11** (the *N-tert*-butyloxycarbonyl [BOC] derivative of *d*-threonine) was also prepared during this program; it played a curious and important role in the rapid development of aztreonam.

Thus, when Richard B. Sykes and his colleagues in the Microbiology Department discovered, developed, and began to use the screen that uncovered the monobactams, expectation and interest, relevant experience, and significant resources were at hand to exploit their discovery.

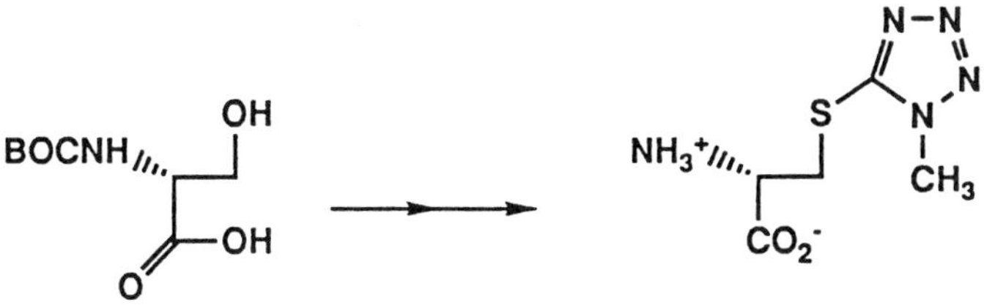

BOCNH OH OH O → NH_3^+ S N N N N CH_3 CO_2^-

11 **12**

The Discovery of Aztreonam

The Screen

The most difficult task in the search for new antibiotics is "dereplication": the identification of a microorganism as a replicate of a previously described antibiotic producer and, most important, the identification of an antibiotic as a replicate of one previously described. As more and more antibiotics were described, this task became the central issue of new antibiotic screening. In the early 1970s our primary screen was a simple one: in vitro activity against a Gram-negative bacterium. The high percentage of active cultures encountered in similar screens provoked a search for more selective screening criteria in laboratories throughout the world.

At Squibb, that search first centered on the discovery of a selective screen for β-lactam antibiotics. The safety, efficacy, and commercial success of β-lactams provided a powerful incentive. The continued emergence of strains resistant to then-available β-lactams provided additional incentive. A third incentive was the familiarity of the β-lactam field to Squibb scientists: Richard Sykes, the discoverer of the basis of the selective screen, had been associated with the β-lactam field since his graduate research (*11*).

In addition to selectivity, the ideal screening methodology should be sensitive and adaptable to a simple plate assay so that high throughput (>1,000 organisms/day) and bioautography can be achieved. The advantage of high sensitivity is obvious: new structures can be detected at low concentrations. High throughput allows *all* encountered organisms to be screened without applying any preconceived criteria for admission to the screen. Finally, the use of the screen in a bioautographic mode allows rapid "dereplication" by thin-layer chromatography and electrophoretic comparison to the handful of reference, naturally occurring β-lactams.

Sykes conceived a selective screen early in 1978, and he and his colleagues rapidly developed it to the point that it fit all of these criteria. It was sensitive: it could be used to detect β-lactams at concentrations that were too low to inhibit bacterial growth. It was specific: the only exception noted was the discovery of antimicrobially active β-lactones such as **13** (*12*). It was easily adaptable to a plate assay and capable of high throughput: through the use of the blob technique, multiple organisms could be screened on one agar plate. It could be used in a bioautographic mode: replicate chromatographic or electrophoretic separation, followed by bioautography on the screening organism for β-lactamase induction and on other indicator organisms that were sensitive or resistant to reference β-lactams, provided a powerful dereplication tool. The screen was simple enough to be used by isolation chemists to inform the development of isolation protocols and rapid enough that two sequential experiments could be performed daily; the second experiment, using the specific information obtained from the first, could be completed and fully analyzed the same day. Finally, it was very rugged: the chemist, W. L. "Larry" Parker, who isolated the first monobactam at Squibb, would routinely tuck plates under his arm, take them home, and incubate them on top of his furnace. After the evening news, the chromogenic reagent could be added and the results read and used to plan the next day's experiments.

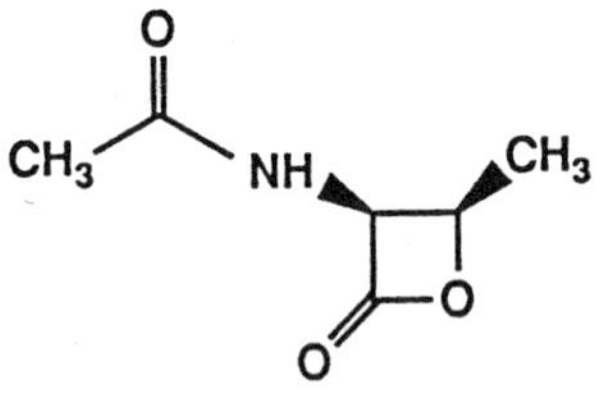

13

The basis for this screen has been described (*13*). It relies on the ability of β-lactams to induce the formation of β-lactamase in a strain of *Bacillus*, and the ability to detect β-lactamase with a chromogenic cephalosporin substrate.

The Soil Sample

The discovery of penicillin followed a chance occurrence—the observation of a lytic zone around a colony of a mold that had contaminated an agar plate of a Gram-positive bacterium (*14*). The discovery of the monobactams, at both Takeda Chemical Industries and Squibb, was the result of subjecting soil microorganisms to selective screening methodology. The Takeda screen used mutants that were hypersensitive to β-lactams (*15*).

Squibb's screen was first used with the usual sources of organisms and with the usual organisms (Actinomycetes). The powerful dereplication capability of the screen allowed us to assign the activities encountered in this way to known classes of β-lactams; after several months, no new classes were encountered. To validate this conclusion, an organism that produced an activity that we had assigned as a "carbapenem" was subjected to isolation. After several attempts, an intensive effort (by W. C. Liu) that featured long hours of chromatographic work at 5 °C gave a few milligrams of concentrated activity. The results of ^{1}H NMR spectroscopy allowed the definitive conclusion that the activity was indeed due to the carbapenem class. No further work with this organism was attempted; we believed that the novel insight we sought would not be found in the exact structure of an additional carbapenem.

At this point, Richard Sykes made a momentous decision: to look for novel β-lactams in novel microorganisms. Since the discovery of penicillin, all β-lactams had been discovered as products of the Actinomycetes. Whatever the origin of the soil sample, the techniques used to isolate individual organisms for admission to the screen were usually designed around this fact. Because media and isolation protocols often favored the growth of these organisms, it was no accident that novel β-lactam nuclei (e.g., clavulanic acid and thienamycin) had been discovered in their culture filtrates. To avoid this well-trodden path, Sykes decided to screen soil bacteria for the presence of β-lactams.

At that time (1978), no β-lactam with a penicillinlike mechanism of antibacterial activity had been reported to occur in bacteria, although bacteria were known to produce many antibiotics. The selectivity of β-lactamase induction, however, would preclude the isolation of any antibiotics but the desired β-lactams. The sensitivity of the screen would allow the detection of agents at concentrations far less than those needed to inhibit growth. Finally, the high throughput of the screening protocol would permit the detection of a "one-in-a-million" occurrence. These facts convinced Richard Sykes that a serious effort to search for novel β-lactam structures in bacteria was warranted. This decision at first affected only the screening group; however, once the first ring of induction was noted around a bacterial blob culture, more and more colleagues were enlisted in the venture, and eventually a large part of the discovery, development, regulatory, manufacturing, and marketing resources of the company was engaged. From the beginning, the project assumed a special significance across the entire company. This is best typified by a button that John Talley commissioned early in 1980: a red heart on a white background that proclaimed "I Love Monobactams". The importance of this company-wide enthusiasm for the rapid development of Azactam cannot be overstated.

Curiously, although our soil screening program regularly received soil samples from many parts of the globe, the bacterium that produced the first monobactam isolated at Squibb was a native of the New Jersey Pine Barrens. It was collected on a routine field trip by members of the screening group and brought back to Squibb's Lawrenceville laboratories exactly 50 years after Fleming noticed the lytic zone around his colony of *Penicillium notatum*. The rapid development of aztreonam that followed the observation of an induction zone around a colony of this purple bacterium (5 years to the Azactam NDA filing) owed much to the knowledge accumulated in the intervening 50 years.

SQ 26,180

The first lead delivered to the isolation group from the induction screen was designated EM 5117. The EM designation was used to identify important cultures that distinguished themselves in early screening studies; these numbers were assigned by Edward Myers, the eponymous "EM", a longtime member of the Squibb new antibiotics effort. EM 5117 was assigned to Larry Parker, a member of the isolation group who had performed his doctoral research under the direction of R. B. Woodward. The information that accompanied EM 5117 was extremely exciting: it was both an inducer of and stable to β-lactamase; it was electrophoretically mobile toward the cathode at pH 1.8; it was an antibacterial agent; it was the product of a purple, Gram-negative bacterium (*Chromobacterium violaceum*); and—we were all convinced—it was a novel β-lactam.

Parker devised an efficient isolation scheme (*16*) based on ion-pair extraction and an equally efficient purification method based on several chromatographic steps that preceded crystallization and recrystallization. Cursory examination of the results of elemental analysis ($C_6H_9N_2SO_6K$) and infrared and NMR spectroscopy, coupled with our conviction, based on the induction screen, that EM 5117 was a β-lactam of penicillinlike absolute configuration, allowed only two structures to be considered. Alkylsulfamic acid (**14**) and azetidinone-1-sulfonate (**15**) were the only possible candidates. There was literature precedent for *N*-acylazetidinones (*17*), but there was no report of an azetidinone-1-sulfonate at this time.

Initial interpretation of a single piece of chemical evidence—a positive test for secondary sulfamic acids (*18*)—led us to consider **14** as a likely

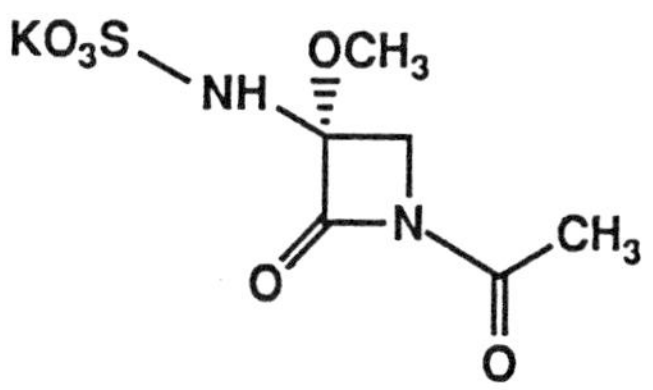

14 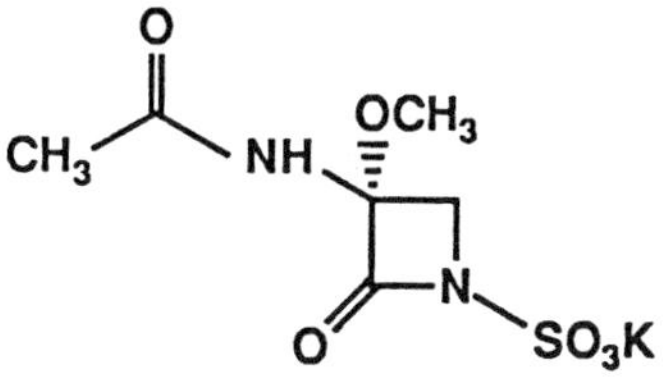 **15**

structure. However, the excellent correspondence of infrared and proton NMR data for this substance and typical cephamycins made **15** much more probable. This was further substantiated by the similarity in chemical shifts of the side-chain amide carbonyl. The chemical instability of EM 5117 provided a reasonable explanation of these contradictory results. Hydrolysis of **15**, which was accompanied by a loss of activity, should lead to the sulfamic acid **16**. If this hydrolysis occurred during the sulfamic acid test, EM 5117 would appear to contain a secondary sulfamic acid moiety. This interpretation allowed us to be convinced by the spectroscopic evidence, and we assigned structure **15** to EM 5117. It was entered into the Squibb Chemical Registry System as SQ 26,180; as mentioned previously, the absolute configuration was assigned on the basis of our belief in the strict enantiospecificity of β-lactamase induction.

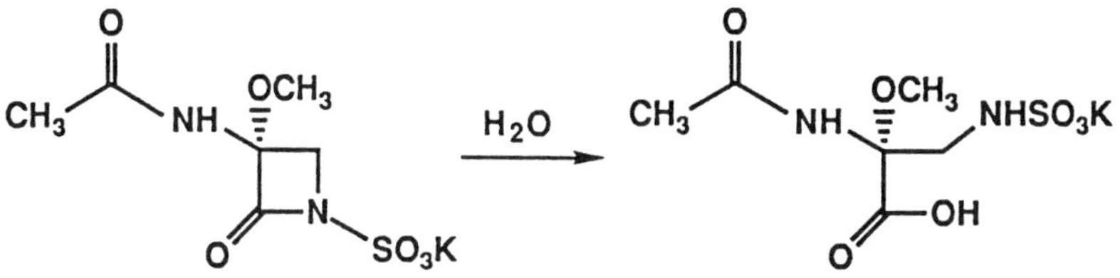

H_2O

15 (SQ 26,180) **16**

Interestingly, no azetidinone-1-sulfonic acid had been reported prior to the assignments made at Takeda and Squibb to naturally occurring monobactams. The corresponding "acid chlorides" (azetidinone-1-sulfonylchlorides such as **17**) were well-known in the chemical literature as the products of cycloadditions of olefins and chlorosulfonylisocyanate (Scheme 1). Reaction of isobutene, for example, provides the 4,4-dimethyl derivative **17** (*17*) in 70% yield. Hydrolysis of this material under acidic or neutral conditions provides the unsubstituted azetidinone **19** (*17*) by preferential N–S bond cleavage, rather than monobactam **18**. One possible explanation for this behavior is indicated in Scheme 1.

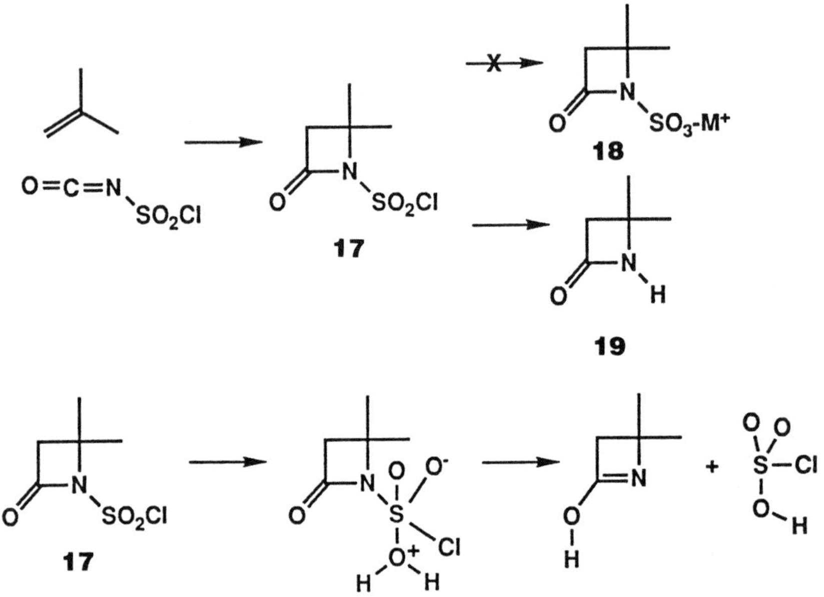

O=C=N
SO$_2$Cl
SO$_2$Cl
17
SO$_3$-M$^+$
18
H
19
17
Cl
H H
H
Cl
H

Scheme 1

If hydrolysis of these readily available acid chlorides had proceeded normally, 2-azetidinone-1-sulfonic acids would have been discovered and reported by synthetic chemists working in the polymer field in the 1960s. The accompanying description of their reactivity would surely have suggested the synthesis of the corresponding 3-acylamino-derivatives to β-lactam chemists long before their recognition as the products of Gram-negative bacteria in 1979.

Once the structural assignment was made, a number of things were perceived to be of utmost importance. The screening group should redouble its already considerable effort in surveying bacteria for additional novel β-lactams. The isolation group should redouble its efforts to isolate and identify other leads that the screen had already uncovered. The biochemical group involved in characterizing potential carboxypeptidase–transpeptidase-β-lactamase inhibitors should examine the interaction of SQ 26,180 with these enzymes and with penicillin-binding proteins of Gram-positive and Gram-negative bacteria. The biochemical group that supported screen development should examine the biosynthesis of SQ 26,180. The β-lactam chemists perceived a like number of imperatives. Side chains that were particularly effective in the cephamycin series should be formally (or actually) exchanged for the acetyl moiety of SQ 26,180. Nonmethoxylated monobactams should be prepared so that their activity and β-lactamase stability could be compared with the same properties of SQ 26,180. In light of our tentative conclusions concerning the mechanism of action (penicillin-like) and β-lactam activation (by electron withdrawal of the sulfonate moiety), the addition of substituents with varied steric and electronic properties at the 4-position seemed very important. Finally, the search for alternative groups that, like the sulfonate moiety, would activate the β-lactam bond of a monocyclic azetidinone was viewed as an equally important goal.

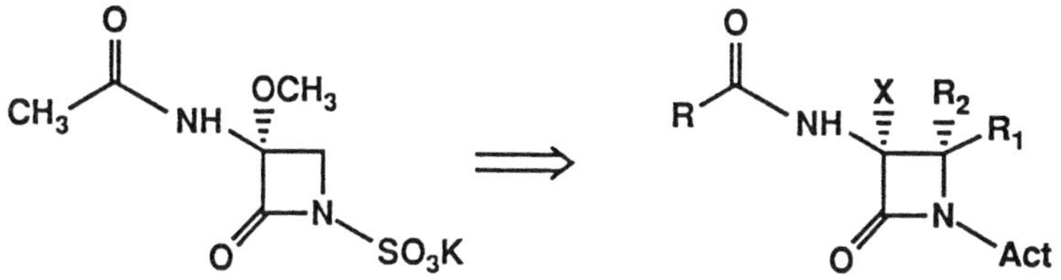

R = moieties suggested by penicillin, cephalosporin, and cephamycin

X = CH_3O, H

R_1, R_2 = any moiety that was synthetically accessible

Act = negatively-charged, electron-withdrawing moiety

Side-Chain Analogues of SQ 26,180

Three lines of synthetic investigation were begun in order to access side-chain analogues of SQ 26,180. One relied on the cephamycin precedent of acyl exchange developed by the Merck group (*19*); however, all attempts to effect conversion of **20** to **22** failed. A likely explanation is the chemical instability of methoxylated monobactams to β-lactam hydrolysis; this instability would be exacerbated in intermediate **21** by the presence of an additional electron-withdrawing substituent and would lead to β-lactam, rather than imide, hydrolysis.

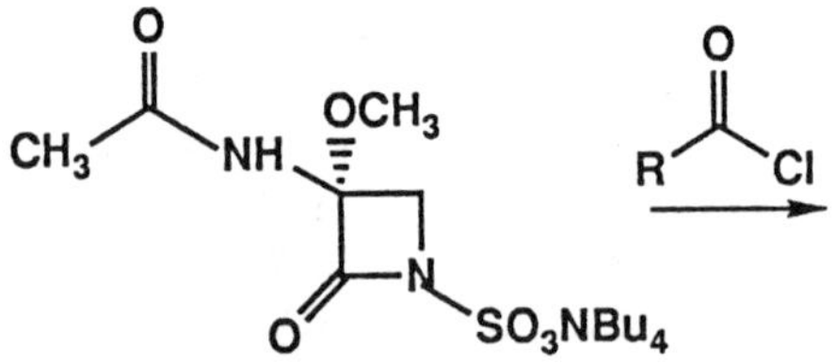

20
(SQ 26,180 tetra-*n*-butylammonium salt)

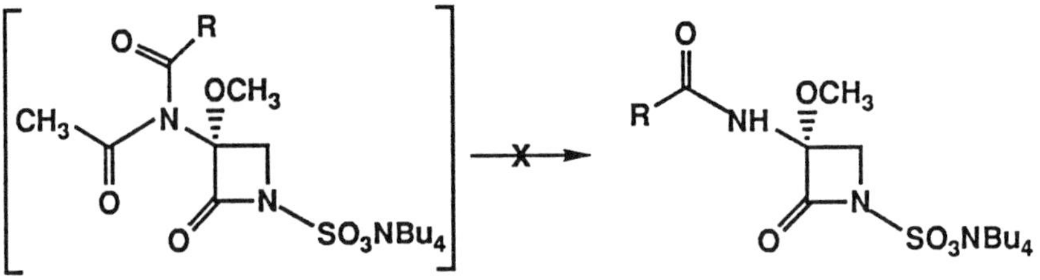

21

22

Another line of investigation was based on our own work in the cephamycin field, which involved reaction of iminochlorides with bidentate nucleophiles that removed the side chain (*20*). In the event, this approach proved impossible because of a curious alternative reaction (Scheme 2): fragmentation of the imino chloride **23** to acetonitrile and an alkyl chloride. All attempts to convert SQ 26,180 (as its tetra-*n*-butylammonium salt **20**) to imino chloride **23**, followed by reaction with 2-mercaptoethylamine, failed to produce deacylation product **24**. Instead, an amphoteric compound was produced that was eventually recognized as **26** (*21*). Evidence for this pathway to **26** included the direct NMR observation of the production of acetonitrile.

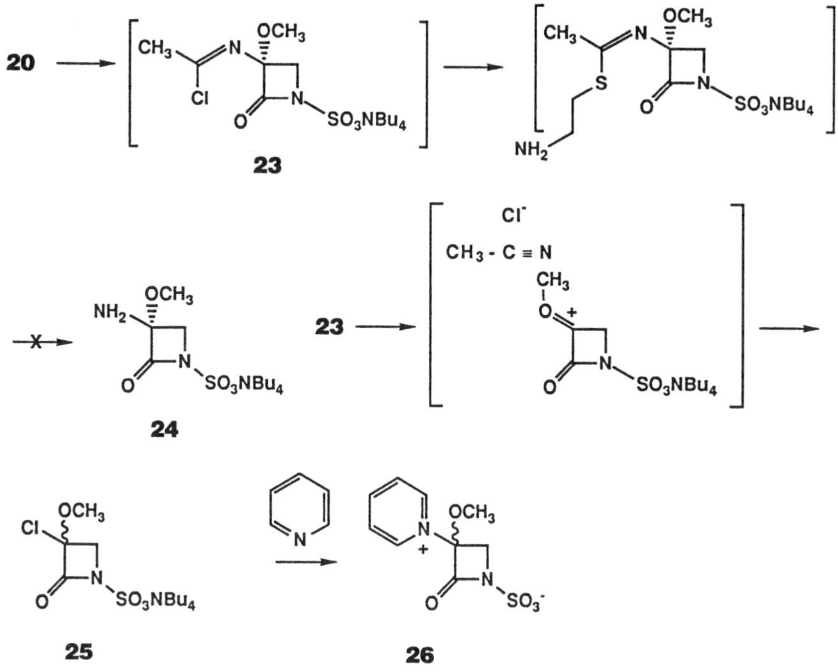

Scheme 2

The third synthetic approach was initially frustrated by our inability to sulfonate the N1-unsubstituted azetidinone **27**. Azetidinone **27** was prepared by the precedented route (*22–24*) illustrated in Scheme 3. The specific rotation of **27** supported our assignment of cephamycinlike absolute configuration to SQ 26,180; however, initial attempts to convert **27** to SQ 26,180 and provide a final substantiation of our structural assignment failed. In retrospect, this failure can be attributed to the propensity of methoxylated monobactams to undergo β-lactam hydrolysis.

The isolation research group, rather than the synthetic group, provided the first side-chain analogue of SQ 26,180 and allowed a comparison of their antimicrobial activity. The second lead (EM 5220) delivered to the isolation group as a novel β-lactam produced by the Gram-negative bacterium *Gluconobacter oxydans* differed from SQ 26,180 in that it was neutral (by electrophoretic mobility) at pH 1.8, although at higher pH it was negatively

Scheme 3

charged. Biological characterization of the activity and initial spectroscopic characterization of the pure substance strongly indicated another methoxylated azetidinone-1-sulfonate. Detailed interpretation of spectra and chemical evidence led to the assignment of structure **28** to EM 5220 (25) and eventual designation as SQ 26,445. SQ 26,445 proved to be identical to sulfazecin, one of two diastereomeric side-chain isomers reported by Imada and the Takeda group as products of the Gram-negative bacterium *Pseudomonas*.

The publication in *Nature* (15) of the paper describing sulfazecin (**29**) and isosulfazecin (**30**) had little impact for us because we were already aware of the Takeda group's discovery of monobactams. In 1980 we became aware of a Takeda patent application that described the synthesis of methoxylated and nonmethoxylated analogues of SQ 26,180. After a careful reading of this application, we decided to continue with our high-priority effort, for the patent did not disclose the line of investigation (4-substitution) we believed, at the time, to be the most fruitful.

The comparison of antimicrobial activity of SQ 26,180 and SQ 26,445 (Table 1) provided useful information despite the relatively poor activity of both. It established that side-chain modification would result in modified antimicrobial activity. Extrapolating from penicillin, cephalosporin, and cephamycin precedent, we believed this would be the case; however, it was reassuring to have proof.

We eventually succeeded in developing a general synthesis of racemic analogues of SQ 26,180. As illustrated in Scheme 4, the approach utilized the benzyloxycarbonyl group to facilitate introduction of the required methoxyl

Table 1. Antimicrobial Activity of SQ 26,180 and SQ 26,445 Compared

Organism	*SC No.*	*SQ 26,180*	*SQ 26,445*
Staphylococcus aureus	1276	50	>100
Staphylococcus aureus	2400	50	>100
Escherichia coli	8294	>100	50
Escherichia coli	10857	>100	100
Klebsiella aerogenes	10440	>100	50
Proteus mirabilis	3855	>100	12.5
Proteus vulgaris	9416	>100	6.3
Enterobacter cloacae	8236	>100	50
Serratia marcescens	9783	25	25
Pseudomonas aeruginosa	9545	3.1	50
Pseudomonas aeruginosa	8329	50	100

NOTE: Activities are given as minimum inhibitory concentration (MIC) in micrograms per milliliter at 10^4 colony-forming units (cfu).

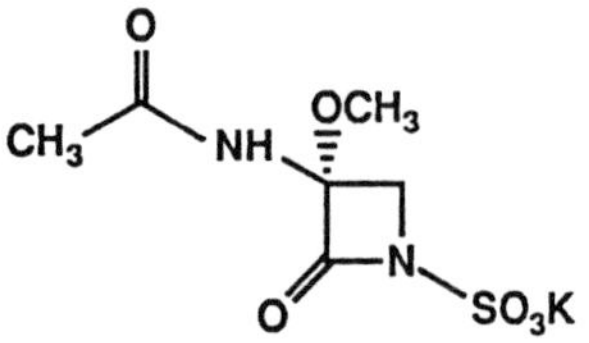

15 (SQ 26,180)

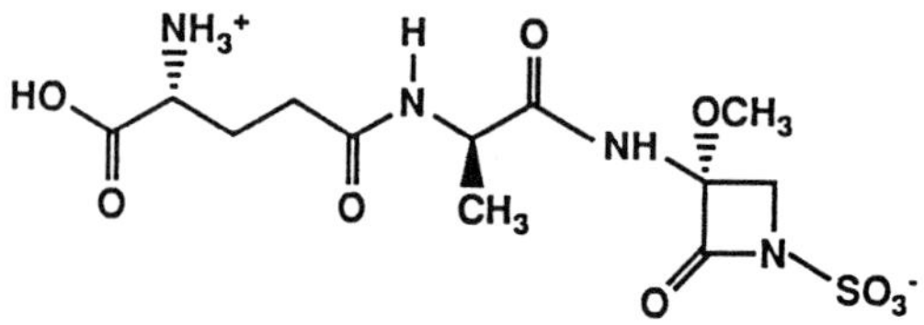

28 (SQ 26,445)

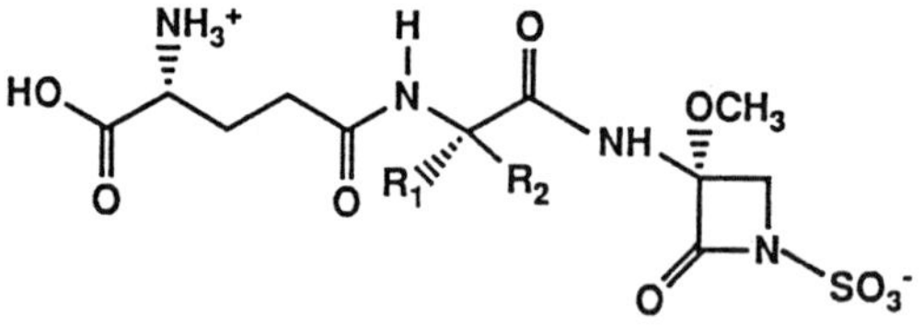

	R_1	R_2	
29	H	CH_3	**(Sulfazecin)**
30	CH_3	H	**(Isosulfazecin)**

group. Reaction of **31** with two equivalents of sodium hypochlorite (Clorox) provided the isolable dichloro derivative (**32**). Exposure of **32** to lithium methoxide, followed by exposure to sodium bisulfite, gave racemic 3-methoxyazetidinone (**35**) via acylimine (**33**) and *N*-chloroazetidinone (**34**). Sulfonation of **35** was possible with procedures that were initially developed and demonstrated with **31** (the unmethoxylated parent of **35**).

Sulfonation of **31** with pyridine–sulfur trioxide complex in dichloromethane-dimethylformamide (DMF) (1:1), followed by solvent removal, gave a quantitative yield of the pyridinium salt of **37**. However, it proved much more useful to quench this solution in bicarbonate buffer and use ion-pair extraction to isolate azetidinone-1-sulfonates as tetra-*n*-butylammonium salts. This protocol, developed by William H. Koster, greatly facilitated the transition from cephalosporin chemistry to monobactam chemistry. The salt **37** was the key intermediate in our synthesis of nonmethoxylated monobactams.

A second procedure involved reaction of the *N*-trimethylsilylazetidinone (**38**) with trimethylsilyl chlorosulfonate to give, via trimethylsilyl (TMS)

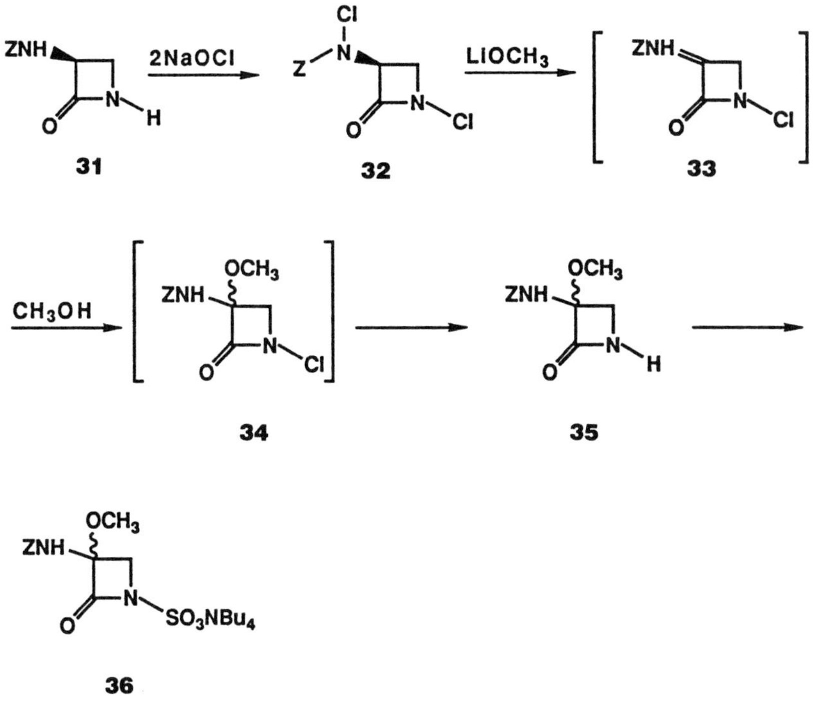

(Z is $PhCH_2OCO$ or benzyloxycarbonyl)

Scheme 4

ester (**39**), the potassium salt (**40**) after quench. This procedure, though successful with **35** as well as **31**, was not successful with **27**. Presumably, the corresponding intermediate **41** was unstable with respect to β-lactam hydrolysis. The same fate apparently met pyridinium salt **42**, generated from **27** with pyridine–sulfur trioxide complex. Under the conditions in which it was formed, the β-lactam bond underwent rapid hydrolysis.

A third procedure, and the most general one that allowed sulfonation of all classes of N1-unsubstituted azetidinones, was the simplest: reaction of the azetidinone, in DMF, with a DMF solution of $DMF{\bullet}SO_3$ complex. This solution could be prepared by reaction of trimethylsilyl chlorosulfonate with DMF and concentration to remove trimethylsilyl chloride. The generation of azetidinone 1-sulfonates as "free acids" (such as **43**) in DMF was generally and surprisingly successful. In this solvent, **43** most likely was stabilized by

ZNH, A, B, or C, ZNH, N, H, O, N, SO_3NBu_4, O

31 → **37**

A: $N \cdot SO_3$ (pyridine), DMF/CH_2Cl_2 → $KHCO_3$ → Bu_4NHSO_4, CH_2Cl_2

B: $TMSCl/Et_3N$, CH_2Cl_2 → $TMSOSO_2Cl$ → $KHCO_3$ → Bu_4NHSO_4, CH_2Cl_2

C: $DMF \cdot SO_3$ → $KHCO_3$ → Bu_4NHSO_4, CH_2Cl_2

conversion to the salt (**44**) or to a hydrogen-bonded equivalent that imparted substantial negative character to the sulfonate group.

Application of this procedure to **27** provided a formal total synthesis of SQ 26,180 (*16*). The correspondence of spectroscopic and physical properties of a sample prepared in this way with one isolated from *Chromobacterium violaceum* finally proved the assignment of structure and absolute configuration that we had made. The comparison was facilitated by a simple conversion of tetrabutylammonium salts to their potassium salts using potassium perfluorobutanesulfonate in acetone. The potassium salt could be isolated as a crystalline solid with this procedure, either directly or after dilution with ether, and recrystallized or chromatographed on a macromolecular styrene–divinylbenzene copolymer (HP-20) in water for final purification. These simple protocols (use of tetrabutylammonium salts in organic solvents, conversion to potassium salts in acetone–ether, and chromatographic purification on HP-20) greatly facilitated much of the synthetic work that we subsequently undertook.

Tetra-*n*-butylammonium salt (**36**) provided access to a variety of side-chain analogues of SQ 26,180 by the sequence illustrated in Scheme 5. Acylation of **45** with acetyl chloride provided a sample of racemic SQ 26,180 (**46**; $R = CH_3$) before we had developed the DMF SO_3 sulfonation; therefore,

31 → ZNH ... N–TMS (38) $\xrightarrow[-\ TMSCl]{TMSOSO_2Cl}$ [ZNH ... N–SO_3TMS] (39)

$\xrightarrow{KHCO_3}$ ZNH ... N–SO_3K (40) $\xrightarrow{Bu_4NCl}$ 37

27 → [CH_3C(O)NH, OCH_3 ... N–SO_3TMS] (41) → β-lactam hydrolysis

27 → [CH_3C(O)NH, OCH_3 ... N–SO_3^- pyridinium H+] (42) → β-lactam hydrolysis

this was the first demonstration that our assignment of structure to SQ 26,180 (exclusive of absolute configuration) was correct (*21*).

We prepared a number of racemic methoxylated monobactams and compared them to cephamycin counterparts that contained the same side chain and a variety of 3′-substituents. In general, the activity of these monobactams was less than that of their cephamycin counterparts (Table 2).

$$\mathrm{TMSOSO_2Cl + DMF \xrightarrow{DMF} TMSCl + DMF{\cdot}SO_3 \xrightarrow[-\,TMSCl]{vacuum}}$$

$$\mathrm{DMF{\cdot}SO_3/DMF} + \mathbf{31} \longrightarrow \mathbf{43}$$

$$\mathbf{43} \rightleftharpoons \mathbf{44} \longrightarrow \mathbf{37}$$

More important for structure–activity work, the chemical stability of these compounds was very poor. As was true of SQ 26,180, their stability at pH 7 was less than at lower pH. The data in Table 3 illustrate the poor stability of SQ 26,180 and the comparatively worse stability of its phenylacetyl counterpart, SQ 26,522 (**46**; R = $PhCH_2$). Coupled with antimicrobial activity no better than that of cephamycin counterparts, this poor stability forced us to explore other avenues to find a monobactam that could compete with the best of contemporary β-lactam antibiotics.

Before we abandoned 3-substitution, we checked the cephalosporin precedent and prepared the ethoxy (**47b**), butoxy (**47c**), and methyl (**47d**) analogues (phenylacetyl side chain) for comparison with the methoxylated parent (**47a**). As the data in Table 4 indicate, cephamycin precedent was followed: the compounds were virtually inactive. The relatively poor antimicrobial activity of these side-chain analogues of SQ 26,180, and their remarkably poor stability, could have led to reduced interest in and priority for the monobactam program. Based on cephamycin and cephalosporin precedent, nonmethoxylated analogues of SQ 26,180 were predicted to be less stable to β-lactamase (26). In addition, they were predicted to be less stable to chemical hydrolysis (27). We were able to show, very early in the program, that nonmethoxylated monobactams were much more stable to chemical hydrolysis. They were certainly stable enough to support clinical development. It took considerable effort, however, to overcome the lack of β-lactamase stability of the initial nonmethoxylated monobactams that we prepared.

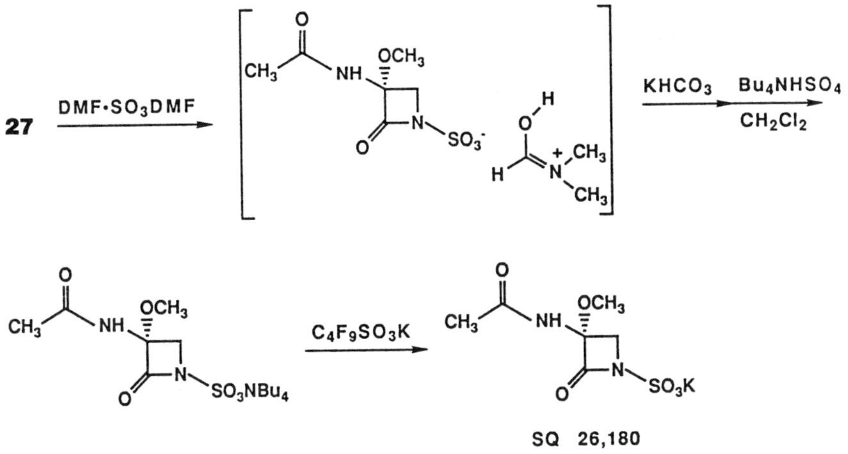

27 $\xrightarrow{DMF \cdot SO_3DMF}$ [CH_3 NH OCH_3 O N SO_3^- ; O H H $\overset{+}{N}$ CH_3 CH_3] $\xrightarrow[CH_2Cl_2]{KHCO_3 \quad Bu_4NHSO_4}$

CH_3 NH OCH_3 O N SO_3NBu_4 $\xrightarrow{C_4F_9SO_3K}$ CH_3 NH OCH_3 O N SO_3K

SQ 26,180

ZNH OCH_3 O N SO_3NBu_4 **36** $\xrightarrow[Pd/C]{H_2}$ [NH_2 OCH_3 O N SO_3NBu_4 **45**] $\xrightarrow{RCOCl}$

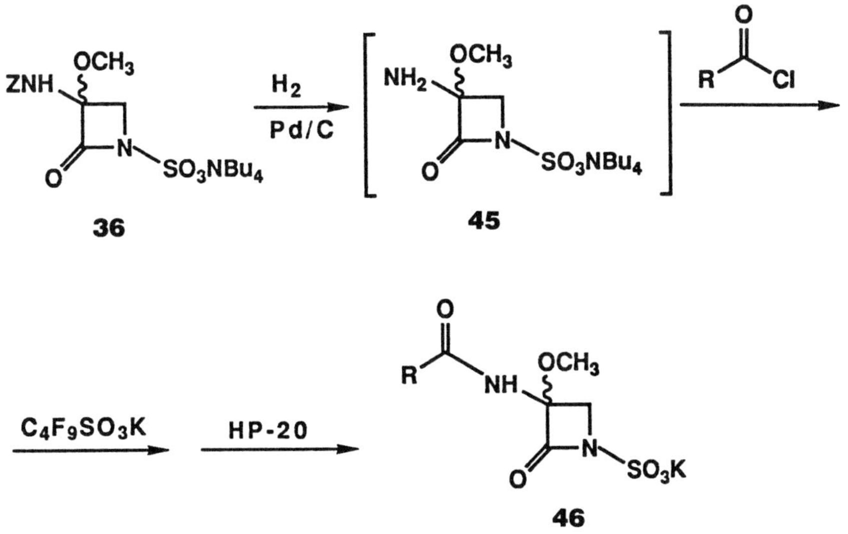

$\xrightarrow{C_4F_9SO_3K}$ $\xrightarrow{HP\text{-}20}$ R NH OCH_3 O N SO_3K **46**

Scheme 5

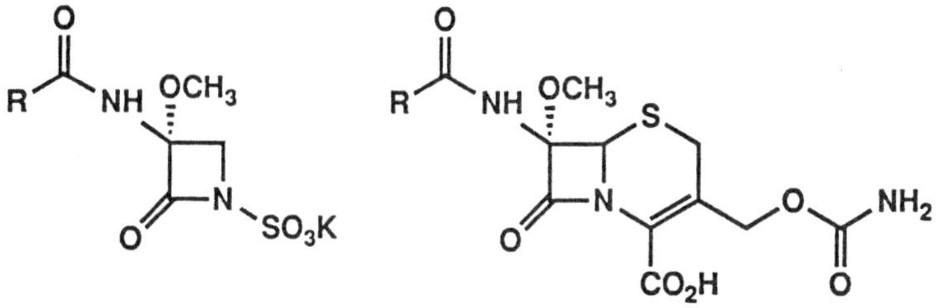

SQ 26,559 and cefoxitin: R =

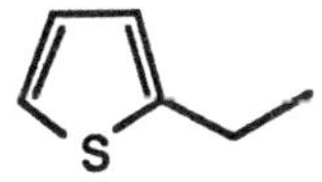

SQ 26,554 and cefmetazole: R =

Table 2. Comparison of Antimicrobial Activity of SQ 26,559, SQ 26,554, and Their Cephamycin Counterparts

Organism	*SC No.*	*SQ 26,559*	*Cefoxitin*	*SQ 26,554*	*Cefmetazole*
Staph. aureus	1276	6.3	1.6	25	0.4
Staph. aureus	2400	12.5	3.1	50	0.8
E. coli	8294	12.5	12.5	50	1.6
E. coli	10857	12.5	3.1	50	0.8
K. aerogenes	10440	25	12.5	50	3.1
Prot. mirabilis	3855	25	3.1	>100	1.6
Prot. vulgaris	9416	50	3.1	>100	1.6
Ent. cloacae	8236	50	>100	25	100
Ser. marcescens	9783	25	25	25	12.5
Ps. aeruginosa	9545	25	50	25	50
Ps. aeruginosa	8329	>100	>100	>100	>100

NOTE: Activities are given as MIC in micrograms per milliliter at 10^4 cfu.

Table 3. Stability of SQ 26,180 and Its Phenylacetyl Counterpart, SQ 26,522, at Different pH Values

	SQ 26,180		*SQ 26,522*		
Time (h)	*pH 5*	*pH 7*	*pH 3*	*pH 5*	*pH 7*
0	100	94	100	100	100
1	104	77	103	82	93
2	105	39	105	44	47
4	101	15	106	18	18
24	97	0	84	1	1

NOTE: Stability is given as percent compound remaining as determined by high-performance liquid chromatography.

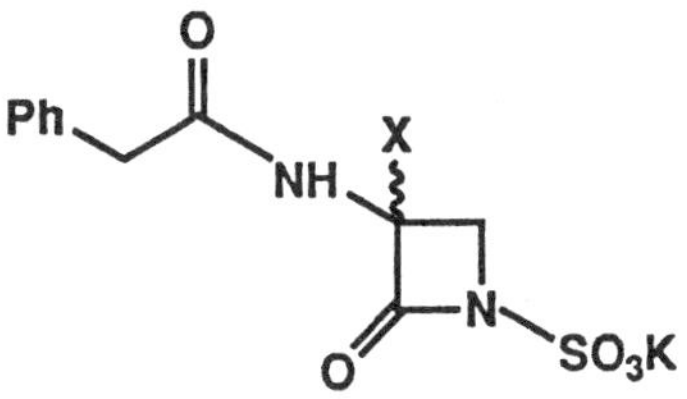

No.	X
47a	CH_3O
47b	CH_3CH_2O
47c	$CH_3(CH_2)_3O$
47d	CH_3

Table 4. Comparison of Antimicrobial Activity of 3-Methoxy, 3-Ethoxy, 3-*n*-Butoxy, and 3-Methyl Monobactams

Organism	*SC No.*	47a	47b	47c	47d
Staph. aureus	1276	6.3	25	>100	>100
Staph. aureus	2400	6.3	25	>100	>100
E. coli	8294	25	>100	>100	>100
E. coli	10857	25	>100	>100	>100
K. aerogenes	10440	50	>100	>100	>100
Prot. mirabilis	3855	50	>100	>100	>100
Prot. vulgaris	9416	25	>100	>100	>100
Ent. cloacae	8236	>100	>100	>100	>100
Ser. marcescens	9783	50	>100	>100	>100
Ps. aeruginosa	9545	>100	12.5	>100	>100
Ps. aeruginosa	8329	>100	>100	>100	>100

NOTE: Activities are given as MIC in micrograms per milliliter at 10^4 cfu.

Nonmethoxylated Analogues of SQ 26,180

The first attempts to synthesize nonmethoxylated monobactams were confounded by our unfamiliarity with the chemistry and properties of these ionic species. 3*S*-3-Phenylacetylamino-2-azetidinone (**49**) was prepared following literature precedent (Scheme 6).

Desulfurization of penicillin G, first described in the penicillin monograph, gave azetidinone **48**, which was converted to **49**. Reaction of **49** with pyridine•SO_3 complex, followed by HP-20 chromatography, provided the first nonmethoxylated monobactam evaluated in our laboratories. Appropriately, it contained the same side chain (phenylacetyl) as the first β-lactam (penicillin G) studied at Squibb. Its activity is compared with that of penicillin G and SQ 26,180 in Table 5. We were excited to find a compound with increased activity relative to SQ 26,180 that approached the spectrum

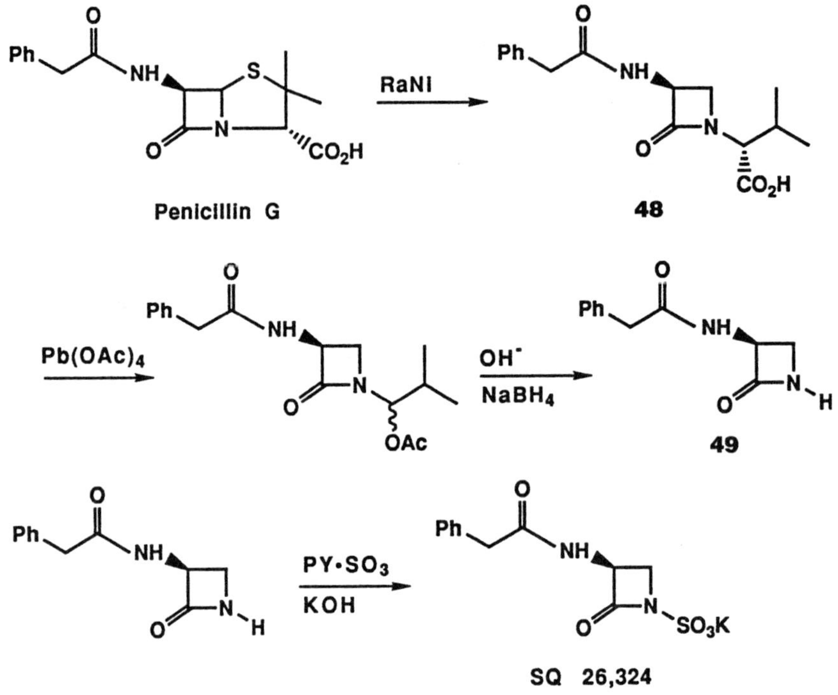

Scheme 6

Table 5. Comparison of Antimicrobial Activity of SQ 26,180, SQ 26,324, and Penicillin G

Organism	*SC No.*	*SQ 26,180*	*SQ 26,324*	*Penicillin G*
Staph. aureus	1276	50	3.1	<0.05
Staph. aureus	2400	50	3.1	3.1
E. coli	8294	>100	50	50
E. coli	10857	>100	25	12.5
K. aerogenes	10440	>100	50	25
Prot. mirabilis	3855	>100	50	6.3
Prot. vulgaris	9416	>100	50	6.3
Ent. cloacae	8236	>100	>100	>100
Ser. marcescens	9783	25	100	>100
Ps. aeruginosa	9545	3.1	100	50
Ps. aeruginosa	8329	50	>100	>100

NOTE: Activities are given as MIC in micrograms per milliliter at 10^4 cfu.

of penicillin G. Though modest at best, the activity of SQ 26,324 was very encouraging.

Table 6 compares the antimicrobial activity of SQ 26,324 and its methoxylated counterpart, SQ 26,522. This first comparison of a methoxylated and nonmethoxylated monobactam was extremely encouraging, but not because the antimicrobial activity had increased dramatically (it hadn't) or because the β-lactamase stability had survived removal of the 3-methoxyl group (it hadn't). Rather, we were encouraged because the chemical stability of SQ 26,324 was quite obviously better than that of SQ 26,180. This stability was subsequently highlighted when a number of its more water-insoluble side-chain analogues were purified by recrystallization from boiling water.

In order to fully evaluate the potential for nonmethoxylated monobactams, we sought a general synthesis that avoided the individual preparation and sulfonation of 3-acylamino-2-azetidinones. Hydrogenolysis of tetrabutylammonium salt **37** and acylation (dicyclohexylcarbodiimide–HOBt) of **50**, in situ, with the side-chain acid provided a wide variety of monobactams for biological evaluation. Most of these were isolated as their potassium salts (**51**) (using exchange with potassium perfluorobutanesulfonate) and purified by reverse-phase chromatography on HP-20.

This general synthesis allowed us to prepare radiolabeled monobactams to directly examine their interaction with the enzymes (β-lactamase and transpeptidase) that mediated the interaction between bacteria and β-lactams. It also allowed us to search for "mbp's" (monobactam-binding proteins), different from "pbp's" (penicillin-binding proteins), that might contribute to

Table 6. Comparison of Antimicrobial Activity of SQ 26,324 and Its Methoxylated Counterpart, SQ 26,522

Organism	*SC No.*	*SQ 26,324*	*SQ 26,522*
Staph. aureus	1276	3.1	6.3
Staph. aureus	2400	3.1	6.3
E. coli	8294	50	50
E. coli	10857	25	25
K. aerogenes	10440	50	50
Prot. mirabilis	3855	50	50
Prot. vulgaris	9416	50	50
Ent. cloacae	8236	>100	>100
Ser. marcescens	9783	100	100
Ps. aeruginosa	9545	100	100
Ps. aeruginosa	8329	>100	>100

NOTE: Activities are given as MIC in micrograms per milliliter at 10^4 cfu.

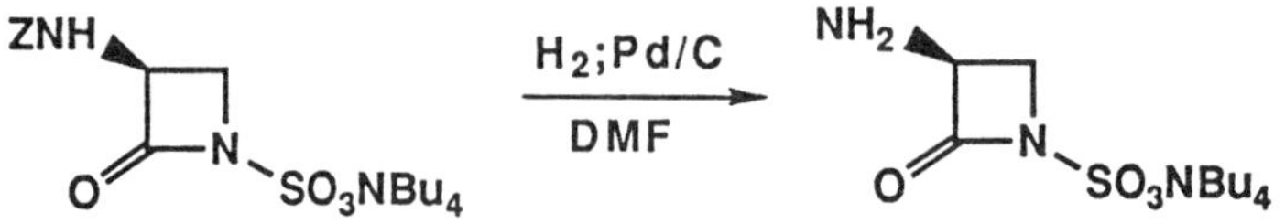

ZNH, H_2;Pd/C, DMF, NH_2, N, SO_3NBu_4, O

37 50

RCO_2H / DCC, HOBt; 1) $C_4F_9SO_3K$ 2) HP-20 chromatography

R, O, NH, N, SO_3K

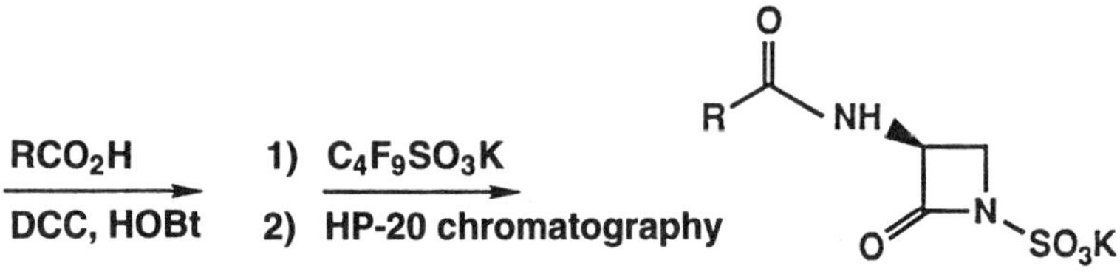

51

their antimicrobial activity. The penicillin G and cefotaxime analogues **52** and **53** were prepared by coupling the appropriate radiolabeled acid with **50**, as described in the previous paragraph. Detailed investigation of the interaction of **52** and **53** with bacterial enzymes and cell walls led to the conclusion that monobactams were typical β-lactams and shared a penicillin-like mechanism of action (*28, 29*). No special "mbp's" were uncovered utilizing **52** and **53** as probes; the second ring of penicillins and cephalosporins, therefore, provided no special interactions over those possible with their monocyclic counterparts. This finding led to the expectation of similar absolute specificity for bacterial cell-wall enzymes and an attendant lack of mammalian toxicity.

We prepared a variety of nonmethoxylated monobactams and compared them with their methoxylated parents. Examples were found in which the activity of the nonmethoxylated analogue was poorer than, similar to, or much greater than that of its methoxylated counterpart (Table 7). As

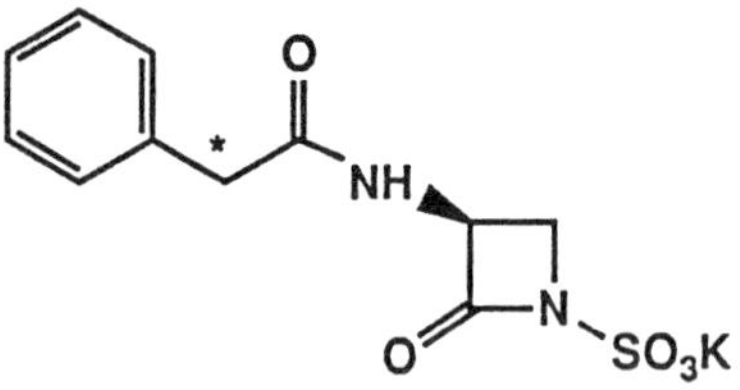

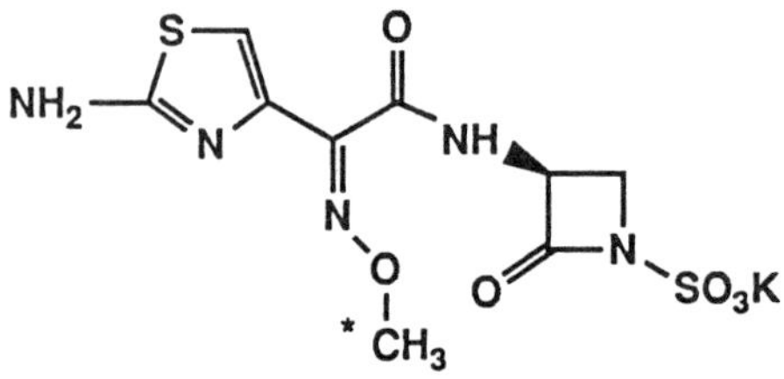

O, NH, N, SO_3K; S, NH_2, N, O, CH_3

52 $* = {}^{14}C$ 53

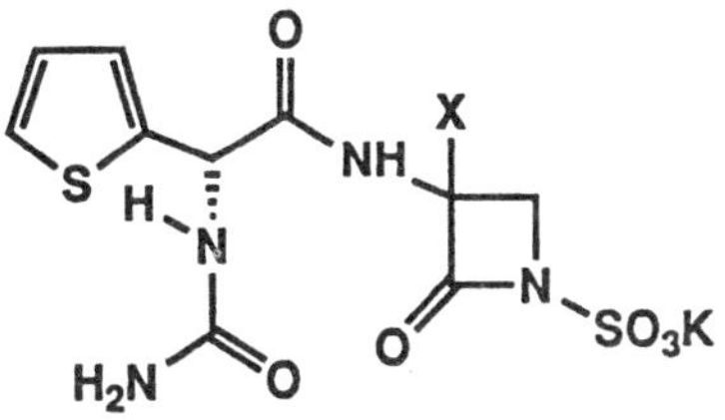

54

No.	X
54a	H
54b	CH_3O

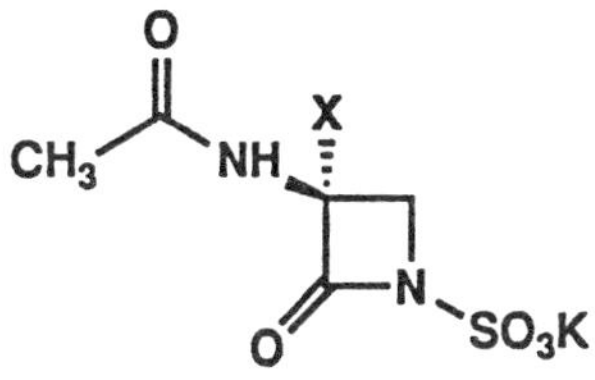

55

No.	X
55a	H
55b	CH_3O

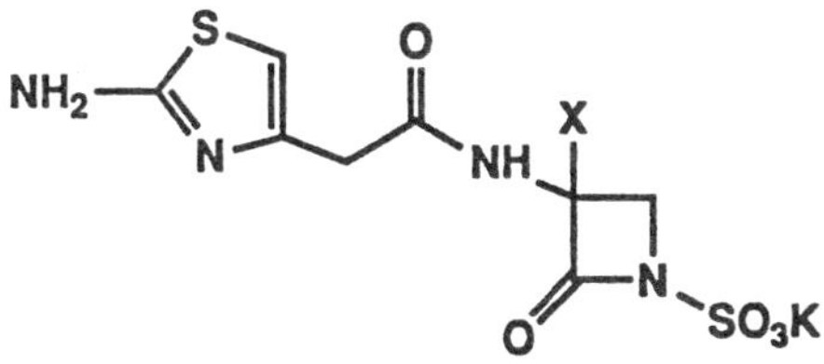

56

No.	X
56a	H
56b	CH_3O

Table 7. Comparison of Antimicrobial Activity of Various Nonmethoxylated Monobactams and Their Methoxylated Parents

Organism	*SC No.*	54a	54b	55a	55b	56a	56b
Staph. aureus	1276	12.5	6.3	50	50	6.3	50
Staph. aureus	2400	12.5	6.3	100	50	6.3	50
E. coli	8294	100	3.2	>100	>100	0.8	50
E. coli	10857	50	3.2	>100	>100	0.4	12.5
K. aerogenes	10440	100	3.2	>100	>100	0.8	50
Prot. mirabilis	3855	50	3.2	>100	>100	0.8	12.5
Prot. vulgaris	9416	50	3.2	>100	>100	0.8	25
Ent. cloacae	8236	100	12.5	>100	>100	0.8	>100
Ser. marcescens	9783	25	6.3	25	25	25	50
Ps. aeruginosa	9545	12.5	12.5	3.1	3.1	1.6	>100
Ps. aeruginosa	8329	100	100	50	50	25	>100

NOTE: Activities are given as MIC in micrograms per milliliter at 10^4 cfu.

discussed previously, the chemical instability of methoxylated monobactams led us to emphasize their nonmethoxylated counterparts in subsequent work. A wide array of side chains were incorporated; none were found to confer high activity on monobactams without prior penicillin–cephalosporin precedent. Two series emerged as most interesting: α-ureidoarylacetyl compounds, such as SQ 81,491, and α-oximino-4-amino-2-thiazolylacetyl compounds, such as SQ 81,377. Many compounds were prepared in both series. A number of the ureido compounds displayed high serum binding, low urinary recovery, and poor ED_{50} values in mice; however, the major problem with both series was illuminated by the secondary screening data (Table 8). The lack of activity at high inoculum (10^6 colony-forming units) for the β-lactamase-producing strains (TEM+, P99+, Kl+) was evident in both series of monobactams. Although this inoculum effect was also present with cefoperazone, it was absent with cefotaxime for the clinically important TEM lactamase. This apparent lack of β-lactamase stability could be confirmed in formal studies (*30*) with isolated enzyme preparations and was extremely disappointing. As discussed previously, the methoxylated monobactams, though β-lactamase-stable, were too unstable to simple hydrolysis at pH 7 to be useful clinically. If nonmethoxylated monobactams (which were chemically stable) could not be rendered stable to β-lactamase by the appropriate choice of side chain, then the prospects for clinical development of a monobactam of any kind seemed remote.

Because of the singular antimicrobial activity of SQ 81,377 in comparison with the other compounds studied (Table 9), and because of cephalosporin precedent, we decided to prepare additional analogues of SQ 81,377 in the hope of increasing β-lactamase stability. The series of analogues reported in Table 10 (secondary screen) is typical of the data we accumulated. The isobutyric acid moiety of SQ 81,402 (**59f**), common to ceftazidime, imparted a degree of β-lactamase stability, especially to the clinically important TEM enzyme, and antipseudomonal activity. However, when compared with cefotaxime and ceftazidime (Table 11), these compounds were still inferior. No additional modification of the side chain provided a compound equivalent to these benchmark cephalosporins. It remained for the third line of investigation (4-substitution) to overcome this critical lack of β-lactamase stability.

4-Substituted Monobactams

When we first decided to investigate 4-substituted monobactams, there were many descriptions of 3-amino-2-azetidinones with 4-heteroatom-linked

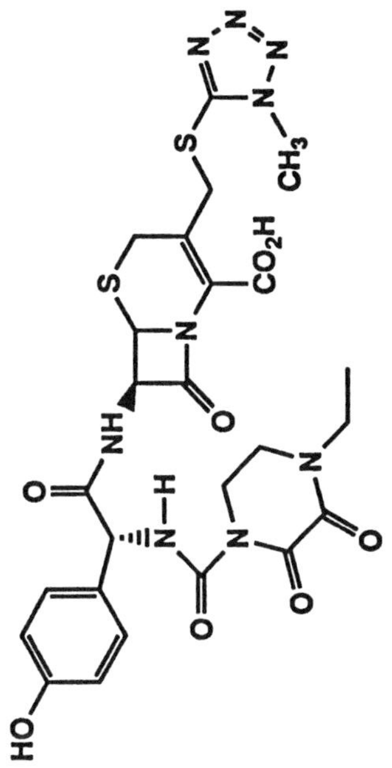

Cefoperazone

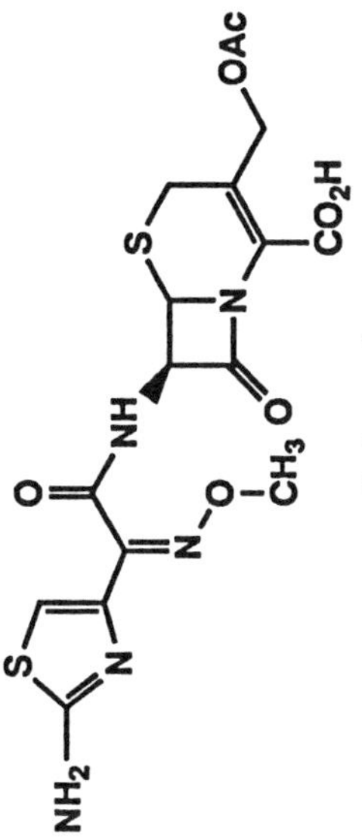

Cefotaxime

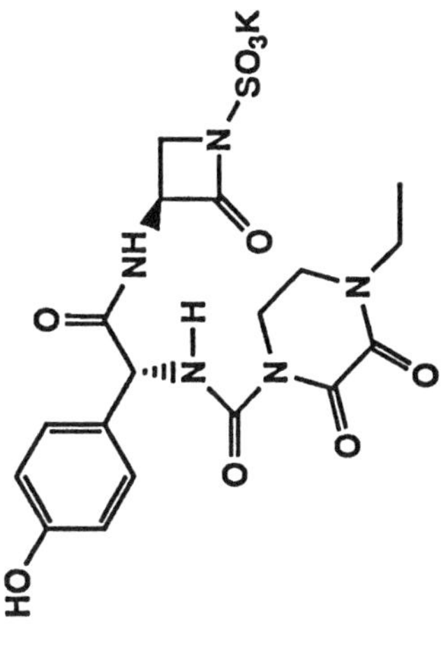

SQ 81,491

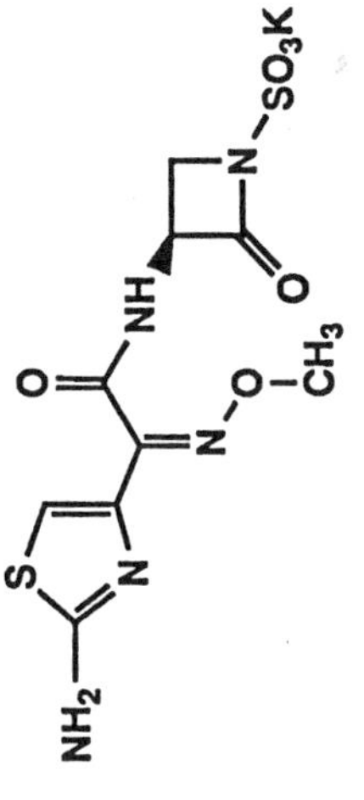

SQ 81,377

Table 8. Comparison of Antimicrobial Activity (β-Lactamase-Producing Strains) of Monobactams and Their Cephalosporin Counterparts

Organism	*SC No.*	*SQ 81,491*	*Cefoperazone*	*SQ 81,377*	*Cefotaxime*
E. coli TEM+	10404	>100	>100	>100	<0.05
E. coli TEM−	10439	0.4	0.4	1.6	0.1
Ent. cloacae P99+	10435	>100	>100	>100	>100
Ent. cloacae P99−	10441	0.8	0.4	1.6	0.4
K. aerogenes K1+	10436	>100	>100	>100	100
K. aerogenes K1−	10440	6.3	0.2	1.6	0.2
Ps. aeruginosa	8329	>100	25	25	>100
Ps. aeruginosa	9545	0.8	0.8	12.5	3.1

NOTE: Activities are given as MIC in micrograms per milliliter at 10^6 cfu.

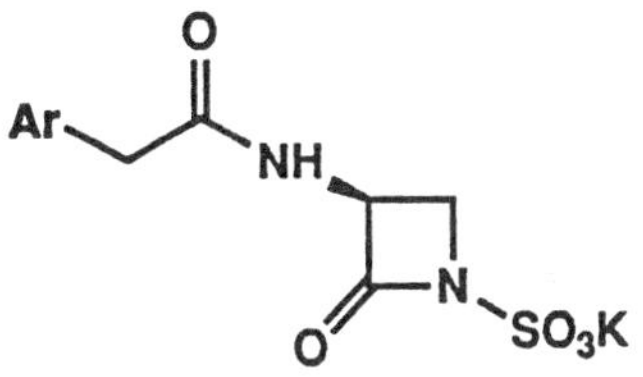

57 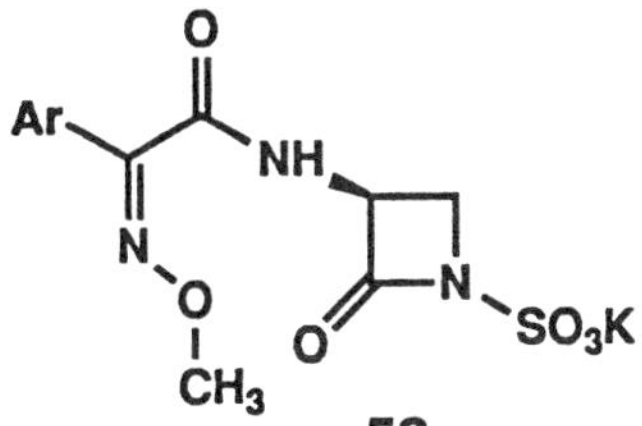 **58**

No.	Ar	No.
57a	Phenyl	**58a**
57b	2-Thienyl	**58b**
57c	2-Amino-4-thiazolyl	**58c**

Table 9. Comparison of Antimicrobial Activity of Aryclacetylomonobactams and Their (*Z*)-α-Methoxyamino Counterparts

Organism	*SC No.*	**57a**	**58a**	**57b**	**58b**	**57c**	**58c**
Staph. aureus	1276	3.1	6.3	6.3	12.5	12.5	6.3
Staph. aureus	2400	3.1	25	6.3	12.5	12.5	6.3
E. coli	8294	50	100	100	25	>100	0.8
E. coli	10857	25	25	50	6.3	>100	0.4
K. aerogenes	10440	50	>100	100	25	>100	0.8
Prot. mirabilis	3855	50	12.5	100	25	>100	0.8
Prot. vulgaris	9416	50	25	100	3.1	>100	0.8
Ent. cloacae	8236	>100	>100	100	100	>100	0.8
Ser. marcescens	9783	100	>100	12.5	25	6.3	25
Ps. aeruginosa	9545	100	12.5	12.5	25	50	1.6
Ps. aeruginosa	8329	>100	>100	>100	>100	>100	25

NOTE: Activities are given as MIC in micrograms per milliliter at 10^4 cfu.

substituents (exemplified by sulfone **60**) (*31*). These compounds had been obtained by degradation of penicillin, normally by cleavage of the sulfur-quaternary carbon bond. Our research plan, however, called for preparation of 4-methyl compounds as benchmarks by which to evaluate additional steric and electronic effects of 4-substituents. Early in 1979, the simplest precedented approach appeared to be 2 + 2 achiral cycloaddition (*32*) to prepare racemic azetidinones such as **61** and **62** with 4-substituents that could be derived from the original carboxylate ester or arylvinyl moiety, respectively. Indeed, this approach was initially utilized by the Takeda group to arrive at carumonam.

59

No.	R	No.	R
59a	CH_3	**59d**	CH_2CO_2H
59b	CH_3CH_2	**59e**	CH_3CHCO_2H
59c	$(CH_3)_2CH$	**59f**	$(CH_3)_2CO_2H$

Table 10. Comparison of Antimicrobial Activities (β-Lactamase-Producing Strains) of SQ 81,377 Analogues (Secondary Screen)

Organism	*SC No.*	**59a**	**59b**	**59c**	**59d**	**59e**	**59f**
E. coli TEM+	10404	25	100	100	3.1	6.3	3.1
E. coli TEM−	10439	1.6	0.4	1.6	1.6	100	0.4
Ent. cloacae P99+	10435	>100	>100	>100	>100	>100	>100
Ent. cloacae P99−	10441	1.6	0.8	3.1	0.8	0.8	0.4
K. aerogenes K1+	10436	>100	>100	>100	>100	>100	>100
K. aerogenes K1−	10440	0.8	0.4	6.3	>100	>100	0.8
Ps. aeruginosa	8329	>100	>100	>100	>100	>100	>100
Ps. aeruginosa	9545	100	50	1.6	>100	>100	1.6

NOTE: Activities are given as MIC in micrograms per milliliter at 10^6 cfu.

This situation changed significantly in the summer of 1979 with the communication (33) by Marvin Miller and his group at Notre Dame of the utility of the hydroxamate moiety in preparing chiral 3-acylamino-2-azetidinones. This methodology, based on facile generation of a nucleophilic anion from its parent *O*-alkyl hydroxamate ester, proved to be the linchpin around which Miller's group developed significantly improved access to many biologically important compounds. The original paper described the se-

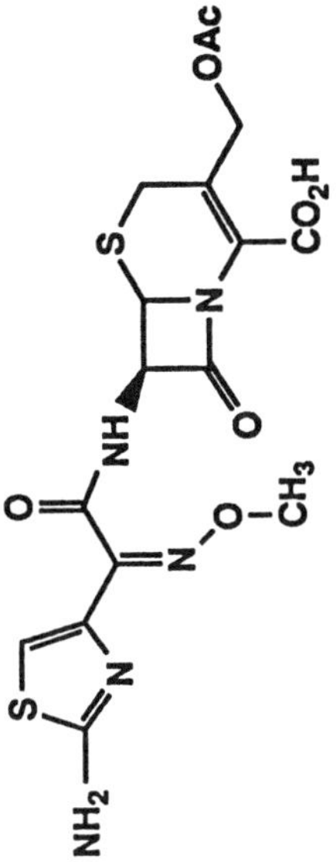

59a (SQ 81,377)

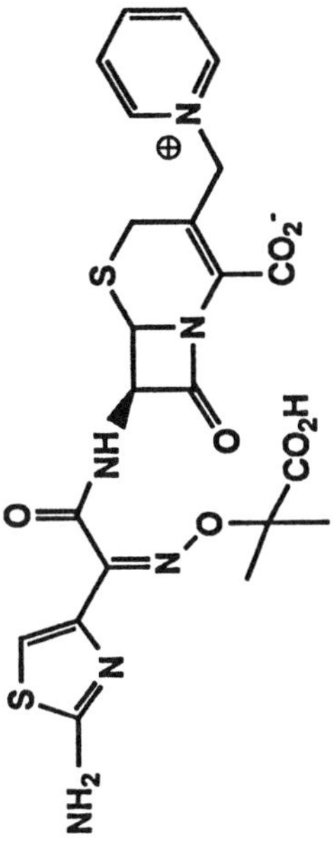

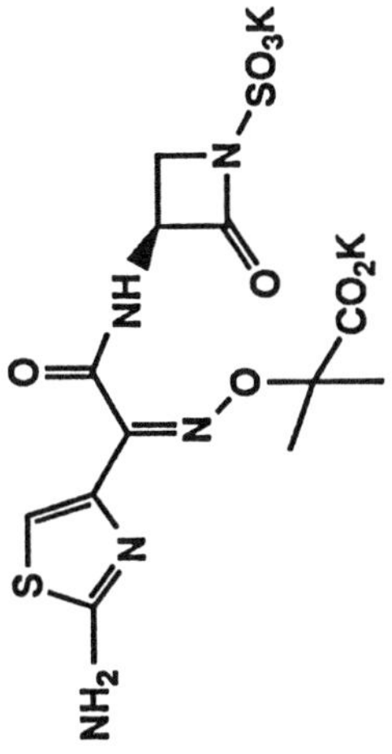

59f (SQ 81,402)

Cefotaxime

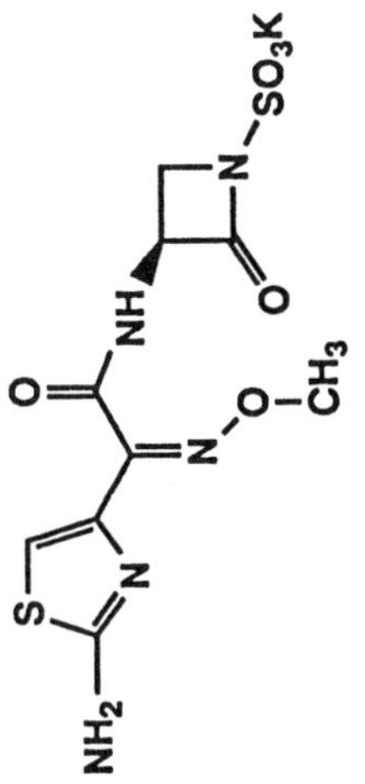

Ceftazidime

Table 11. Comparison of Antimicrobial Activities (β-Lactamase-Producing Strains) of Monobactams and Their Cephalosporin Counterparts

Organism	*SC No.*	*SQ 81,377*	*Cefotaxime*	*SQ 81,402*	*Ceftazidime*
E. coli TEM+	10404	>100	<0.05	3.1	0.4
E. coli TEM−	10439	1.6	0.1	0.4	0.4
Ent. cloacae P99+	10435	>100	>100	>100	100
Ent. cloacae P99−	10441	1.6	0.4	0.4	0.4
K. aerogenes K1+	10436	>100	100	>100	1.6
K. aerogenes K1−	10440	1.6	0.2	0.8	0.1
Ps. aeruginosa	8329	25	>100	>100	1.6
Ps. aeruginosa	9545	12.5	3.1	1.6	0.8

NOTE: Activities are given as MIC in micrograms per milliliter at 10^6 cfu.

quence in Scheme 7 to chiral 3-acylamino-2-azetidinones (**64**; R = BOC) via reduction of the *N*-hydroxyazetidinones (**63**). It also described the use of *l*-threonine as starting material to prepare 4α-methylazetidinone (**65**).

Using an analogous scheme, we were able to convert *d*-serine to the enantiomeric (3*R*) monobactams **66** and **67**. These compounds, as predicted, were devoid of activity. This finding added to the evidence that monobactams, despite their novel structure, depended only on a penicillinlike mode of action for their activity. The finding that **67** was inactive at 100 Mg/mL while its 3*S* enantiomer was active at <0.05 Mg/mL is testament to the stereoselectivity of the Miller chemistry and the bacterial enzymes involved.

As reported by Miller, treatment of hydroxymate **68** with Ph_3P–diethyl azodicarboxylate gave azetidinone **65** in excellent yield. Hydrogenolysis of **65** produced the *N*-hydroxyazetidinone **69**. Reduction with titanium trichloride (with the use of ammonium acetate buffer to eliminate the need to control pH by portionwise addition of base) gave the 4α-methyl derivative **70** (Scheme 8). The BOC-protecting group was exchanged for benzyloxycarbonyl protection by treatment with trifluoroacetic acid (TFA) followed by benzyl chloroformate. Sulfonation provided the 4α-methyl analogue **72** of the key intermediate (**37**) we were using to prepare 4-unsubstituted monobactams. The chemistry of **72** proved to be analogous to that of **37**; it provided simple access to 4α-methylmonobactams such as **73**.

The first comparisons we evaluated involved the 4α-methyl counterparts. The data in Table 12 provided little impetus to proceed; however, our plan involved using both 4-methyl compounds as benchmarks. We therefore

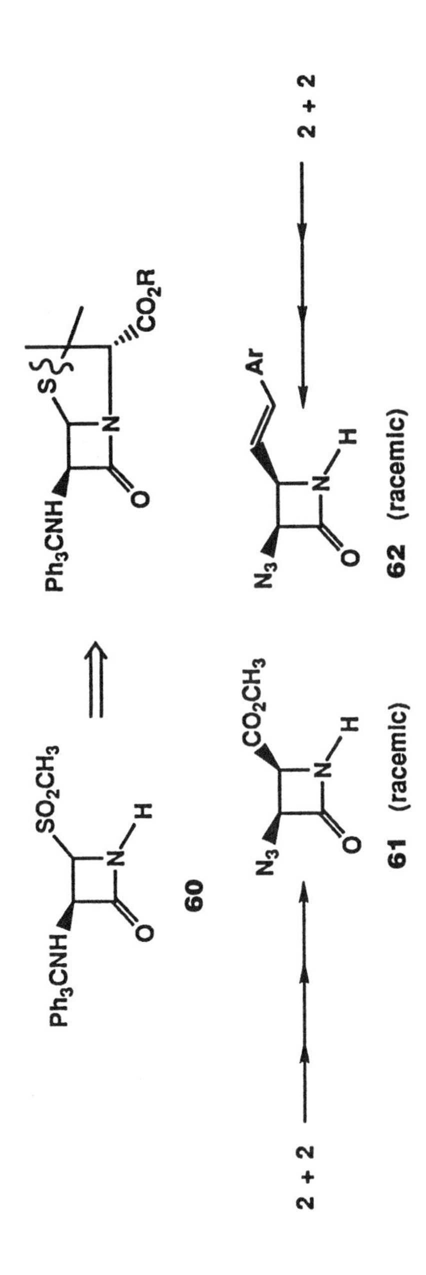
Ph$_3$CNH
SO$_2$CH$_3$
60
CO$_2$R
2 + 2
N$_3$
CO$_2$CH$_3$
61 (racemic)
Ar
62 (racemic)

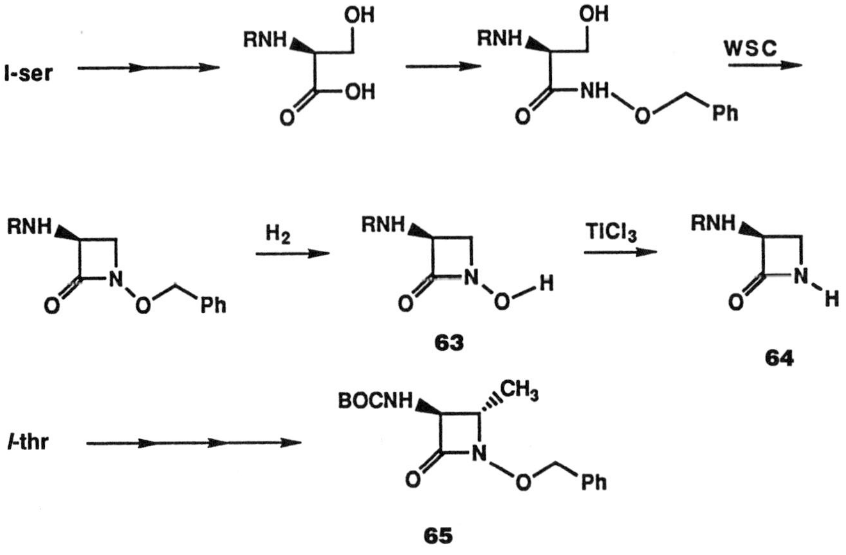

l-ser → RNH (OH, OH, O) → RNH (OH, NH, O, Ph, O) WSC →

RNH (O, N, O, Ph) H_2 → RNH (O, N, O, H) **63** $TiCl_3$ → RNH (O, N, H) **64**

l-thr → BOCNH (CH_3, O, N, O, Ph) **65**

Scheme 7

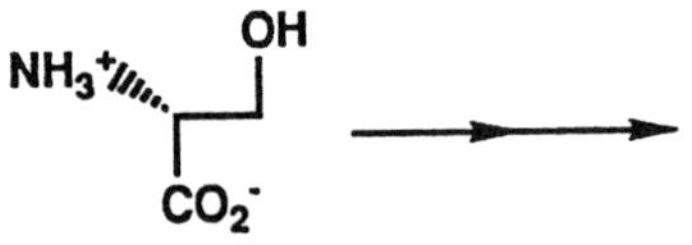

NH_3^+ (OH, CO_2^-) →

***d*-serine**

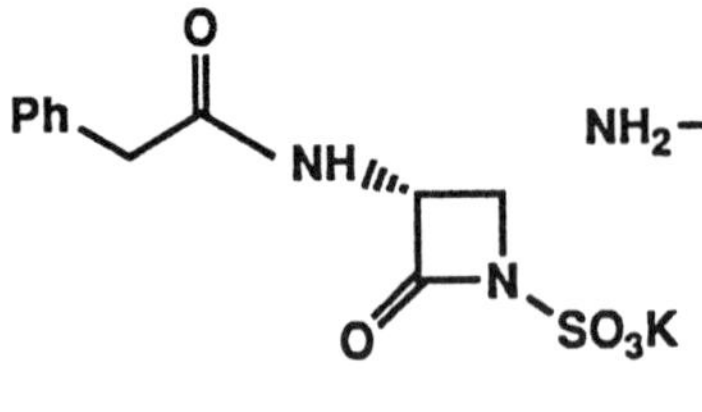

Ph (O, NH, O, N, SO_3K) **66**

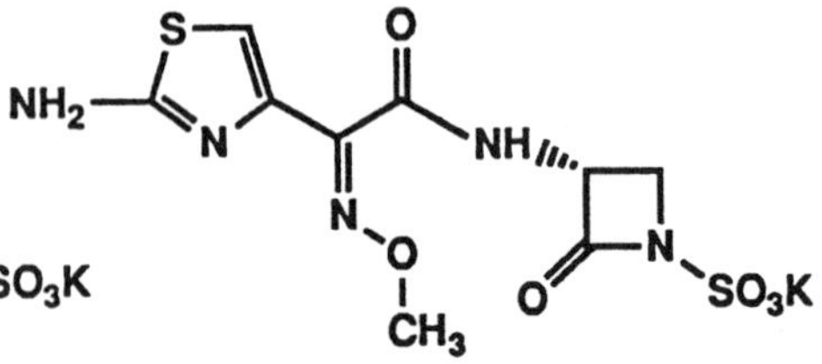

NH_2 (S, N, O, N, O, CH_3, NH, O, N, SO_3K) **67**

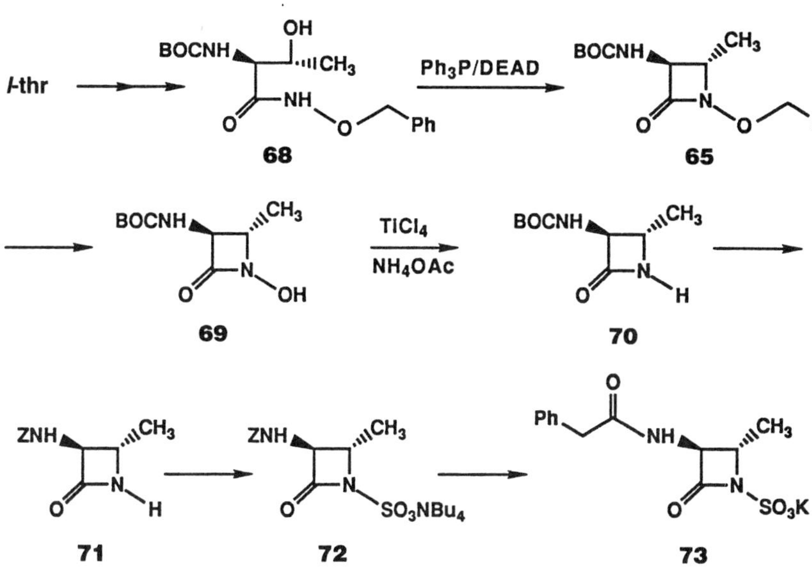

Scheme 8

converted *d,l*-allothreonine to racemic 4β-methylazetidinone **76**. (We had no *l*-allothreonine, we were confident that the result would predict the chiral case after division by 2, and we did not want to wait to obtain the answer.) Conversion of **76** to the phenylacetyl analogue **77** completed the first triad of 3-acylaminomonobactams that included the unsubstituted parent as well as its 4α- and 4β-methyl counterparts. Again, the results in Table 13 provided only a hint of the ultimate benefit of 4-substitution. The β-lactam frequencies (CCl_4 solution) of these compounds as their tetra-*N*-butylammonium salts (4-unsubstituted, 1755 cm^{-1}; 4-α-methyl, 1760 cm^{-1}; and 4β-methyl, 1765 cm^{-1}) indicated slightly increased activation of the β-lactam bond and provided some encouragement to continue.

By this time, we had evaluated enough side-chain analogues to know that "aminothiazole oximes" were preferred; therefore we prepared methoximes **79b** and **79c** for comparison to SQ 81,377 (**79a**). The results (Table 14) were electrifying, precipitating a celebration in Richard Sykes' conference room on the day the primary screen of **78b** was read. Data from the secondary screen were just as encouraging. The increased β-lactamase stability that the screen implied was soon confirmed by formal β-lactamase stability studies. The compounds proved to be active in vivo at levels

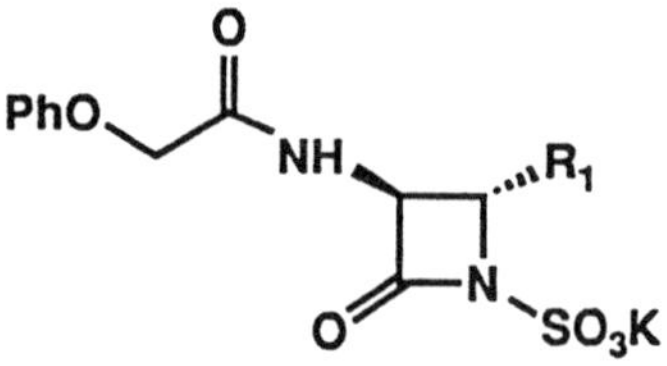

No.	R_1	No.
74a	H	75a
74b	CH_3	75b

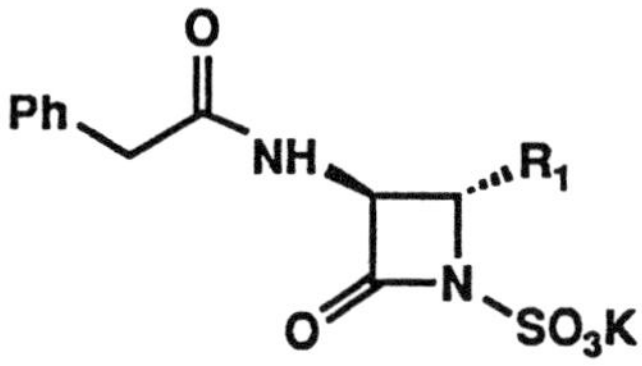

Table 12. Comparison of Antimicrobial Activities of 4-Unsubstituted Monobactams and Their 4α-Methyl Counterparts

Organism	*SC No.*	74a	74b	75a	75b
Staph. aureus	1276	6.3	100	3.1	50
Staph. aureus	2400	3.1	100	3.1	50
E. coli	8294	>100	>100	50	100
E. coli	10857	>100	>100	25	50
K. aerogenes	10440	>100	>100	50	100
Prot. mirabilis	3855	>100	>100	50	100
Prot. vulgaris	9416	>100	>100	50	100
Ent. cloacae	8236	>100	>100	>100	>100
Ser. marcescens	9783	>100	>100	100	50
Ps. aeruginosa	9545	>100	>100	100	50
Ps. aeruginosa	8329	>100	>100	>100	>100

NOTE: Activities are given as MIC in micrograms per milliliter at 10^4 cfu.

predicted by their minimum inhibitory concentrations, and their pharmacokinetics were normal. In the few days that it took to obtain this data we became convinced that an excellent clinical candidate would soon emerge from this line of investigation.

Some weeks earlier we had reached a decision to evaluate SQ 81,402 in depth, and our colleagues in Germany (the Squibb Institute had drug discovery, chemical development, and toxicology facilities at the site of its Regensburg manufacturing plant), who prepared most of the side-chain analogues, were in the middle of producing 50 g of SQ 81,402. We quickly prepared its 4α- and 4β-methyl analogues for comparison. SQ 26,726 (the dipotassium salt of aztreonam) and SQ 26,743 (its racemic 4β-methyl

d,l-allothreonine   **76** (racemic)   **77** (racemic)

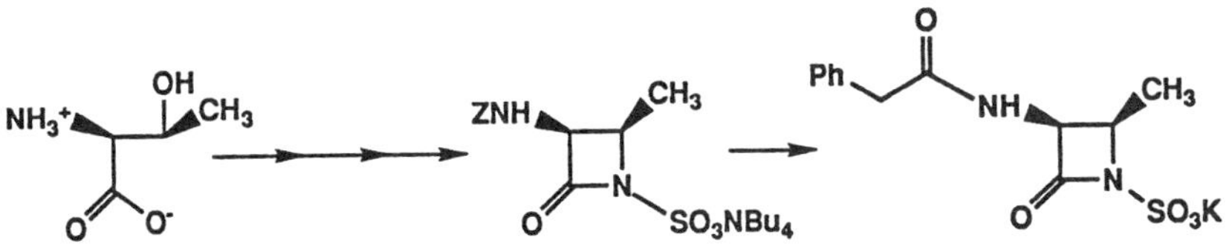

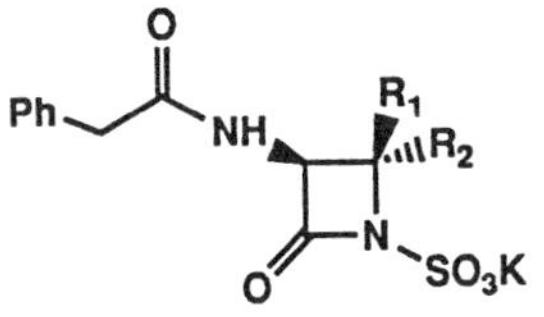

No.	R_1	R_2
78a	H	H
78b	H	CH_3
78c	CH_3	H

78

Table 13. Comparison of Antimicrobial Activities of SQ 26,324 and Its 4α- and 4β-Methyl Analogues

Organism	*SC No.*	78a	78b	78c
Staph. aureus	1276	3.1	50	12.5
Staph. aureus	2400	3.1	50	25
E. coli	8294	50	100	50
E. coli	10857	25	50	25
K. aerogenes	10440	50	100	>100
Prot. mirabilis	3855	50	100	>100
Prot. vulgaris	9416	50	100	>100
Ent. cloacae	8236	>100	>100	>100
Ser. marcescens	9783	100	50	>100
Ps. aeruginosa	9545	100	50	>100
Ps. aeruginosa	8329	>100	>100	>100

NOTE: Activities are given as MIC in micrograms per milliliter at 10^4 cfu.

counterpart) were superior to their 4-unsubstituted parent (Table 15), as predicted by the methoxime triad just discussed. We made an exhaustive comparison of SQ 81,402, SQ 26,726 and SQ 26,917 (the chiral compound derived from *l*-allothreonine) and decided to evaluate SQ 26,726 in the clinic.

This decision was made for a variety of reasons. The most important was the timely evaluation of the interaction of monobactams and humans: by that time we were expending our total drug discovery effort in antiinfectives (close to 100 individuals) in evaluating the interaction of monobactams with bacteria, mice, rats, and monkeys. We needed to confirm our predictions of

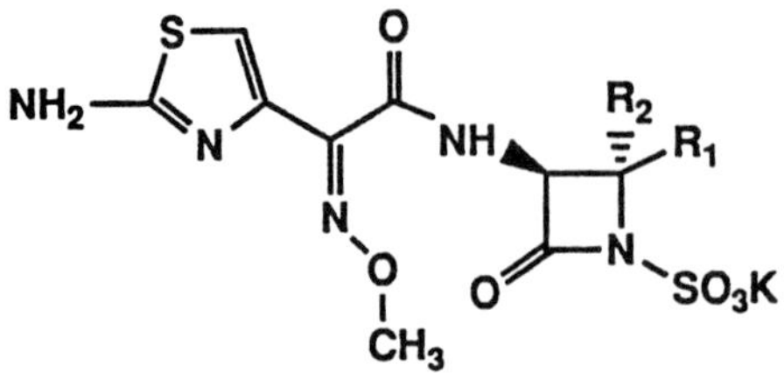

79

No.	R_1	R_2
79a	H	H
79b	H	CH_3
79c	CH_3	H

Table 14. Comparison of Antimicrobial Activities of SQ 81,377 (79a) and Its 4α- and 4β-Methyl Analogues

Organism	*SC No.*	**79a**	**79b**	**79c**
Staph. aureus	1276	6.3	100	50
Staph. aureus	2400	6.3	50	25
E. coli	8294	0.8	0.1	0.1
E. coli	10857	0.4	<0.05	<0.05
K. aerogenes	10440	0.8	0.2	0.2
Prot. mirabilis	3855	0.8	<0.05	0.1
Prot. vulgaris	9416	0.8	<0.5	<0.05
Ent. cloacae	8236	0.8	0.1	0.2
Ser. marcescens	9783	25	0.2	0.8
Ps. aeruginosa	9545	1.6	0.4	0.8
Ps. aeruginosa	8329	25	>100	100

NOTE: Activities are given as MIC in micrograms per milliliter at 10^4 cfu.

the safety and efficacy of monobactams in humans, predictions that had been based on comparison of this data with similar data on penicillins and cephalosporins.

The decision to advance SQ 26,726 to clinical study proved a milestone in the monobactam program at Squibb. SQ 26,726 became the benchmark by which we evaluated all other monobactams. It became the focus of intense effort by diverse groups in the company, many of them outside the Institute for Medical Research. For those involved in the program from its beginning, it brought both satisfaction and a sense of urgency: satisfaction, because what had begun as an idea—novel β-lactams from novel sources—now had a finite realization, and urgency, because, having made the decision, we wanted to implement it as soon as possible. The knowledge that at least one other β-lactam group was involved in monobactam research added to the sense of urgency.

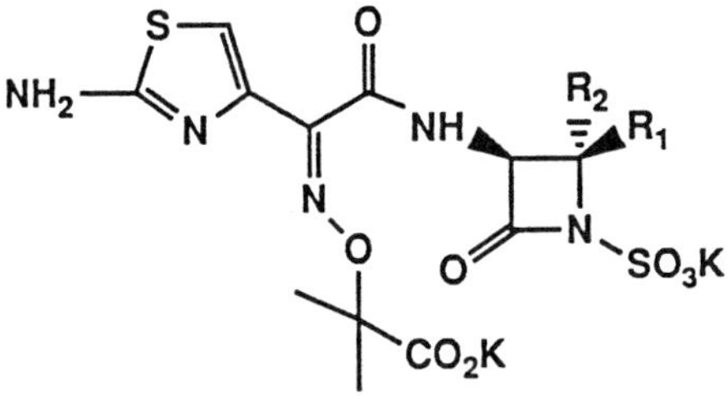

SQ No.	R_1	R_2
81,402	H	H
26,726	H	CH_3
26,743	CH_3	H (racemic)
26,917	CH_3	H

Table 15. Comparison of Antimicrobial Activities of SQ 81,402 and Its 4α- and 4β-Methyl Analogues

Organism	*SC No.*	*SQ 81,402*	*SQ 26,726*	*SQ 26,743*
E. coli TEM+	10404	3.1	0.4	0.1
E. coli TEM−	10439	0.4	0.4	0.4
Ent. cloacae P99+	10435	>100	100	>100
Ent. cloacae P99−	10441	0.4	0.4	0.4
K. aerogenes K1+	10436	>100	>100	6.3
K. aerogenes K1−	10440	0.8	0.2	0.4
Ps. aeruginosa	8329	>100	50	>100
Ps. aeruginosa	9545	1.6	0.8	0.4

NOTE: Activities are given as MIC in micrograms per milliliter at 10^6 cfu.

That implementation would eventually involve most of the development resources of the company. Before it could begin, however, synthetic access had to be improved. Large amounts of SQ 26,726 would be needed for toxicology, formulation development, further microbiological evaluation, and, we hoped, clinical study. Existing synthetic methodology would not support the rapid development that was expected.

Synthetic Access to Aztreonam

The scheme that was used to prepare SQ 26,726 for the first time (Scheme 9) relied on benzyloxycarbonyl protection. When the scheme was designed, we had little experience with the chemistry of monobactams and chose the most conservative approach: deprotection of the benzyloxycarbonylamino moiety could be accomplished under mild neutral conditions. There was no liability in the case of 4-unsubstituted compounds derived from 6-aminopenicillanic acid; however, our original application of the Miller hydroxamate chemistry involved initial BOC protection (A in Scheme 9) of threonine, deprotection (B) of the BOC moiety and replacement (C) with the "required" benzyloxycarbonyl group, and final deprotection (D) by hydrogenolysis prior to side-chain incorporation. The side-chain introduction also involved removal (E) of benzhydryl protection.

Protocols that were expected to confound rapid scale-up into our New Brunswick pilot plant included the following:

two hydrogenolyses (lack of suitable equipment)
water-soluble carbodiimide (WSC) (availability and cost)
Ph_3P-DEAD (chromatographic removal of Ph_3PO)
$TiCl_3$ reduction (large volumes required)
two TFA deprotections (availability of TFA)
HP-20 chromatography (large volume, cost)

In addition, isolation of SQ 26,726 from aqueous solution by evaporation to dryness followed by trituration with an organic solvent was not a simple protocol to envision in large scale. We had recognized most of these drawbacks before the synthesis of SQ 26,726 and had begun an effort to improve synthetic methodology well before the identification of a clinical candidate. We eventually found solutions for all of these problems; none would cause any significant delay in the development program.

Initial attempts to further improve synthetic access to monobactams were based on the knowledge of their acid stability (acylated monobactams were stable to TFA deprotection). We realized that the BOC protecting group might provide a direct route to 3-aminomonobactams that should be stable under acidic deprotection conditions. Sulfonation of **70** provided the tetrabutylammonium salt **80**. Treatment of a methylene chloride solution of **80** with formic acid gave the zwitterion SQ 26,771 as a stable crystalline solid.

Acylation of SQ 26,771 could be accomplished as readily as in situ acylation of its tetrabutylammonium salt. An important aspect of this

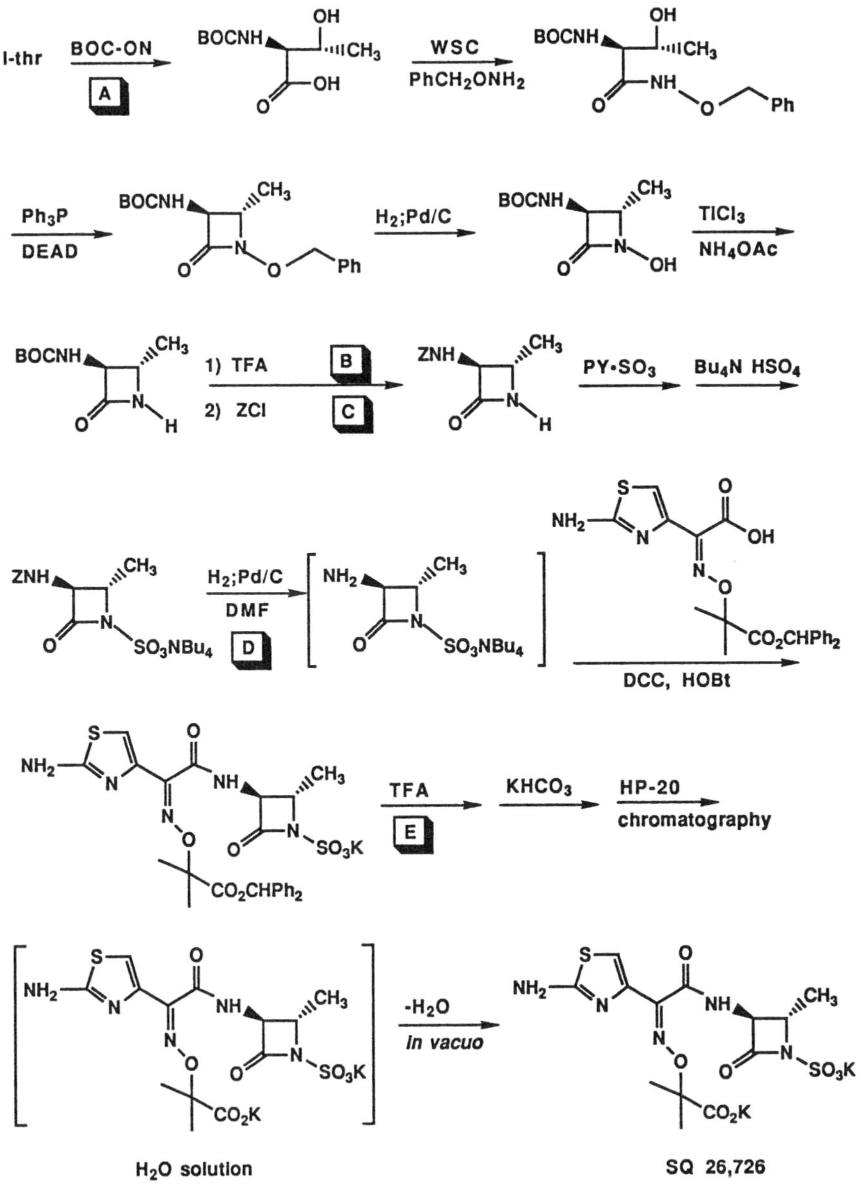

l-thr BOC-ON A

WSC $PhCH_2ONH_2$

Ph_3P DEAD

H_2;Pd/C

$TiCl_3$ NH_4OAc

1) TFA B

2) ZCl C

PY•SO_3

$Bu_4N\ HSO_4$

H_2;Pd/C DMF D

DCC, HOBt

TFA E

$KHCO_3$

HP-20 chromatography

-H_2O *in vacuo*

H_2O solution

SQ 26,726

Scheme 9

70 → **80** → **SQ 26,771**

finding was the ease with which SQ 26,771 could be isolated in highly pure form. No purification of SQ 26,771 was necessary after any of the subsequent, more facile routes to it that we developed: its quality remained essentially constant. This feature was a great asset as we changed the chemistry to access SQ 26,771 and the chemistry that converted it to aztreonam. It also simplified the logistics of our early development work and facilitated subsequent regulatory review. To appreciate how much this seemingly simple finding contributed to the rapid development of aztreonam, consider that, whereas penicillin was discovered in 1928, 6-APA (and, subsequently, the semisynthetic derivatives based on it) only became available more than 30 years later.

The second improvement involved replacement of the two-step deprotection of an *N*-benzyloxyazetidinone to a one-step protocol involving *N*-methoxyazetidinones. Methoxyamine functioned equally well in the Miller cyclization (**81** → **83**) methodology; and drawing on the analogy to reduction of α-alkoxyketones with sodium in ammonia, which generates ketone enolates, David Floyd predicted that *N*-methoxyazetidinones such as **82** would be reduced to the corresponding azetidinone anion **83**. In the absence of a proton source, this anion was expected to survive the reduction conditions; quenching with an appropriate acid should provide the N1-unsubstituted azetidinone **70**. This reduction proved to be facile (*34*) and amenable to pilot-plant scale-up. The reduction of **82** to **70** was eventually carried out on a multikilogram scale. In the earliest stages of our scale-up efforts this was important because it precluded the need for large-scale hydrogenolysis equipment, which we did not have at the time.

These two modifications, with an additional one, facilitated the first pilot-plant synthesis of aztreonam. The sequence is illustrated in Scheme 10 with the third improvement—replacement of the Mitsunobu cyclization by a two-step protocol involving mesylate formation and subsequent cyclization. Though adding a step, the sequence was facile and obviated the need for chromatographic removal of Ph_3PO; intermediates **82** and **84** were nicely crystalline, a property that also facilitated scale-up. One of our development colleagues (LeRoy High) even overcame the need for high dilution in the conversion of **84** to **82** by the simple expedient of adding multiple charges of **84** to a refluxing slurry of K_2CO_3 in acetone. Azetidinone **82** proved to be stable to these conditions for many hours.

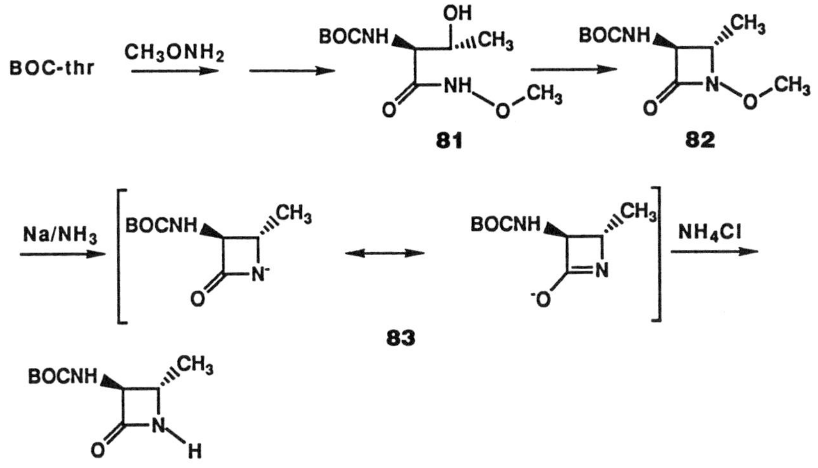

The first kilogram-amounts of aztreonam were prepared using this methodology. We were concerned, however, about the length of the sequence and the availability and cost of some of the key reagents (BOC-ON, methoxyamine, WSC). Soon after we assigned the structure of EM 5117 as a 3-acylamino-2-oxoazetidine-1-sulfonic acid, we began to speculate about the biosynthesis of monobactams. Extrapolating from the work of Townsend on the nocardicin series, we believed that serine was the ultimate precursor of the nucleus of EM 5117. Extrapolating from cephamycin precedent, we also believed that the methoxy group was added last to a preformed monobactam. Subsequent work proved these conclusions to be correct (35).

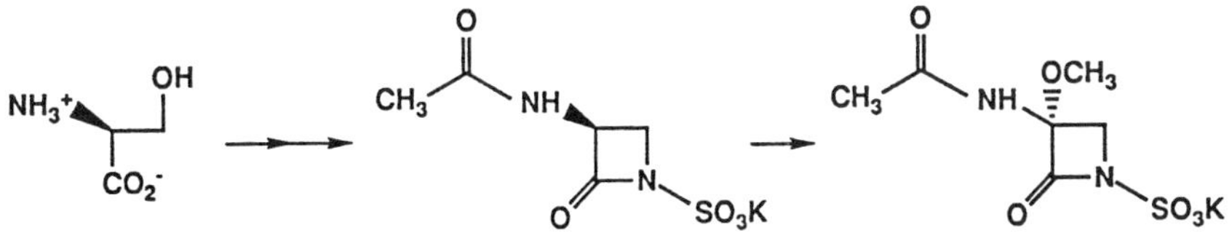

The key question that we sought to answer (on paper and experimentally) was the mode of cyclization. This was of interest for several reasons, including the very practical one of modifying the biosynthetic process to a biomimetic synthesis. Among several hypotheses, we favored the simple one illustrated in Scheme 11. The individual chemical reactions seemed either

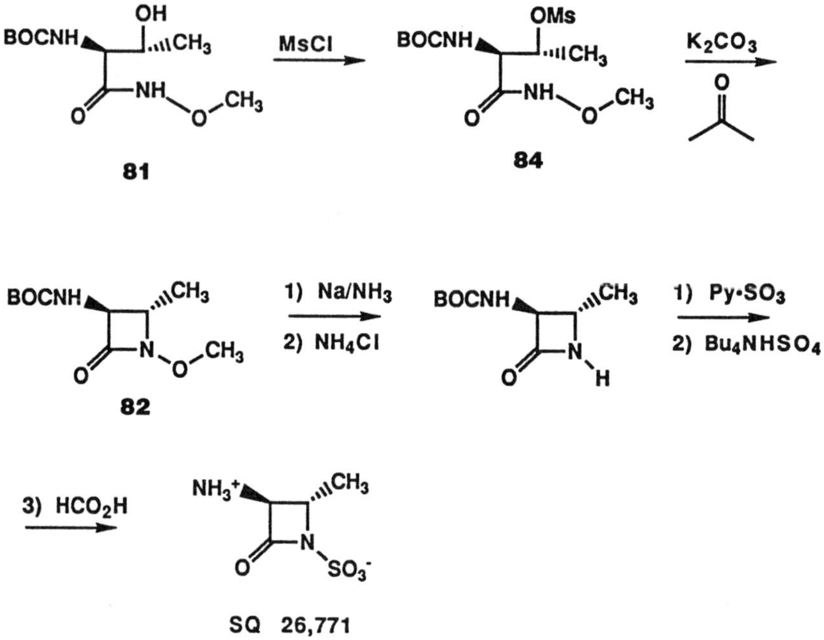

BOCNH, OH, CH3, NH, O, CH3, O — 81 → MsCl → BOCNH, OMs, CH3, NH, O, CH3, O — 84 → K_2CO_3, acetone

82 → 1) Na/NH_3 2) NH_4Cl → BOCNH, CH3, N–H → 1) $Py \cdot SO_3$ 2) Bu_4NHSO_4 3) HCO_2H → NH_3^+, CH3, N, SO_3^-

SQ 26,771

Scheme 10

l-ser → RNH, OH, NH_2, O → RNH, OSO_3^-, N–H, SO_3^- → RNH, OSO_3^-, N^-, SO_3^- → RNH, N, SO_3K

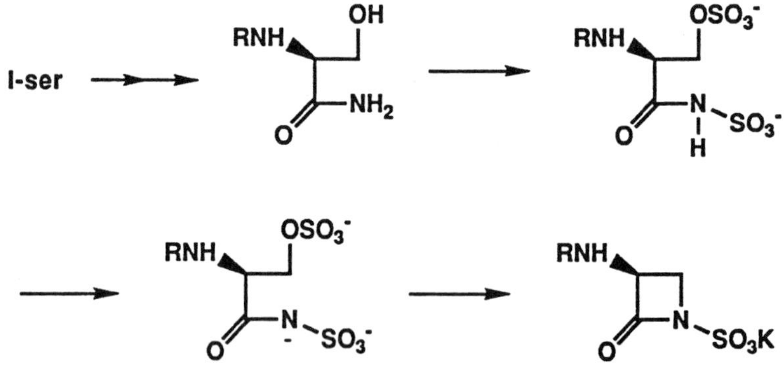

Scheme 11

well precedented or quite plausible (simple acylsulfamates were known (*36*) to form dianions in water, for example) and—an important feature—were open to experimental scrutiny.

The first test of this hypothesis involved β-chloroalanine, which was converted to its benzyloxycarbonyl derivative **85**. After much experimentation, a protocol (tetrahydrofuran solution, WSC, tetrabutylammonium sulfamate) was found that converted **85** to the potential cyclization substrate **86**. Treatment of **86** with sodium hydride in DMF provided azetidinone **31** that was identical to an authentic sample in all respects, including magnitude of optical rotation. This exciting result (*37*) provided the impetus to design and exhaustively explore more practical routes to SQ 26,771 based on this cyclization methodology. Among many other potential attributes, the direct construction of the monobactam nucleus without protecting groups was viewed as sufficient incentive.

The difficulty in coupling *Z*-serine with sulfamic acid was magnified in the case of BOC-threonine (**87**); we were unable to find conditions for this transformation of **87** to **88**. Because sulfonation of simple amides was precedented (*38*), we decided to look at this obvious alternative construction. We chose to use the mesylate of BOC-L-threonineamide as a sulfonation substrate. This choice avoided O-sulfonation, precluded the need for O-protection, and provided the necessary leaving group (demonstrated with hydroxamate methodology). Threonine could be simply converted (Scheme 12), via unisolated methyl ester and amide, to BOC-threonineamide (**89**) and its mesylate (**90**), both of which were nicely crystalline.

Sulfonation of **90** with pyridine–sulfur trioxide did not proceed; therefore the use of more active sulfonating agents was explored (Scheme

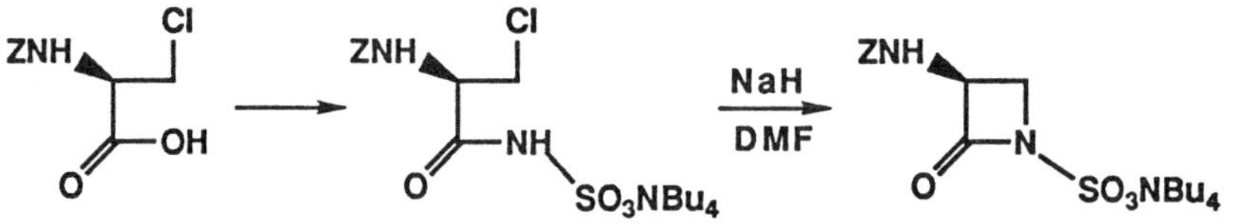

ZNH, Cl, OH, O — **85** → ZNH, Cl, NH, SO_3NBu_4, O — **86** —NaH / DMF→ ZNH, N, SO_3NBu_4, O — **31**

BOCNH, OH, CH_3, OH, O — **87** —X→ BOCNH, OH, CH_3, NH, SO_3NBu_4, O — **88**

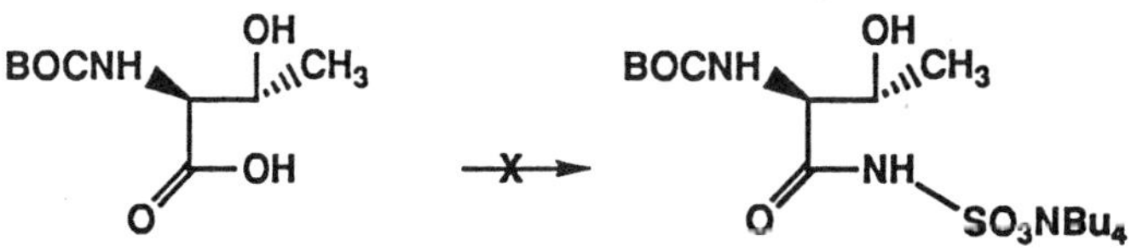

13). 2,6-Lutidine-sulfur trioxide, which required cooling below room temperature, or 2-picoline-sulfur trioxide, which required reflux in methylene chloride, converted **90** to sulfamate **91** in high yield; ion-pair extraction provided the tetrabutylammonium salt **92**. After considerable effort, we found that potassium carbonate was an effective reagent for the cyclization of **92** to **80**. Eventually the need for isolation of **92** was eliminated. The

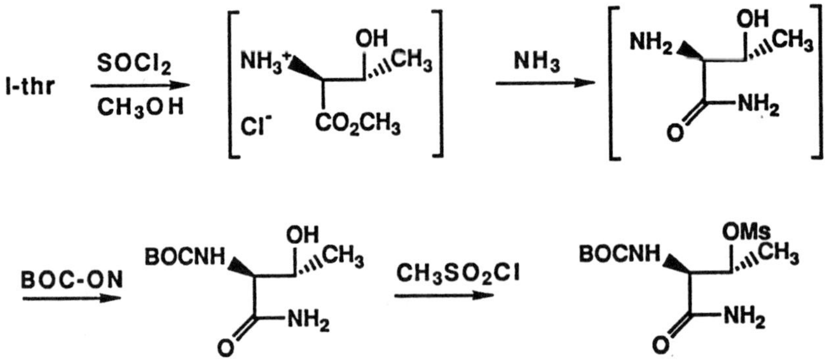

l-thr $\xrightarrow[CH_3OH]{SOCl_2}$ [NH_3^+ Cl^- OH CH_3 CO_2CH_3] $\xrightarrow{NH_3}$ [NH_2 OH CH_3 NH_2 O] $\xrightarrow{BOC\text{-}ON}$ BOCNH OH CH_3 NH_2 O **89** $\xrightarrow{CH_3SO_2Cl}$ BOCNH OMs CH_3 NH_2 O **90**

Scheme 12

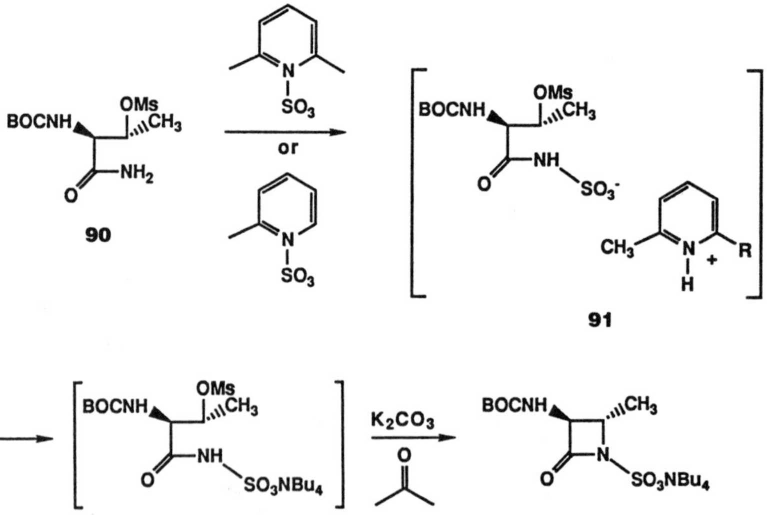

BOCNH OMs CH_3 NH_2 O **90** $\xrightarrow{\text{2,6-lutidine-}SO_3 \text{ or 2-picoline-}SO_3}$ [BOCNH OMs CH_3 NH SO_3^- O; CH_3 N^+ H R] **91** $\longrightarrow$ [BOCNH OMs CH_3 NH SO_3NBu_4 O] **92** $\xrightarrow[\text{acetone}]{K_2CO_3}$ BOCNH CH_3 N SO_3NBu_4 O **80**

Scheme 13

sulfonation solution (in methylene chloride) was quenched in potassium bicarbonate solution, refluxed to effect cyclization, cooled, and subjected to ion-pair extraction. Concentration of the methylene chloride solution and addition of formic acid led to the precipitation of SQ 26,771 in good overall yield.

Our last concern (the cost and availability of BOC-ON or an alternative BOC-transfer reagent) became moot when the corresponding sequence with Z-protection was demonstrated to lead to the crystalline tetrabutylammonium salt **95** (Scheme 14). Hydrogenolysis of **95**, followed by protonation, provided SQ 26,771 in very high yield on large scale. The complete sequence to SQ 26,771 that supported the clinical development phase of aztreonam is illustrated in Scheme 14. The facile isolation (and concomitant purification) of **93**, **94**, and **95** by crystallization and the facile deprotection of **95** by hydrogenolysis (by then we had developed this capability at pilot-plant scale) provided simple and efficient access to SQ 26,771 in the necessary time frame.

While this work was going on, the need for large amounts of SQ 26,726 forced us to consider improvements to the last part of the sequence and to the isolation and purification of SQ 26,726. Recognizing that purification by reversed-phase chromatography and the in vivo use of a dipotassium salt would prove to be obstacles, we sought better alternatives. The close link we had established between the isolation research group and the synthetic research group, which jointly occupied one side of a floor in the Lawrenceville laboratory, was a great advantage in all of the monobactam work. It had an immediate benefit when we asked Larry Parker, who had isolated SQ 26,180, to look at this question. Because we could not divert any SQ 26,726 at this stage from microbiological evaluation, Larry worked with its enantiomer. This compound had inadvertently become available when we first attempted to make the β-methyl analogue of SQ 26,726. In our haste to evaluate this compound, we had unknowingly taken BOC-*d*-threonine (**96**) rather than the desired BOC-*d,l*-allothreonine (**98**) from the refrigerator shelf and subjected it to the sequence illustrated in Scheme 9. HP-20 purification provided a dipotassium salt that was submitted for thorough analytic examination after the usual cursory laboratory characterization. At that time, this examination included bioautography after chromatographic and electrophoretic separation. This methodology, developed and refined over many years by Octavian Kocy, was a tremendous benefit to several decades of β-lactam work at Squibb (*39*). Nevertheless, we were surprised when Kocy reported that the "β-methyl analogue" was microbiologically inactive and, curiously, had the same chromatographic and electrophoretic behavior as SQ 26,726.

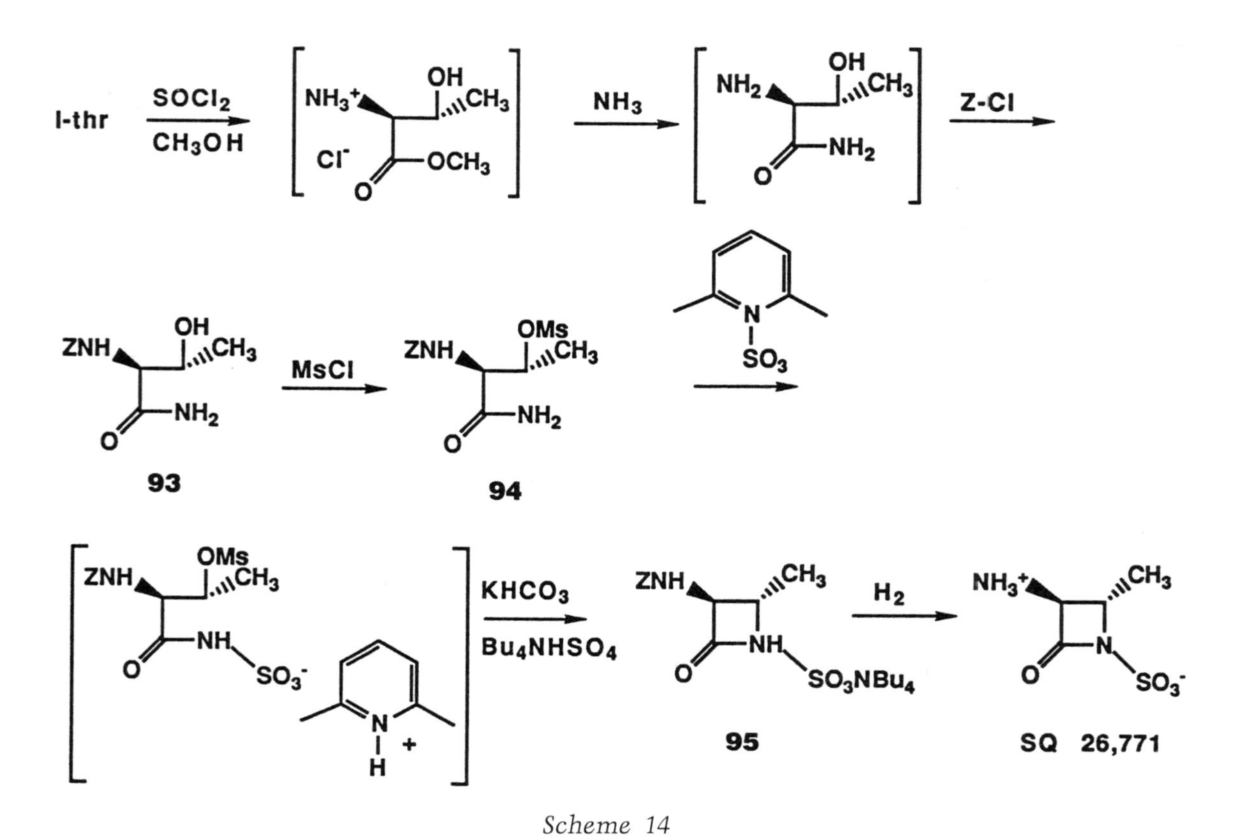
l-thr
SOCl2
CH3OH
NH3
Z-Cl
MsCl
93
94
KHCO3
Bu4NHSO4
95
H2
SQ 26,771

Scheme 14

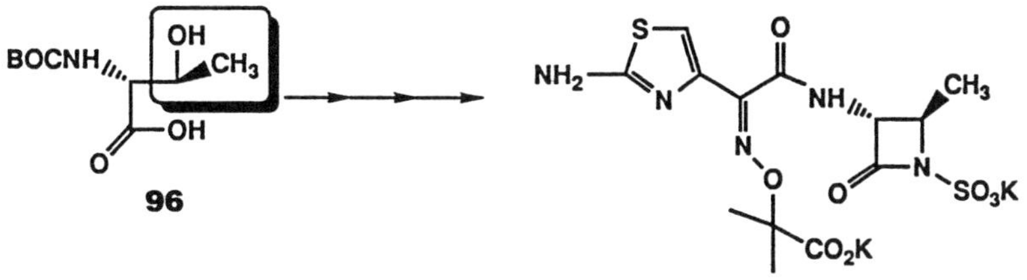

96

97

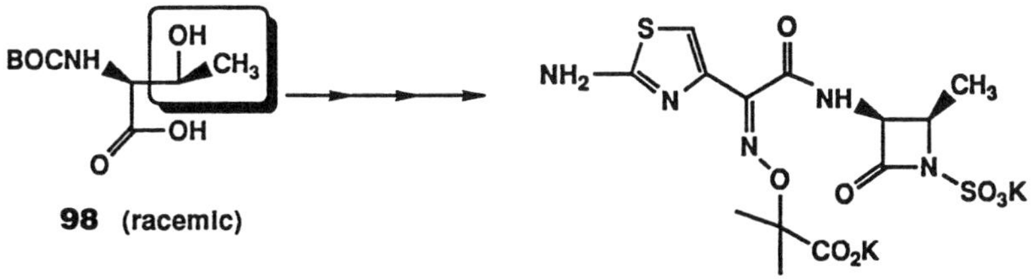

98 (racemic)

SQ 26,743 (racemic)

Close inspection of laboratory notebooks revealed the unfortunate original choice of **96**. Remembering the saying attributed to R. B. Woodward—"the only good model is the enantiomer"—we gave the sample of "enantio" SQ 26,726 (**97**) to his student, Larry Parker. Parker soon discovered that it crystallized from aqueous solution at low pH as the zwitterion **99**, and that this provided a convenient and useful purification that eliminated the need for chromatography. This crystalline material was assigned the internal registry number SQ 26,775. When some of the precious correct enantiomer was provided to Larry, it was converted (as predicted) to its zwitterion (**100**) and eventually assigned the designation SQ 26,776. Larry also showed that the benzhydryl ester precursor of SQ 26,776 could be crystallized as its zwitterion. These two observations, coupled with the facile synthesis developed for SQ 26,771, set the stage for rapid development of SQ 26,776. However, much remained to be done to get from SQ 26,776 to Azactam.

Azactam

As supplies of SQ 26,726 and then SQ 26,776 became available, extensive in vitro and in vivo comparison was made to reference antibacterial agents. These studies defined SQ 26,776 as a highly active, β-lactamase-stable agent

97 →

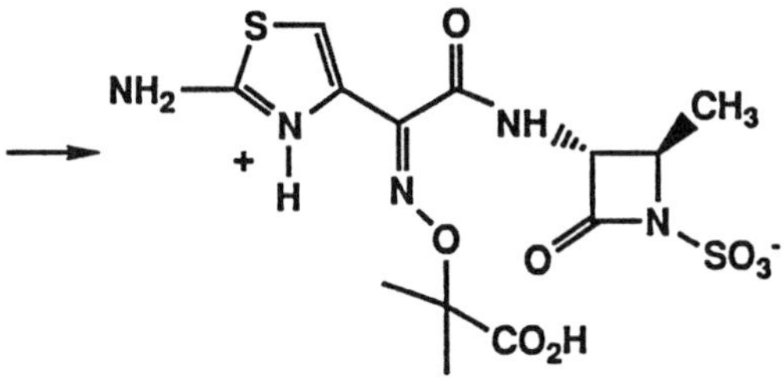

99 (SQ 26,775)

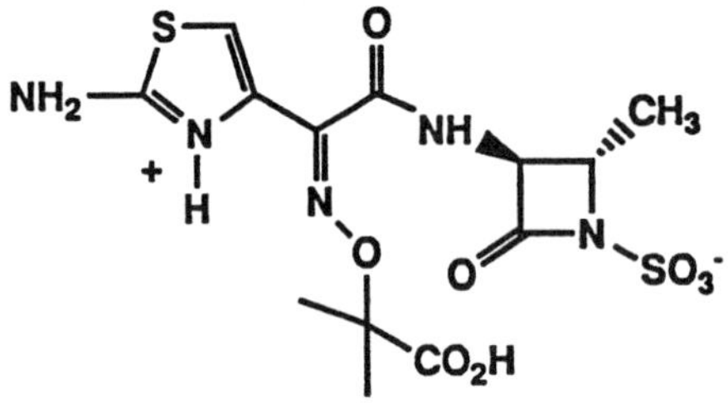

100 (SQ 26,776)

with specific activity against aerobic Gram-negative bacteria, including *Pseudomonas*, and essentially no anti-Gram-positive activity. Low oral absorption was not viewed as a drawback because much of the therapy associated with severe infections is administered parenterally. It did require, however, that we develop a sterile preparation for clinical use.

Richard Sykes and his colleagues designed and executed a program that convincingly demonstrated a potential niche for this agent in the chemotherapy of serious Gram-negative infections. These findings were first discussed openly outside Squibb in the summer of 1981 at a symposium entitled "Azthreonam, a Synthetic Monobactam," held at the XII International Congress of Chemotherapy (Florence, Italy, 23 July 1981). Azthreonam (azetidinone threonine monobactam) was originally submitted as a United States Adopted Name (USAN). It was modified to aztreonam, at the request of WHO, to avoid four consecutive consonants that included the *th* sound (-*zthr*-). The 17 papers presented at the symposium were collected as a supplemental volume to the *Journal of Antimicrobial Chemotherapy* (Supplement E to Vol. 8, December 1981). From this beginning, the continued involvement of academic investigators in the aztreonam project facilitated acceptance by internal commercial and marketing colleagues. This experimentation, and the toxicological studies that qualified SQ 26,776 for evaluation in humans, were accomplished with aqueous solutions made by adjusting a suspension of SQ 26,776 to physiologic pH. Such a procedure, which facilitated this early work, could not be used for clinical evaluation or commercial use. It became imperative to define a sterile (and stable) dosage form as quickly as possible.

It had already been demonstrated that SQ 26,776 was not stable at room temperature and was incompatible, because of rapid reaction, with sodium carbonate or arginine (agents that could have been used in blends that would reconstitute with water to the required sterile salt solution). We could find no salt that was crystalline, easily purified, and amenable to facile sterile production. At the same time, initial studies of lyophilized salts (sodium, arginine) showed an unacceptable instability. This situation resulted in a formal request from the pharmaceutics group for additional work on an alternative final form (either examination of more salts or different crystalline forms of SQ 26,776). Fortunately, the experiment that provided the solution had been completed just days before the request was made.

During a search for optimized isolation protocols, a suspension of SQ 26,776 in methanol was allowed to remain in the freezer for some time. When this material was provided to Octavian Kocy for evaluation of its purity, he noted that it had different dissolution characteristics. Close examination revealed a different crystal habit and a modified infrared spectrum in the solid state. The sample was analyzed and shown to be high-quality SQ 26,776. Further examination revealed that the solid was a new crystalline form of SQ 26,776 that differed from the original one in several key aspects. It was relatively free of water of hydration (less than 2%) compared to the original form (ca. 15%) and it was much more stable. Either alone or in combination with arginine, this "beta" form of SQ 26,776 was much more stable than the original "alpha" form. Improved protocols for preparing the beta form were developed quickly, and accelerated stability studies of blends with arginine led to the prediction that a sterile blend of beta-form aztreonam and arginine would be a suitable dosage form. Indeed, Azactam was eventually introduced worldwide as a sterile blend of these components suitable for reconstitution with water and stable over its shelf life at room temperature.

Further development of lyophilization protocols eventually led to a stable preparation of Azactam that precluded the need for sterile beta-form aztreonam and sterile arginine. This lyophile has been registered in many countries as an alternative to the dry-powder blend and represents a more elegant dosage form that is simpler to produce. Ten years after the synthesis of the first kilogram of sterile, beta-form aztreonam in New Brunswick, the initial production run of lyophilized Azactam was made in a newly constructed and FDA-approved facility there. The lyophilization chambers in this state-of-the-art sterile facility are the largest in the world.

With adequate clinical supplies, investigation of the safety and efficacy of Azactam in humans proceeded rapidly and successfully. During 1983–1985, documents requesting authorization to market Azactam were filed in many countries. Much of the clinical data incorporated in these filings was reviewed in 1986 (*40*). The summary statement of the conclusion section is repeated here.

> Aztreonam appears to be potentially very useful as a replacement for many broad-spectrum agents when more directed therapy is necessary, or for aminoglycosides in many situations where antibacterial actively against Enterobacteriaceae and *Pseudomonas* are needed.

Many of the predictions made during extensive in vitro and in vivo laboratory investigation of aztreonam were realized in these clinical studies of Azactam. They included:

1. Microbiologic cure rates of ca. 90% for varied infections involving the Enterobacteriaceae, *Pseudomonas aeruginosa, Neisseria gonorrrhoeae,* and *Hemophilus influenzae.* Aztreonam is highly active against Gram-negative aerobic organisms and stable to most of the β-lactamases that they elaborate.
2. Lack of increased resistance during therapy. Aztreonam does not induce the formation of β-lactamases.
3. A low incidence of diarrhea that is usually attributed to radical changes in microbial gut flora. Aztreonam is inactive against Gram-positive organisms and anaerobes and provides selective decontamination of the intestinal tract).
4. An extremely low incidence of allergic reactions, even in individuals with a previous history of allergies to other β-lactams. Aztreonam does not cross-react with penicillin in vitro.

Azactam is currently marketed in 67 countries, and its annual sales first exceeded $100 million in 1988. In the context of captopril or ranitidine, aztreonam's commercial success has not reached blockbuster proportions. A recent Pharmaceutical Manufacturer's Association (PMA) study revealed that the cost of successful development of a new chemical entity is ca. $230 million. By this benchmark, Azactam can probably be judged successful;

with annual sales of $100 million or more, the company should eventually recoup its R&D investment.

Though we initially thought about a "family" of monobactams, only one other—carumonam, with very similar structure and properties—is marketed, by Takeda in Japan. We have brought other structural types of monocyclic β-lactams to toxicological (SQ 83,360) and, in some cases, clinical investigation (SQ 82,531, Tigemonam). To date none has matched aztreonam's initial promise. A wide range of additional alternatively activated monocyclic azetidinones were studied at Squibb and other laboratories; to date, no others have progressed to clinical study.

The discovery of sulfazecin at Takeda and SQ 26,180 at Squibb has stimulated much thought and inspired much new chemistry. It was the impetus for the development of aztreonam and carumonam; their clinical use provides an additional and unique weapon in the struggle between humans and microbes. Ten years after his initial involvement with aztreonam, Neu (*41*) has provided a clinician's view of the utility of aztreonam:

> At the Columbia-Presbyterian Medical Center, aztreonam often is used as empiric therapy, especially when there is concern about the possibility of resistant organisms, such as *Pseudomonas,* and aminoglycoside toxicity. As empiric therapy, it has been used for Gram-negative coverage in combination with a number of agents, including erythromycin, clindamycin, and vancomycin, in patients with various infections, including respiratory tract infection and abdominal infection. Aztreonam is used as an aminoglycoside replacement, particularly in patients with compromised renal function.
>
> From all of the data presented at this meeting, we can conclude that aztreonam is a safe therapy that provides activity against *Pseudomonas* spp., *E. coli, Klebsiella* spp., and other Gram-negative organisms. It will not damage the liver or the kidneys, will not cause any hematologic problems, will not cross-react with the penicillins, and will not cause seizures. Most importantly, it is effective; over the past five years, clinical failures have not been a problem.

I do not believe we expected more for aztreonam when we prepared it.

Interestingly, two of the major findings of our synthetic work with monobactams could have been derived from inspection of the natural world. We isolated a complex of monobactam antibiotics (EM 5400) from an *Agrobacterium* species (*42*); among them were the methoxylated-nomethoxylated pair **101** and **102**. Study of this pair would have demon-

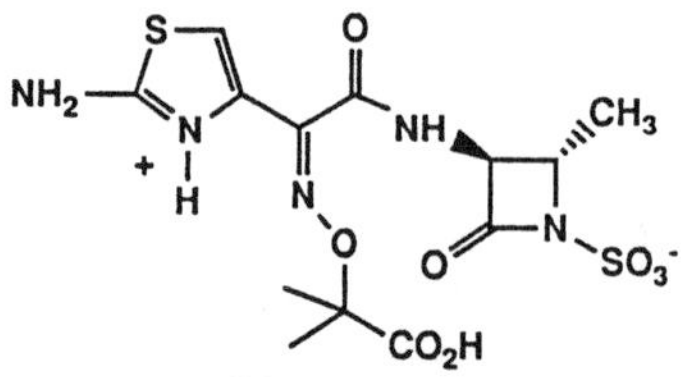

SQ 26,776

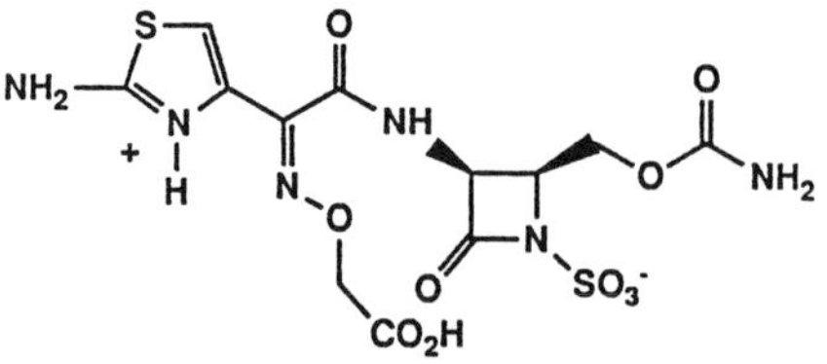

Carumonam

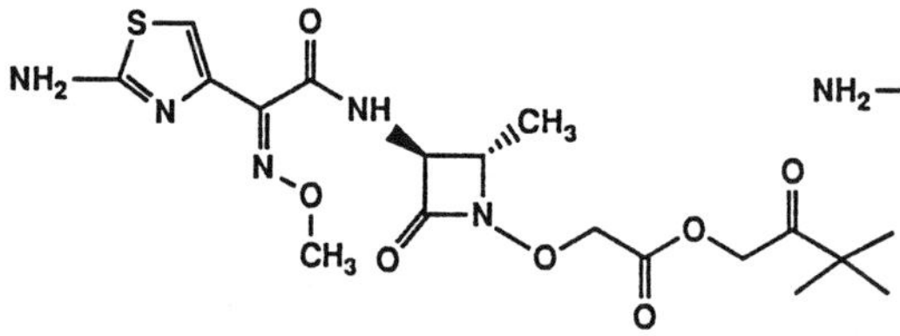

SQ 82,531

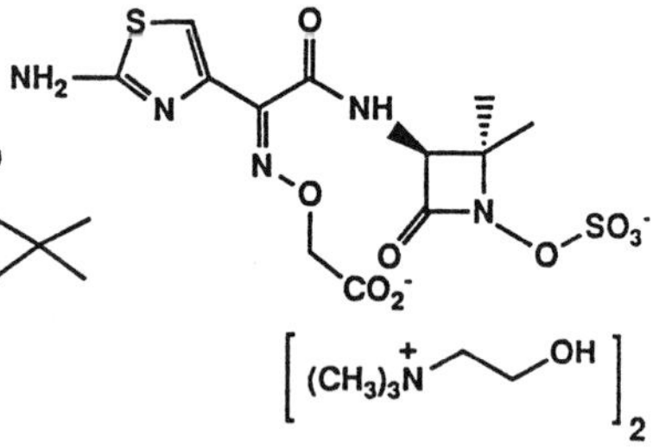

Tigemonam

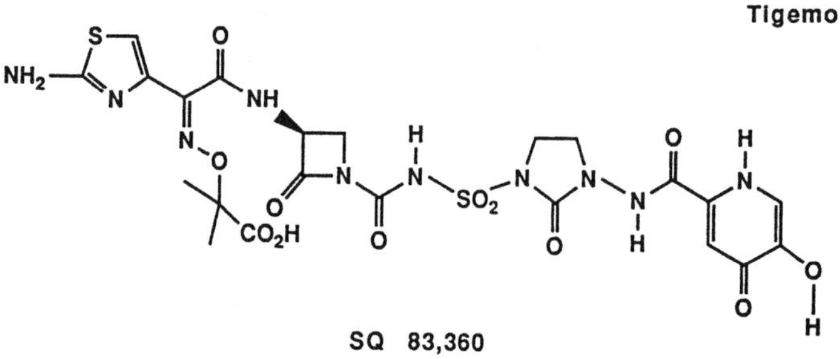

SQ 83,360

strated the better chemical stability of **101**. In 1988, workers at Beecham (*43*) reported the occurrence of the 4β-methylmonobactam **103** in a bacterium isolated from a soil sample collected in Japan. If the Takeda group had collected this organism instead of the one that produced sulfazecin, the discovery and development of aztreonam (if it occurred at all) would have taken a different course from the one described here.

The most useful lesson from the discovery and development of aztreonam may be a general one. The products of life are ubiquitous and highly variable. Even in a mature field like β-lactam research, understanding them continues to provide unique insight. Basic research programs to identify novel and highly selective targets for chemical modulation should consider these products as an important source of structural leads. Efforts to devise assays based on such targets are likely to be rewarded and to inform the discovery process.

No.	X
101	H
102	CH_3O

103

Literature Cited

1. Dolfini, J. E.; Applegate, H. E.; Bach, G.; Basch, H.; Bernstein, J.; Schwartz, J.; Weisenborn, F. L. *J. Med. Chem.* **1971,** *14,* 117–119.
2. Condon, M. E.; Cimarusti, C. M; Fox, R.; Narayanan, V. L.; Reid, J.; Sundeen, J. E.; Hauck, F. P. *J. Med. Chem.* **1978,** *21,* 913–922.
3. Nagarajan, R.; Boeck, L. D.; Gorman, M.; Hamill, R. L.; Higgins, C. E.; Hoehn, M. M.; Stark, W. H.; Whitney, J. G. *J. Am. Chem. Soc.* **1971,** *93,* 2308–2310.
4. Aoki, H.; Sakai, H.; Kohsaka, M.; Konomi, T.; Hosoda, J.; Kubochi, Y.; Iguchi, E.; Imanaka, H. *J. Antibiot.* **1976,** *29,* 492–500.
5. Howarth, J. T.; Brown, A. G. *J. Chem. Soc. Chem. Commun.* **1976,** 266–267.
6. Albers-Schonberg, G.; Arison, B. H.; Hensen, O. D.; Hirshfield, J.; Hoogsteen, K.; Kaczka, E. A.; Rhodes, R. E.; Kahan, J. S.; Kahan, F. M.;

Ratcliffe, R. W.; Walton, E.; Ruswinkel, L. J.; Morin, R. B.; Christensen, B. G. *J. Am. Chem. Soc.* **1978,** *100,* 6491–6499.
7. Woodward, R. B. In *Recent Advances in the Chemistry of β-Lactam Antibiotics*; Elks, J., Ed; Special Publication No. 38; Royal Society of Chemistry: London, 1977; pp 167–180.
8. Van Handuyt, H. W.; Pyckavet, M. *Antimicrob. Agents Chemother.* **1979,** *16,* 109–111.
9. Ondetti, M. A.; Rubin, B.; Cushman, D. W. *Science (Washington, DC)* **1977,** *196,* 441–444.
10. Lynch, J. L.; Neuhaus, F. C. *J. Bacteriol.* **1966,** *91,* 449–460.
11. Richmond, M. H.; Sykes, R. B. *Adv. in Microb. Physiol.* **1973,** *2,* 31–88.
12. Parker, W. L.; Rathnum, M. L.; Liu, W. C. *J. Antibiot.* **1982,** *35,* 900–902.
13. Sykes, R. B.; Wells, J. S. *J. Antibiot.* **1985,** *38,* 119–121.
14. Fleming, A. *Br. J. Exp. Pathol.* **1929,** *10,* 226–236.
15. Imada, A.; Kitano, K.; Kintaka, M.; Muroi, M; Asai, M. *Nature (London)* **1981,** *289,* 590–591.
16. Parker, W. L.; Koster, W. H.; Cimarusti, C. M.; Floyd, D. M.; Liu, W. C.; Rathnum, M. L. *J. Antibiot.* **1982,** *35,* 189–195.
17. Graf, R. *Angew. Chem.* **1968,** 3, 172–182.
18. Feigl, F. *Spot Tests in Organic Analysis*; Elsevier: Amsterdam, 1966; p 306.
19. Karady, S.; Pines, S. J.; Weinstock, L. M.; Roverts, F. E.; Brenner, G. S.; Hoinowski, A. M.; Cheng, T. Y.; Sletzinger, M. *J. Am. Chem. Soc.* **1972,** *94,* 1410–1411.
20. Applegate, H. E.; Cimarusti, C. M.; Slusarchyk, W. A. *J. Chem. Soc. Chem. Commun.* **1980,** 293–294.
21. Cimarusti, C. M.; Applegate, H. E.; Chang, H. W.; Floyd, D. M.; Koster, W. H.; Slusarchyk, W. A.; Young, M. G. *J. Org. Chem.* **1982,** *47,* 179–180.
22. Gordon, E. M.; Chang, H. W.; Cimarusti, C. M.; Toeplitz, B.; Gougoutas, J. Z. *J. Am. Chem. Soc.* **1980,** *102,* 1690–1702.
23. Bose, A. K.; Tsai, M.; Sharma, S. D.; Manhas, M. S. *Tetrahedron Lett.* **1973,** 3851–3852.
24. Kamiya, T. In *Recent Advances in the Chemistry of β-Lactam Antibiotics*; Elks, J., Ed; The Chemical Society: London, 1977; pp 281–94.
25. Liu, W. C.; Parker, W. L.; Wells, J. S.; Principe, P. A.; Trejo, W. H.; Bonner, D. P.; Sykes, R. B. In *Proceedings of the 12th International Congress on Chemotherapy*; Miami, 1981; pp 328–329.

26. Gordon, E. M.; Sykes, R. B. In *Chemistry and Biology of β-Lactam Antibiotics*; Morin, R. B.; Gorman, M., Eds.; Academic: New York, 1982; Vol. 1.
27. Indelicato, J. M.; Wilham, W. L. *J. Med. Chem.* **1974,** *17,* 528–529.
28. Georgopapadakou, N. H.; Smith, S. A.; Cimarusti, C. M. *Eur. J. Biochem.* **1982,** *124,* 507–512.
29. Georgopapadakou, N. H.; Smith, S. A.; Cimarusti, C. M.; Sykes, R. B. *Antimicrob. Agents Chemother.* **1983,** *23,* 98–104.
30. Cimarusti, C. M.; Sykes, R. B. *Med. Res. Rev.* **1984,** *4(1),* 1–24.
31. Eglington, A. J. *J. Chem. Soc. Chem. Commun.* **1977,** 720.
32. Huffman, W. F.; Holden, K. G.; Buckley III, T. F.; Gleason, J. G.; Wu, L. *J. Am. Chem. Soc.* **1977,** *99,* 2352–2353.
33. Mattingly, P. G.; Kerwin, J. F., Jr.; Miller, M. *J. Am. Chem. Soc.* **1979,** *101,* 3983–3985.
34. Floyd, D. M.; Fritz, A. F.; Pluscec, J.; Weaver, E. R.; Cimarusti, C. M. *J. Org. Chem.* **1982,** *47,* 5160–5167.
35. Parker, W. L.; O'Sullivan, J.; Sykes, R. B. *Adv. Appl. Microbiol.* **1986,** *31,* 181–205.
36. Baumgarten, P.; Marggraff, I. *Chem. Ber.* **1931,** *64,* 1582–1588.
37. Cimarusti, C. M.; Applegate, H. E.; Change, H. W.; Floyd, D. M.; Koster, W. H.; Slusarchyk, W. A.; Young, M. G. *J. Org. Chem.* **1982,** *47,* 176–178.
38. Audrieth, L. F.; Sveda, M.; Sisler, H. H.; Butler, M. J. *Chem. Rev.* **1940,** *29,* 49–94.
39. Wojtkowski, P. W.; Dolfini, J. E.; Kocy, O.; Cimarusti, C. M. *J. Am. Chem. Soc.* **1975,** *97,* 5628–5630.
40. *Beta-Lactam Antibiotics for Clinical Use*; Queener, S. F.; Webber, J. A.; Queener, S. W., Eds.; Dekker: New York, 1986; pp 593–640.
41. Neu, H. C. *Am. J. Med.* **1990,** *88* (3C), 1–43.
42. Parker, W. L.; Rathnum, M. L. *J. Antibiot.* **1982,** *35,* 300–305.
43. Box, S. J.; Brown, A. G.; Gilpin, M. L.; Gwynn, M. N.; Spear, S. R. *J. Antibiot.* **1988,** *41,* 7–12.

Ganciclovir

Julien P. H. Verheyden
Syntex Discovery Research

*Patience et longueur de temps font plus que force et rage.**

The story of ganciclovir could be told in a few words. Paraphrasing a famous Roman general I could say, "It was conceived, it was synthesized, it was efficacious". Indeed, this drug did not result from the optimization of a lead compound after a careful structure–activity relationship study had elaborated the best possible structure. It is very difficult to optimize a nucleoside analogue antiviral agent because the nucleoside itself is not the antiviral agent. It is the corresponding triphosphate that will inhibit the viral polymerase and thus prevent viral replication. In order for the nucleoside to be converted to the antiviral entity, it must be a substrate for three successive and different kinases; then the resulting triphosphate must be selective enough to inhibit the viral enzyme without affecting the cellular polymerases. If a compound fails in an antiviral bioassay it may be because it is a poor substrate for any one of the kinases or the viral polymerase, and it is not possible to know exactly where the compound fails without investing substantial biochemical effort. As an illustration of this difficulty, we and many others have tried to improve the properties of ganciclovir by synthesizing many analogues, but to date none have matched its potency while being less toxic.

*A French proverb my father often quoted.

2523–0/93/0299$07.75/0

Ganciclovir arose from a long-term research project devoted to exploring the antimicrobial properties of 4′-substituted nucleosides. Although substitutions at all other positions of the ribose ring had been extensively studied, the 4′-position was unexplored. This proved to be a very fruitful position to modify. Thus, I will take the reader through the long journey that led to the conception of ganciclovir, a journey like a cross-country hike in the Sierra that brings you often to a pass where you realize there are other valleys to cross and more hills to climb before you reach your destination. The starting point was my interest in a paper by N. K. Kochetkov and A. I. Usov (*1*), abstracted in *Index Chemicus,* that described a novel one-step synthesis of iodinated sugars, which I applied to nucleosides. The 5′-deoxy-5′-iodinated nucleosides obtained by this method led to the synthesis of decoyinine, which is a member of the 4′,5′-unsaturated nucleoside family. The availability of these unsaturated nucleosides in turn permitted the synthesis of nucleocidin, the first 4′-fluorinated nucleoside isolated from nature. The synthesis of nucleocidin led to an investigation of 4′-substituted nucleosides in general, including the 4′-hydroxymethyl nucleosides. When acyclovir came to the forefront of the antiviral field, other workers (*2–4*) who had independently conceived of ganciclovir regarded it as a secoguanosine, an extension of the work done at Burroughs Wellcome Company, whereas I thought of this compound as a 4′-substituted acyclovir, an extension of my work at Syntex Research.

5′-Deoxy-5′-Iodonucleosides

In 1961, I came as a postdoctoral fellow to the newly formed Syntex Institute for Molecular Biology in Palo Alto, California. The institute had been set up to explore the emerging field of molecular biology. It was made up of three departments: biology, enzymology, and chemistry, and it was unusual in the industry in that research was our main goal. Klaus Pfitzner, another postdoctoral fellow, was studying the reaction mechanism of the now well-known Pfitzner–Moffatt oxidation (*5*) and needed for his study 3′-deoxynucleosides, which were prepared by reduction of the corresponding 3′-iodonucleoside. For this reason, my attention was drawn to a paper by Kochetkov and Usov (*1*) that described a one-step iodination of suitably protected carbohydrates using a reagent developed by Rydon and co-workers (*6, 7*) 10 years earlier. I decided to see if the use of this reagent could be extended to the preparation of iodonucleosides in one step. The classic method involved two steps: tosylation, followed by nucleophilic displacement of the resulting tosylate with sodium iodide. Kochetkov and Usov had used benzene at reflux as the solvent for the reaction. However, because 2′,3′-*O*-isopropylideneuridine was completely insoluble in this solvent, I used the newly introduced solvent dimethylformamide (DMF). Because the conversion in benzene at reflux described by Kochetkov and Usov took many hours, I set up my reaction in DMF and at room temperature to be checked at 1, 2, 4, 8, and 24 h. To my delight, the reaction was complete after 1 h. In fact, I was later able to show that the reaction is over in a few minutes and that the 5′-deoxy-5′-iodo-2′,3′-*O*-isopropylideneuridine could be isolated in 96% yield (Figure 1) (*8, 9*). I then extended this reaction successfully to a series of pyrimidine nucleosides and showed that secondary hydroxyls could also be iodinated as long as they were not part of a cis glycol (*10*). With the exception of inosine, the purine nucleosides gave the corresponding N^3,5′-cyclo derivatives in DMF as the only products. Inosine gave also a small amount (7.8%) of the desired 5′-iodo derivative (*10*).

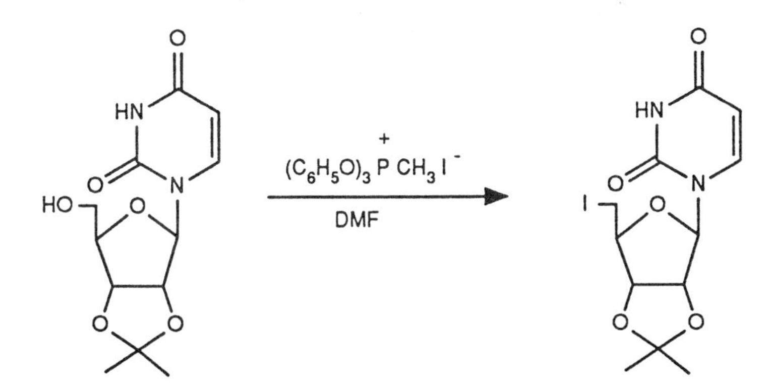

Figure 1. Iodination of pyrimidine nucleoside.

Later on, however, S. Dimitrijevich was able to obtain the desired 5′-deoxy-5′-iodopurine nucleoside by changing the solvent of the reaction from the dipolar aprotic solvent DMF to a nonpolar solvent such as dichloromethane or tetrahydrofuran. Indeed, DMF is known to solvate anions such as iodide, reducing their nucleophilicity and thus favoring attack by the nitrogen (at position 3) of the purine ring on carbon 5′ of the intermediate phosphonium (path *a* in Figure 2) over the attack by the iodide ion (path *b*). With this approach, various 5′-iodopurine nucleosides could be isolated in good yields (*11*).

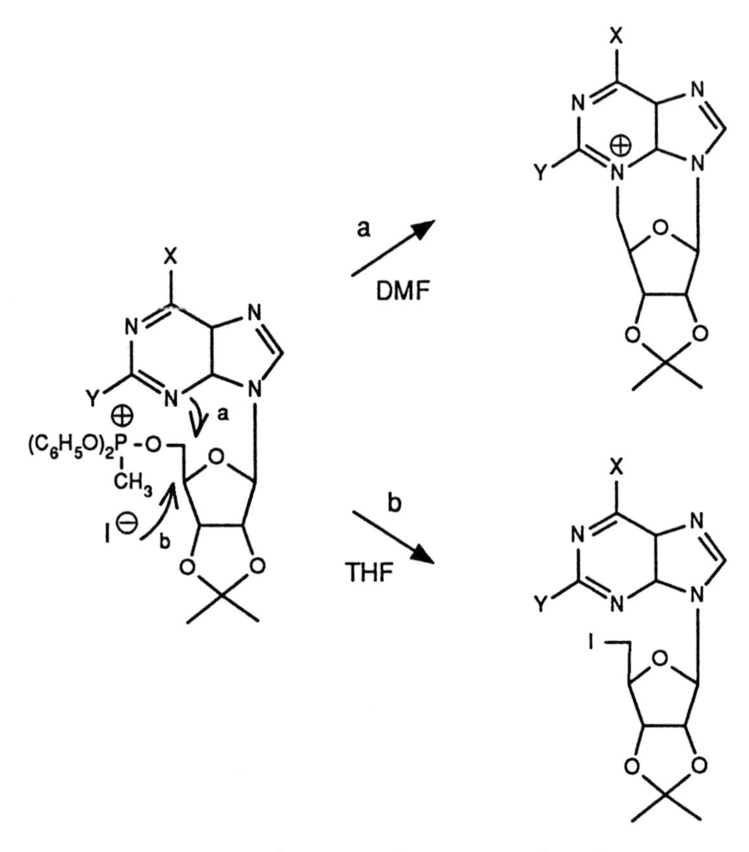

Figure 2. Iodination of purine nucleoside.

4′,5′-Unsaturated Nucleosides

In 1964, Hoeksema et al. (*12*) revised the structure of decoyinine to that derived from psicofuranine (Figure 3). This new structure had a furanose ring and an unusual exocyclic vinyl ether function. In 1966, Hough and Otter (*13*) showed that the reaction used by Helferich and Himmen (*14*) to transform 6-deoxy-6-iodopyranosides into 6-deoxy-5,6-didehydro pyranosides was also applicable to the synthesis of 5-deoxy-4,5-didehydro furanose derivatives. By that time I had accumulated a large collection of 5′-deoxy-5′-iodonucleosides and was eager to search for new compounds with potent antimicrobial activity. Decoyinine appeared to be an interesting target. 2′,3′-

Di-*O*-acetyl-5′-deoxy-5′-iodouridine was reacted with silver fluoride and pyridine, and, after removal of the protecting groups, a nucleoside containing a 4′,5′-double bond (*15*) was isolated, similar in structure to decoyinine (Figure 4).

This work was extended to other pyrimidine nucleosides, and I was able to isolate the 4′,5′-unsaturated derivatives of cytidine, 2′-deoxycytidine, and

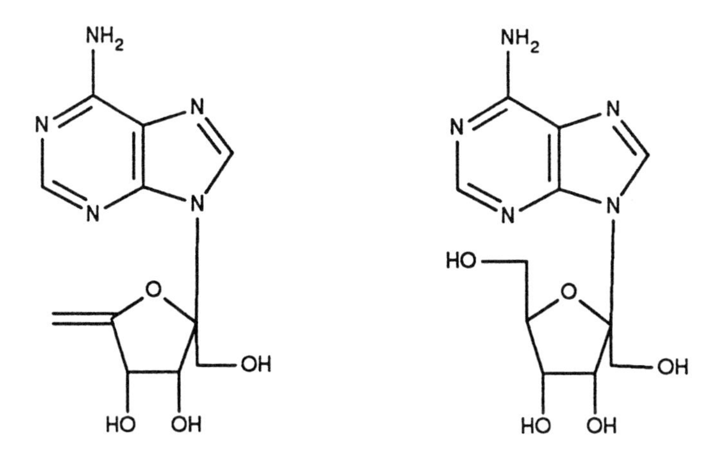

Figure 3. Decoyinine and psicofuramine.

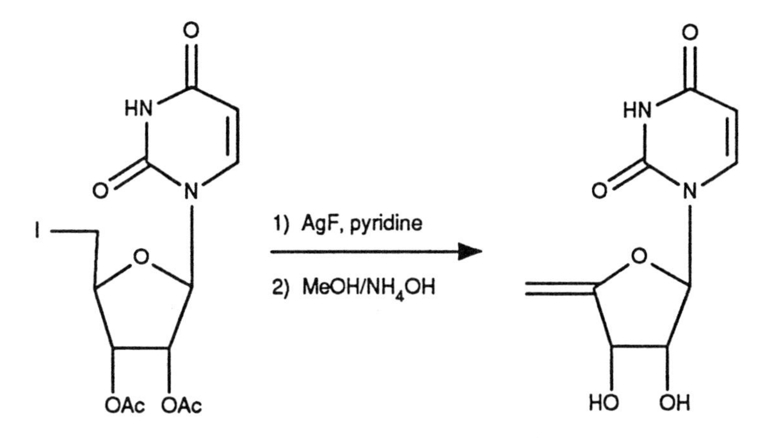

Figure 4. Dehydrohalogenation of a 5′-deoxy-5′-iodouridine derivative.

thymidine by dehydrohalogenation of the corresponding 5′-deoxy-5′-halonucleoside using either silver fluoride in pyridine, 1,5-diazabicyclo[4.3.0]-non-4-ene (DBN), or potassium *t*-butoxide in DMF (*16*).

Using the experience acquired on simple nucleosides, J. Smejkal embarked on the synthesis of decoyinine; E. Prisbe later finished the synthesis and extended it to various analogues (*17*). McCarthy et al. (*18*) had synthesized this antimicrobial starting from psicofuranine, but I desired a more flexible synthesis that would allow us to attach various heterocyclic bases to the glycosidic moiety of decoyinine. The initial target was a suitably protected and activated 6-deoxy-6-iodopsicofuranose that could be easily condensed with a variety of heterocyclic bases. Dehydrohalogenation would then give us the desired unsaturated nucleoside. Toward this goal, fructose was converted into 1,2-3,4-di-*O*-isopropylidenepsicofuranose in four steps. This compound was iodinated and the isopropylidene group replaced by benzoyl groups. Treatment with hydrogen bromide in dichloromethane gave our desired target: 1,3,4-tri-*O*-benzoyl-6-deoxy-6-iodo-D-psicofuranosyl-bromide, which was condensed with N^6-benzoylchloromercuriadenine to give the desired β-psicofuranosylnucleoside in 49% yield. Debenzoylation and dehydrohalogenation were achieved by treatment with sodium methoxide in methanol, and the resulting product was shown to be identical in every respect to an authentic sample of decoyinine (Figure 5). The cytosine and the 1,2,4-triazole 3-carboxamide analogues of decoyinine were then prepared via a similar approach (*17*).

4′-Fluoronucleosides

In 1969 Morton et al. (*19*) established the structure of nucleocidin, a broad-spectrum antibacterial and antitrypanosomal agent (*20, 21*), as 4′-fluoro-5′-*O*-sulfamoyladenosine. Nucleocidin was of great interest to us because it was the first naturally occurring nucleoside having a fluorine atom in its glycosidic moiety; moreover, the fluorine was located at the 4′-position, a position that had not been manipulated by nucleoside chemists. The previous year Hall and Manville (*22*) had shown that iodine monofluoride reacted with glycals, giving a mixture of 1-fluoro-2-iodo hexoses. Because a glycal and a 4′,5′-unsaturated nucleoside are both vinyl ethers, I reasoned that a 4′-fluoronucleoside (for reviews, see refs. 23 and 24) might be obtained via iodine monofluoride addition across the 4′,5′-double bond.

In 1969, Ian Jenkins joined Syntex Institute as a postdoctoral fellow and started the synthesis of nucleocidin (*25*). Addition of iodine monofluoride to

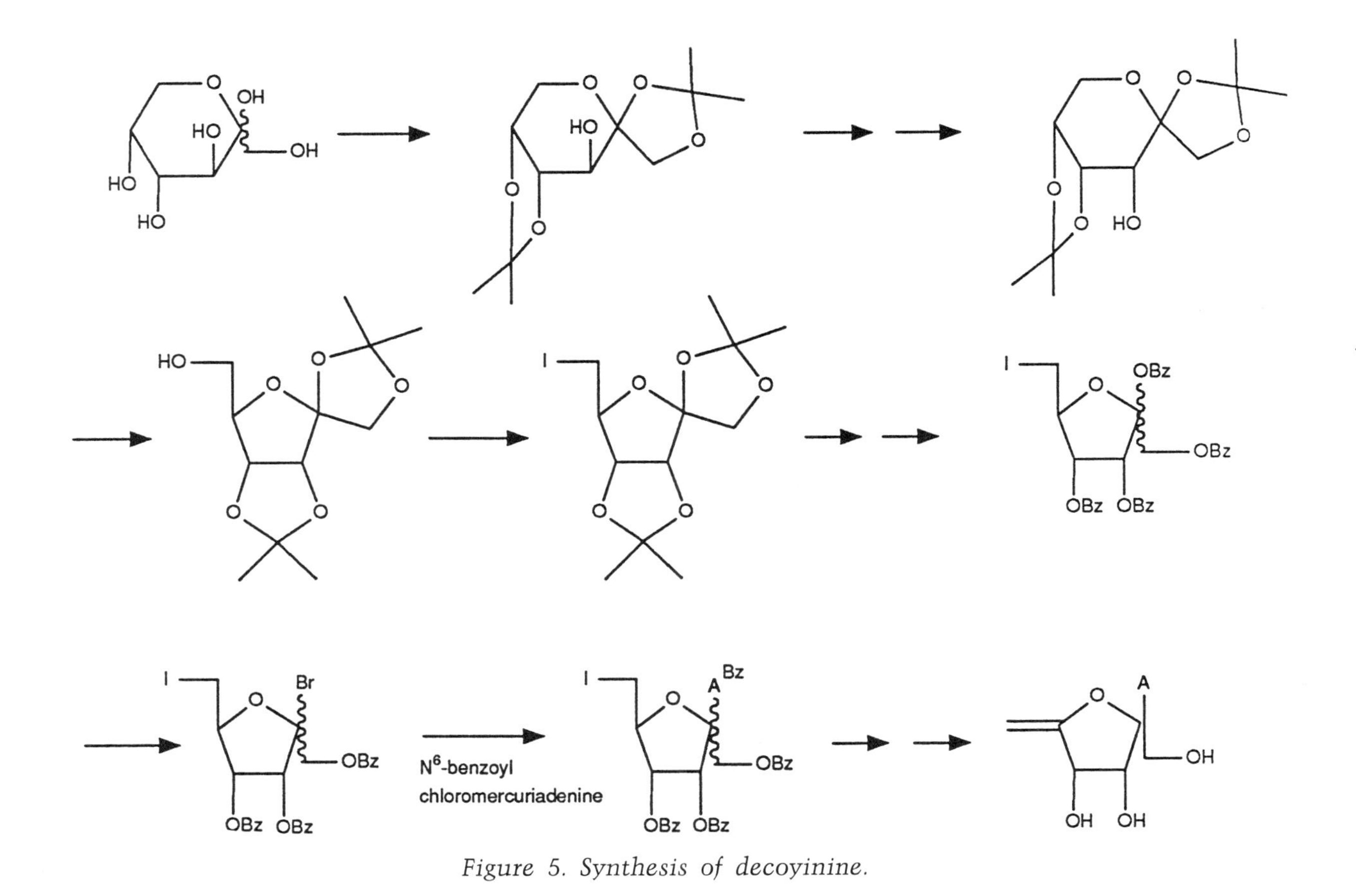

Figure 5. Synthesis of decoyinine.

N^6,N^6-dibenzoyl-9-(5-deoxy-2,3-*O*-isopropylidene-β-D-*erythro*-pent-4-enofuranosyl)adenine gave a 2:1 mixture of isomeric 5′-deoxy-4′-fluoro-5′-iodonucleosides in the β-D-ribofuranosyl and the α-L-lyxofuranosyl configuration. Both isomers were separated by chromatography and debenzoylated. Each isomer was heated in DMF at 100 °C for 20 h, but only one isomer could be converted into an N^3,5′-cyclonucleoside, establishing its ribo configuration. If the introduction of the 4′-fluorine atom had not presented much difficulty, the displacement of the 5′-iodo function became a major challenge. No displacement could be achieved by an oxygen nucleophile even under forcing conditions, owing to the strong deactivating influence of the 4′-fluorine atom. Only azide ion under vigorous conditions gave a 5′-azido-5′-deoxy-4′-fluoro-β-D-ribonucleoside. This successful displacement of the iodine did not immediately solve the problem, because treatment of the azide with nitrosonium tetrafluoroborate (*26*) or deamination of the reduced azide with isoamyl nitrite were not successful in introducing a hydroxyl function at position 5′. We had to resort to photolysis of the azide group (*27*), which converted the azide to a nitrene. Spontaneous isomerization of the nitrene gave an imine, which was then hydrolyzed to an aldehyde; that intermediate was reduced to the desired 4′-fluoro-2′,3′-*O*-isopropylideneadenosine. Esterification with sulfamoyl chloride required activation of the hydroxyl group with hexabutyldistannoxane (*28*). The sulfamoate was obtained in excellent yield. Finally, removal of the isopropylidene group with trifluoroacetic acid gave a sample of nucleocidin that was shown to be identical to the natural product in every respect (Figure 6).

The synthesis of nucleocidin, as described, suffered from the poor reactivity of the iodine at position 5′. In 1971, Bornstein and Skarlos (*29*) described an interesting reagent, lead diacetate difluoride, which added the groups of fluorine and acetoxyl to phenylcyclopropane. This led S. Dimitrijevich to try to add those groups across the 4′,5′-double bond of N^6,N^6-dibenzoyl-9-(2,3-di-*O*-benzoyl-5-deoxy-β-D-*erythro*-pent-4-enofuranosyl)-adenine. The reagent was produced in situ by reacting lead tetraacetate and potassium bifluoride in acetonitrile. Unfortunately, this produced only the 5′-*O*-acetyl-4′-fluoro-α-L-lyxo isomer. However, when iodine monofluoride was added across the 4′,5′-double bond of N^6,N^6-dibenzoyl-9-(2,3-*O*-cyclic-carbonyl-5-deoxy-β-D-*erythro*-pent-4-enofuranosyl)-adenine, only the desired 5′-deoxy-4′-fluoro-5′-iodo-β-D-ribo isomer was produced. This approach could improve the overall yield of the antibiotic substantially and did confirm other studies which compared the stereospecificity of iodine monofluoride addition across the double bond of trichloroethyl 2,3-di-*O*-benzoyl 5-deoxy-β-D-*erythro*-pent-4-enofuranoside and its corresponding 2,3-carbonate. The carbonate gave the ribo adduct as a major product (ratio

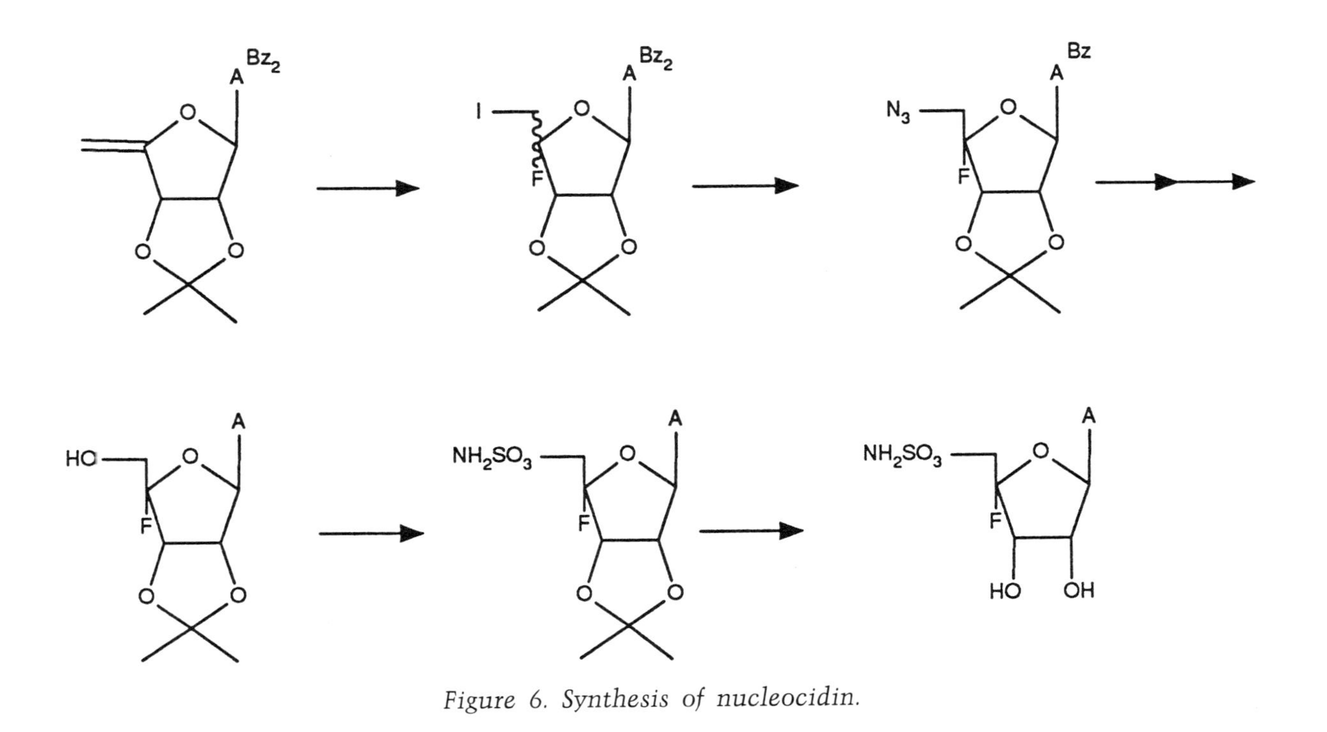

Figure 6. Synthesis of nucleocidin.

6:1), while the dibenzoate gave the lyxo adduct as a main product (ratio 10:1) (J. P. H. Verheyden, J. G. Moffatt, unpublished results).

In 1971, G. Owen, another postdoctoral fellow, made the uridine analogue of nucleocidin (*30*). To our delight the addition of iodine monofluoride occurred stereoselectively and gave only of 5′-deoxy-4′-fluoro-5′-iodo-2′,3′-*O*-isopropylideneuridine from the corresponding 4′,5-unsaturated nucleoside (*31*). This selectivity is likely due to the intermediacy of a 5′-deoxy-5′-iodo-O^2,4′-cyclonucleoside. A similar intermediate was described by Sasaki et al. and later shown to be susceptible to nucleophilic attack (*32*). Displacement of the iodine group by azide again required vigorous conditions. The conversion of the 5′-azido to a 5′-hydroxyl group could be easily achieved using nitrosonium tetrafluoroborate (*26*). The O^2,5′-cyclonucleoside intermediate in this reaction, a result of the attack by the carbonyl group at position 2 of the uracil ring on the 5′-carbonium ion formed initially, was hydrolyzed with acid. Sulfamoylation of 4′-fluoro-2′,3′-*O*-isopropylideneuridine did not require activation by hexabutyldistannoxane as in the synthesis of nucleocidin, but could be achieved in the presence of sulfamoyl chloride and molecular sieves in dioxane. Unfortunately, removal of the isopropylidene group could not be accomplished without partial concurrent hydrolysis of the glycosidic bond, leading to a suboptimal yield (52%) of the uridine analogue of nucleocidin (Figure 7). Attempts to prepare either 4′-fluoroadenosine or 4′-fluorouridine by hydrolysis of the corresponding 2′,3′-*O*-isopropylidene derivative led to substantial cleavage of the glycosidic bond. However, when the 5′-hydroxyl group was esterified by a sulfamoyl or a phosphate group, the protecting group could be removed without release of the heterocyclic base.

4′-Methoxynucleosides

Insofar as a fluorine atom could be introduced at position 4′ of various nucleosides, I decided to see to what extent this position could be modified, and to evaluate the effect of these changes on the antimicrobial properties of these new 4′-substituted nucleosides. A methoxy group is less electronegative than fluorine, more bulky, and should be more stable to base than a fluorine group at this glycosidic position. I therefore attempted the synthesis of 4′-methoxyuridine (*33*) as a model nucleoside. Lemieux and Fraser-Reid (*34*) had provided a precedent in their study of iodomethoxylation of D-glucal triacetate using methanol, iodine, and silver acetate. The hydroxyl groups in position 2′ and 3′ of (5-deoxy-β-D-*erythro*-pent-4-enofuranosyl)uracil needed to be protected with base-labile protecting groups because of the acid

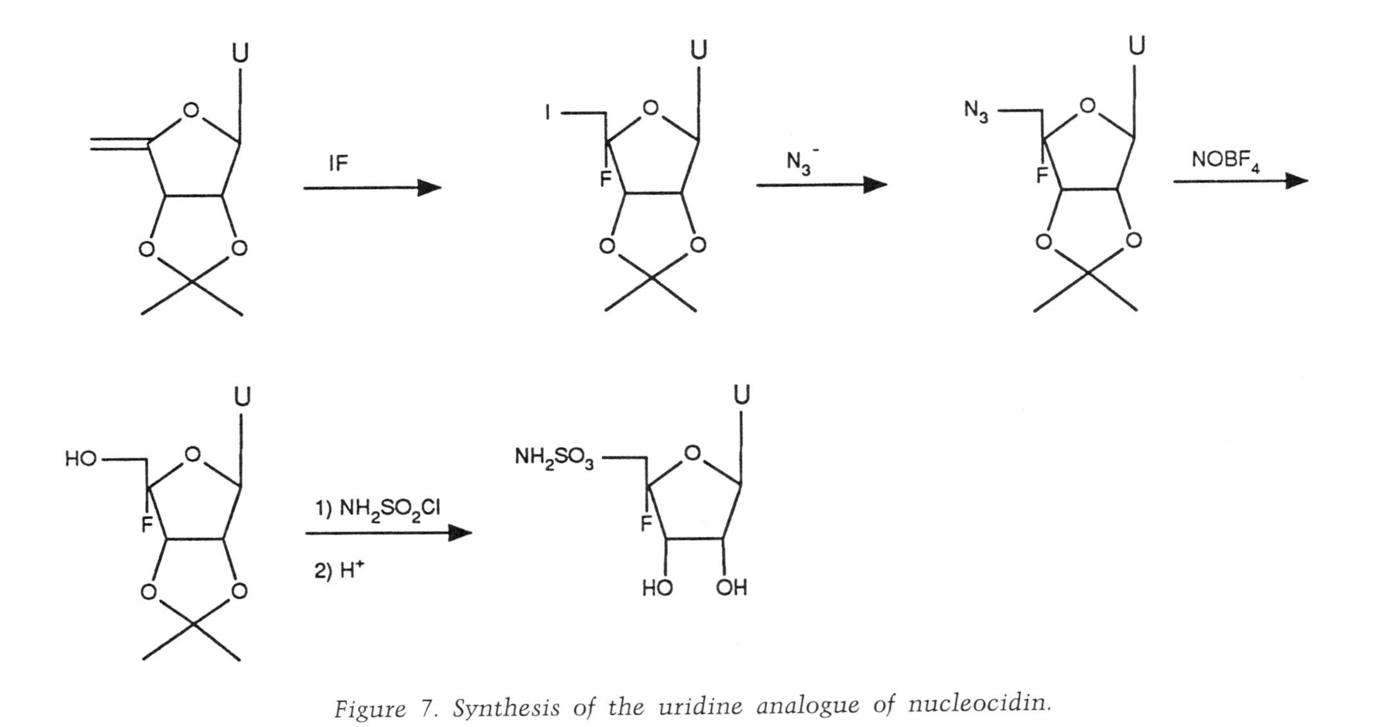

Figure 7. Synthesis of the uridine analogue of nucleocidin.

lability of the 4′-methoxy group. The acetyl group was tried first but rejected because the 3′-*O*-acetyl group participated in the opening of the intermediate 4′,5′-iodonium ion, leading to 3′,4′-acetoxonium ion, which then reacted with methanol to give a mixture of 3′,4′-orthoacetate as well as the desired 5′-deoxy-5′-iodo-4′-methoxy nucleoside.

As mentioned earlier, I had obtained very good results using a 2,3-cyclic carbonate group in the iodine monofluoride addition across the 4,5-unsaturated double bond of trichloroethyl 2,3-*O*-cyclic-carbonyl-5-deoxy-β-D-*erythro*-pent-4-enofuranoside. Accordingly, I decided to investigate this group to protect positions 2′ and 3′ of (5-deoxy-β-D-*erythro*-pent-4-enofuranosyl)uracil. Once again, the addition of iodine and methanol in the presence of lead carbonate to 1-(2,3-*O*-cyclic-carbonyl-5-deoxy-β-D-*erythro*-pent-4-enofuranosyl)uracil gave a single β-D-ribo isomer whose structure was clearly evident from its ^{13}C NMR spectrum. The stereospecificity is probably due to the small size of the carbonate group, thus permitting the unhindered attack by methanol from the α face of the furanose ring. Hydrolysis of the carbonate group followed by reprotection with benzoyl groups gave the 2′,3′-di-*O*-benzoyl-5′-deoxy-5′-iodo-4′-methoxyuridine. Displacement of the iodine group by lithium benzoate in hexamethylphosphorotriamide at 105 °C, followed by removal of all protecting groups with methanolic ammonium hydroxide, gave the desired 4′-methoxyuridine in good yield (Figure 8). A very similar approach was followed by P. Srivastava for the synthesis of 4′-methoxycytidine (for reviews, see refs. 23 and 24). 5′-Deoxy-5′-iodo-4-methylthiouridine was dehydrohalogenated with DBN and protected via reaction with phosgene to give the 4′,5′-unsaturated nucleoside 2′,3′-carbonate. Addition of iodine and methanol gave a single β-D-ribo isomer. Treatment with methanolic ammonium hydroxide, followed by benzoylation, produced the $N^4,O^{2'},O^{3'}$-tribenzoyl-5′-deoxy-5′-iodo-4′-methoxycytidine. Displacement of the 5′-iodo group with lithium benzoate at 120 °C in DMSO, followed by removal of all protecting groups with methanolic ammonium hydroxide, gave the desired 4′-methoxycytidine. This rather lengthy synthesis (14 steps from uridine) was shortened by N. Le-Hong (for reviews, see refs. 23 and 24). Starting with cytidine, the desired 4′-methoxynucleoside was obtained in seven steps in 11% overall yield. The ribofuranosyl configuration was ascertained by ^{13}C NMR.

Richards (35) approached the synthesis of 4′-methoxyadenosine based on the experience acquired during the synthesis of 4′-methoxypyrimidine nucleosides. However, in the purine series, we had to avoid the possible formation of N^3,4′- or N^3,5′-cyclo nucleosides via attack of the N^3 nitrogen of the adenine ring on the intermediate 4′,5′-iodonium ion. This was accomplished by dibenzoylation of the amine function in position 6 of the

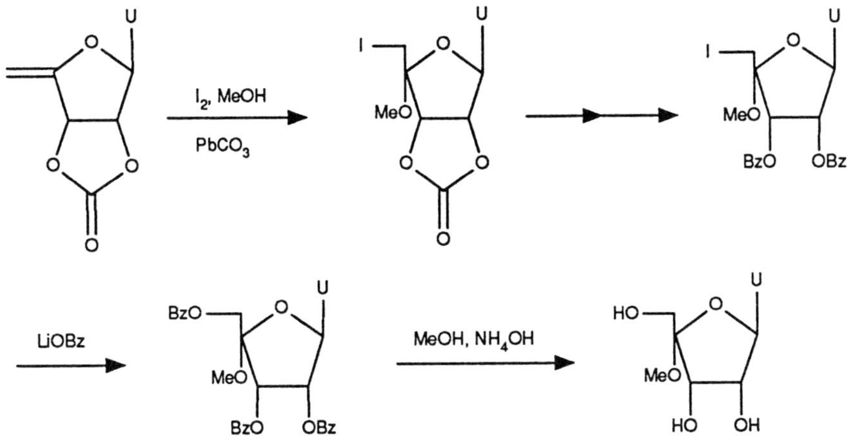

Figure 8. Synthesis of 4-methoxyuridine.

adenine ring. Thus, addition of iodine and methanol to N^6,N^6-dibenzoyl-9-(2,3-*O*-cyclic-carbonyl-5-deoxy-β-D-*erythro*-pent-4-enofuranosyl)adenine gave a single ribo isomer in quantitative yield. Removal of the carbonate protecting group was followed by benzoylation. Displacement of the 5′-iodine atom with lithium benzoate in DMF at 120 °C, followed by removal of all protecting groups with methanolic ammonium hydroxide, gave 4′-methoxyadenosine, whose structure was ascertained by comparing its ^{13}C NMR spectrum with the spectra of the other 4′-methoxynucleosides. 4′-Methoxyadenosine was then protected by forming the corresponding 2′,3′-*O*-ethoxymethylidene derivative, thus permitting selective sulfamoylation at position 5′, after activation with hexabutyldistannoxane. Removal of the very acid-labile protecting group was achieved with 0.5% formic acid without endangering the 4′-methoxy group, and the methoxy analogue of nucleocidin was isolated in good yield.

Other Contributors

The 4′-substituted nucleoside field has been explored by a number of workers. J. A. Secrist et al. (36) have obtained 4′-substituted pyrimidine nucleosides via the corresponding O^2,4′-cyclo nucleoside, prepared from 1-

(5-*O*-acetyl-2,3-*O*-cyclohexylidene-β-D-*erythro*-pent-4-enofuranosyl) uracil. The latter compound resulted from the treatment of the 5′-aldehydo nucleoside with acetic anhydride and potassium carbonate. T. Sasaki et al. have also prepared a 4′-methoxypurine nucleoside, and an O^2,4′-cyclopyrimidine nucleoside starting from the corresponding unsaturated nucleoside (*37, 38*). Recently a 5′-deoxy-4′,5-difluorouridine was synthesized as a prodrug of 5-fluorouracil (*39*), and the 4′-position of carbanucleosides has been modified by a group in France (*40*) and another in England (*41*).

4′-Azidonucleosides

Gougerotin is a dipeptidyl nucleoside antibiotic isolated in 1962 from *Streptomyces gougerotii* by Kanzaki et al. (*42*). Its structure was established as 1-cytosinyl-[4-deoxy-4-(sarcosyl-D-serylamine)-β-D-glucopyrosid] uronamide by J. J. Fox et al. in 1968 (*43*). This antibiotic is part of the 4-aminohexose pyrimidine nucleoside family (*44*). I was intrigued by the structure of these peptidyl nucleosides and wondered if it would be possible to obtain a nucleoside substituted at the 4′-position by an amino acid. As a first step in that direction, the synthesis of 4′-azidocytidine (for reviews, see refs. 23 and 24) was begun by N. Le-Hong in 1974 (N. Le-Hong, J. P. H. Verheyden, J. G. Moffatt, unpublished findings). Addition of iodine azide to N^4-benzoyl-1-(5-deoxy-β-D-*erythro*-pent-4-enofuranosyl)cytosine gave as predicted the corresponding 4′-azido-5′-deoxy-5′-iodo ribonucleoside as the major product (ratio of α-L-lyxo to β-D-ribo = 1:9). Benzoylation followed by displacement of the 5′-iodo function with lithium benzoate at 105 °C in DMF gave the perbenzoylated 4′-azidocytidine. Debenzoylation with methanolic ammonia gave the desired 4′-azidocytidine as a homogeneous foam. This compound had an NMR spectrum consistent with its structure, but a correct elemental analysis could not be obtained. Reduction of the azido group would have given the equivalent of a glycosylamine, which is generally in equilibrium with its open form. Assuming the same equilibrium occurs in 4′-aminonucleoside, rapid cleavage of the glycosidic bond would be anticipated. Thus, it was unlikely that this approach would yield 4′-acylaminonucleoside in high yield. For this and other reasons we did not, at this point, pursue our research on 4′-azidonucleosides. This effort was not lost, as will become clear later on.

4′-Hydroxymethylnucleosides

Having developed methods for the synthesis of nucleosides substituted at the 4′-position with a fluorine atom, a methoxy group, or an azido function, I felt it would be of interest to attach at this position a more stable substituent. A psicofuranosylnucleoside, such as decoyinine, can be considered a 1′-hydroxymethylnucleoside, and thus it became logical to explore the synthesis of 4′-hydroxymethylated nucleosides (for review, see ref. 24). These 4′-substituted nucleosides have many advantages: they are stable to acid and base, and to enzymatic degradation. Moreover, because the asymmetry at position 4′ is lost, we no longer had to contend with mixtures of ribo and lyxo isomers. These compounds were readily accessible via a mixed aldol condensation of nucleoside 5′-aldehydes and formaldehyde followed by Cannizzaro reduction of the resulting hydroxyaldehyde. This approach was based on the substantial amount of experience acquired at Syntex on the preparation of nucleoside-5′-aldehydes via the Pfitzner–Moffatt DMSO/DCC method. Jones had managed to derivatize these rather unstable aldehydes into crystalline 1,3-diphenyl imidazoline derivatives (*45*), and approached the synthesis of 4′-hydroxymethyl nucleosides starting from these preformed nucleosides (*46*). I wanted a more flexible approach and therefore planned the synthesis of a suitably protected and activated 4-hydroxymethylribose onto which any heterocyclic base could be condensed (*47*).

Previous to our work, Schaffer and colleagues (*48*, *49*) had shown that 1,2-*O*-isopropylidene-α-D-xylopentodialdofuranose treated with formaldehyde and aqueous sodium hydroxide was transformed to 4-(hydroxymethyl)-1,2-*O*-isopropylidene-β-L-*threo*-pentofuranose. Preliminary experiments with 1,2-*O*-isopropylidene-α-D-xylo- or α-D-ribopentodialdofuranose had shown that the hydroxyl group in position 3 of the sugar needed to be protected in order to avoid epimerization at this position. This epimerization is due to base-catalyzed reverse aldol cleavage of the C_3-C_4 bond, followed by recyclization of either the starting 3-hydroxy-5-aldehydo sugar or the initial mixed aldol product (*46*, *47*). Therefore, we developed a synthesis starting from 3-*O*-benzyl-1,2-*O*-isopropylidene-α-D-allofuranose, which is readily available from 1,2:5,6-di-*O*-isopropylidene-α-D-glucofuranose. Oxidation with sodium periodate gave the 3-*O*-benzyl-1,2-*O*-isopropylidene α-D-ribo-pentodialdofuranose, which was reacted with formaldehyde and aqueous sodium hydroxide to give a crystalline 3-*O*-benzyl-4-(hydroxymethyl)-1,2-*O*-isopropylidene α-D-*erythro*-pentofuranose in 77% overall yield. Catalytic reduction of the benzyl group followed by acetylation gave the corresponding triacetate. Finally, acetolysis of the 1,2-*O*-isopropylidene group produced the desired intermediate 4-(acetoxymethyl)-1,2,3,5-tetra-*O*-acetyl-D-*erythro*-

pentofuranose. This fully acetylated 4-hydroxymethylribofuranose was then used to prepare a series of 4′-substituted purine and pyrimidine nucleosides. Condensation with 6-chloropurine led to the analogues of adenosine (Figure 9), 6-thio, 6-methylthio, and 6-methoxypurineriboside. We also prepared the analogue of the 5-fluorouracil and the 6-aza-2-thiouracilriboside. Jones et al. (*46*) studied in detail the mechanism of this aldol condensation and prepared the analogues of uridine, cytidine, thymidine, and adenosine. The latter was shown to be identical with the adenosine analogue prepared via the sugar condensation route and therefore confirmed the stereochemical assignments of our series of analogues (*47*).

A Shift of Focus

The various nucleoside analogues described in the preceding pages were submitted to biological screening in order to detect any antiviral, antibacterial, antifungal, or antitumor properties. None showed any useful biological activities. In late 1974, Syntex management, somewhat discouraged by the lack of potential drugs coming out of our synthetic effort, asked us to redirect our attention to fields that were more likely to generate new products. In the ensuing years, a series of potent and selective antiherpetic agents were discovered by other groups: 1-(2-deoxy-2-fluoro-β-D-arabinofuranosyl)-5-iodocytosine (FIAC), (*E*)-5-(2-bromovinyl)-2′-deoxyuridine (BVDU), and, more importantly, 9-[(2-hydroxyethoxy)methyl]guanine (acyclovir) (Figure 10). In view of these developments, Syntex management thought it was appropriate for us to resume our search for an antiviral agent. Because the field of antivirals is vast and the likelihood of finding a broad-spectrum antiviral is remote, I wanted to focus our search. On the advice of our marketing experts, we decided that inhibitors of herpes viruses were good targets. I therefore started reading hundreds of papers on the biology, biochemistry, and antiviral chemotherapy of herpes simplex viruses (HSV). I was amazed by the number of chemical structures claimed to have some activity against HSV. At the beginning of our effort in this new field in 1979, I wanted to have a complete survey of what had been done outside of the nucleoside field because I thought that for an infection due to HSV, a nucleoside would, in the majority of cases, be too toxic for general use. Fortunately, a good review of the field covering 1971-1977 was available (*50*). A literature search updated this review with many more references. My original list included Tromantadine, benzimidazoles, thiosemicarbazones, benzylidene hydantoins, amidines, polyphenols, 2-deoxyglucose, long-chain fatty alcohols, and nonsteroidal antiinflammatory agents. In November

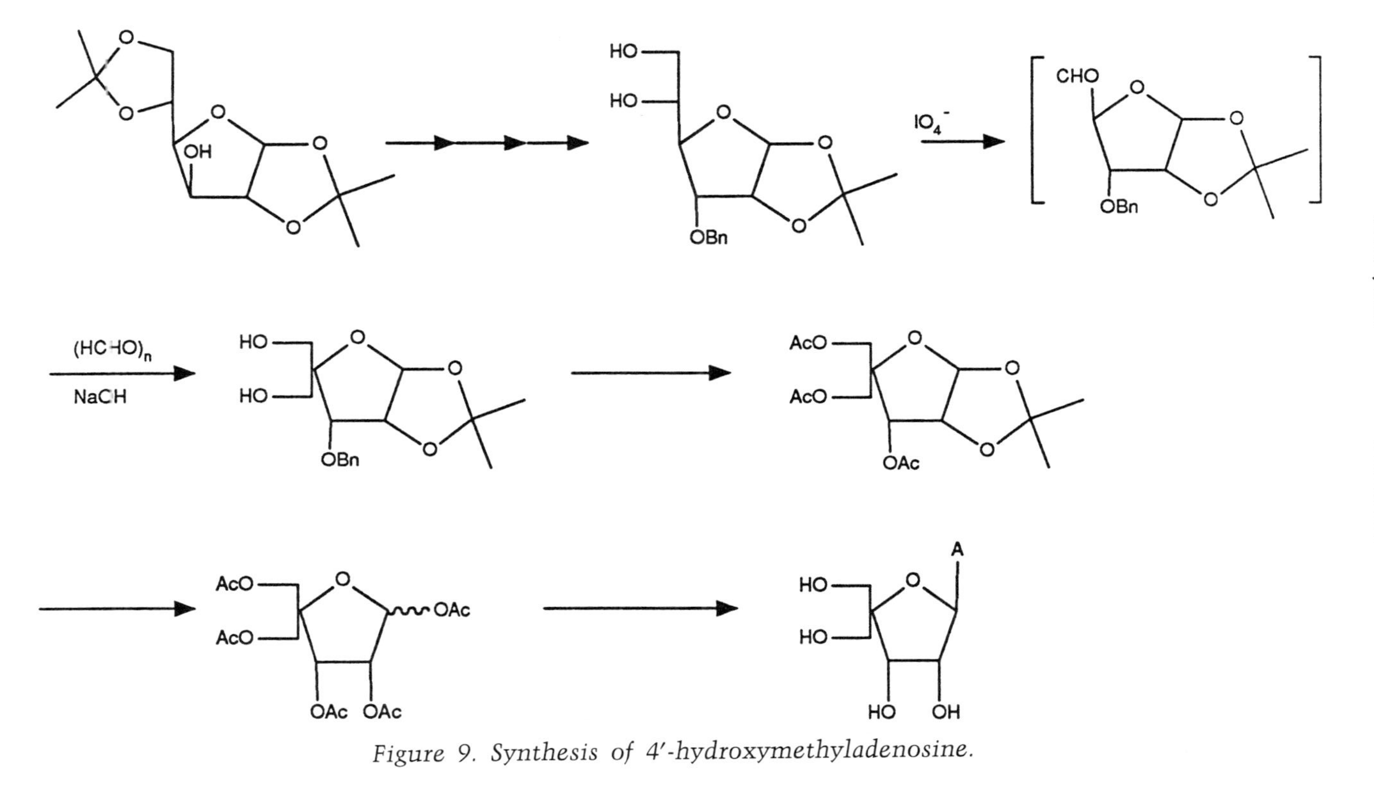

Figure 9. Synthesis of 4′-hydroxymethyladenosine.

1979, our first benzylidene thiohydantoin was submitted for testing. It was the first in a long series of inactive compounds. In February 1980, the first analogue of Tromantadine was tested; again, it was quite inactive. Later we synthesized Tromantadine and found it quite inactive too! We spent some time preparing the carba analogue of 2-deoxyglucose; again, no activity.

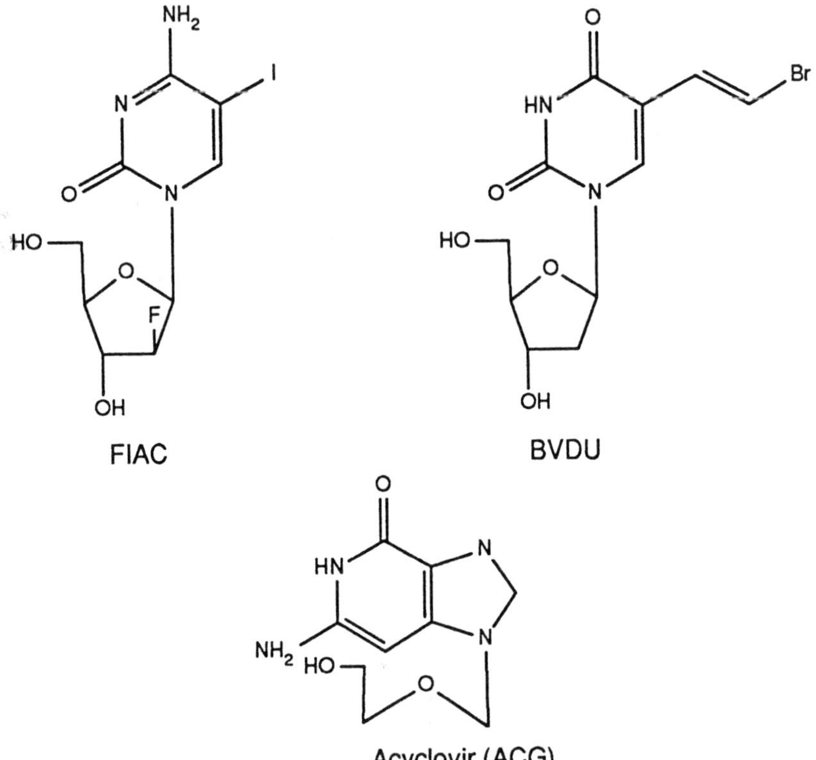

FIAC

BVDU

Acyclovir (ACG)

Figure 10. Structure of various antiherpetic agents.

Synthesis and Development of Ganciclovir

John Martin had been a postdoctoral fellow at Syntex for a year and had joined my group in April 1979 to work on the synthesis of aminoglycoside antibiotics. In the early part of 1980, as an extension of my previous work on 4′-hydroxymethylated nucleosides, I conceived the structure of 4′-hydroxymethylacyclovir, now known as ganciclovir. In a discussion with John, I suggested the synthesis of this compound, and as I wrote the structure on the blackboard, it became clear that this compound should be considered a secoguanosine rather than a 4′-substituted nucleoside. John was a very enthusiastic chemist, eager to learn and highly productive. He went to work immediately. Our first approach started with 1,3-di-*O*-benzoylglycerol, prepared by benzoylation of glycerol. Chloromethylation of the 1,3-dibenzoate gave the 1,2-di-*O*-benzoyl-3-chloro-3-deoxy-glycerol rather than the expected 2-*O*-chloromethyl ether. This product was probably formed via a 2,3-benzoxonium intermediate. In a recent paper by G. Hakimelahi and colleagues (*51*), a similar reaction on 1-*O*-benzoyl-3-chloro-3-deoxy-glycerol was reported, yet the authors did not report the formation of a 1,3-dichloro derivative. However, their chloromethylated product was isolated in only 37% yield.

Obviously, a more stable protecting group was required. The benzyl group was chosen because it would not migrate or be hydrolyzed during the chloromethylation reaction. Epichlorohydrin was reacted with benzyl alcohol in the presence of boron trifluoride etherate to give 1-*O*-benzyl-3-chloro-3-deoxyglycerol, which was treated with sodium hydroxide to form the 2,3-epoxide. A second round of benzylation gave a mixture of 1,3- and 1,2-di-*O*-benzyl glycerols which were separated by chromatography. This synthesis was later improved by reacting epichlorohydrin with sodium benzylate; the resulting mixture of isomers was separated by distillation, yielding the desired 1,3-di-*O*-benzyl isomer in 66% yield (*52*). Chloromethylation of the protected glycerol proceeded in high yield. The resulting chloromethyl ether was reacted with the tris-trimethylsilyl derivative of guanine and gave the expected acyclic nucleoside in 36% yield. Removal of the benzyl protecting group was best achieved with sodium in liquid ammonia, giving the desired hydroxymethylated acyclovir, 9-(1,3-dihydroxy-2-propoxymethyl)guanine (DHPG, RS-21592) (Figure 11).

In July 1980 we submitted 100 mg of product to T. Matthews for biological evaluation. It appeared to be quite active in vitro against HSV-1 and HSV-2. However, a closer look at the Burroughs Wellcome acyclovir patent revealed that the generic claim covered literally millions of variations on acyclovir, including, potentially, our own DHPG. But it was not specifically described or even pointed to anywhere in the patent. We were extremely concerned that we might lose the reward for our effort.

By December 1980, a request for a larger amount of DHPG was made to C. Dvorak, and by early 1981, DHPG had been tested in vivo in a variety of biological systems. In comparison with ACG, FIAC (Figure 10), and PFA (phosphonoformic acid), it appeared to be quite potent. In a systemic HSV-2

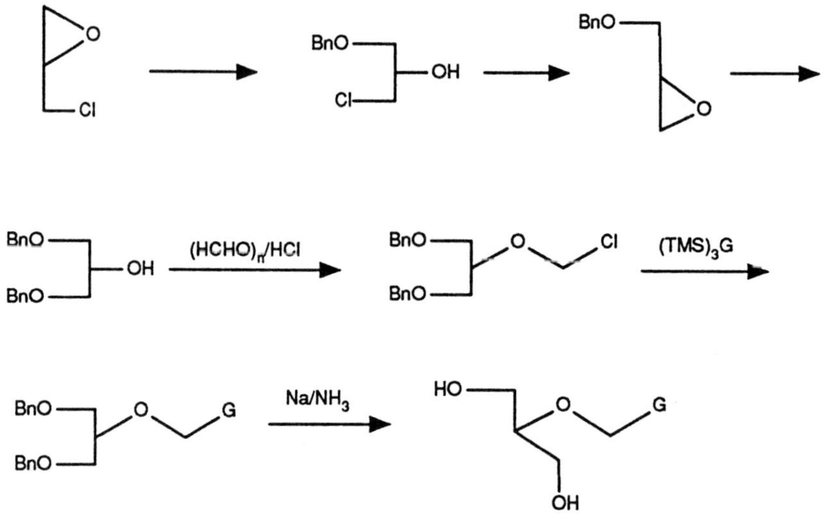

Figure 11. Original synthesis of ganciclovir (DHPG, RS-21592).

infection in mice, it was clearly superior to ACG and FIAC when given subcutaneously 6 h post infection and then once a day for 4 days. When guinea pigs were infected intravaginally with HSV-2 and treated topically (6 h post infection and twice daily for 5 more days), DHPG was vastly superior to ACV and PFA. Further, it soon became evident that DHPG was a broadly active antiherpetic agent having potent activity not only against HSV-1 and HSV-2 but also against varicella-zoster virus (VSV), Epstein-Barr virus (EBV), and cytomegalovirus (CMV). With the emerging epidemic of AIDS, and because most HIV-infected patients seroconvert to CMV, this antiviral activity was an important discovery. On the basis of the tests, in May 1981 we filed a patent to cover DHPG.

We now needed a sensitive method to detect circulating levels of DHPG in serum, so John Martin prepared the hemisuccinate that would be reacted with keyhole limpet hemocyanin and bovine serum globulin in the presence of a carbodiimide to provide an antigen. Once antibodies were raised in rabbits, we would be able to detect the drug via a radioimmunoassay. This was an improvement over our then current analytical method based on high-performance liquid chromatography, which was sensitive only to 100 ng/mL. DHPG was negative in the Ames test but was positive in the sister chromatid exchange assay. In general, the spectrum of toxicity seen, at this point, appeared similar to that reported for acyclovir. To improve the oral bioavailability of DHPG, which was excellent in dogs but mediocre in mice, rats, and monkeys, we started the synthesis of a series of esters of DHPG (53). For this we needed larger amounts of DHPG, and the scale-up was at first not easy. Moreover, initial acute toxicity studies indicated that DHPG was not very toxic; because the toxicity studies needed for regulatory approval require that the high dose show a definite toxic effect, the projected amount of drug required for long-term toxicity studies was large. These concerns put considerable pressure on the group responsible for scale-up. One kilogram was requested by April 1982.

We were still hesitating between various esters of DHPG and DHPG itself as our development compound because some esters appeared more bioavailable. In terms of bioavailability, we wondered if a high level of DHPG in the blood was necessary or even desirable. DHPG is not an antiviral agent in its own right; rather, it is the corresponding triphosphate that inhibits the viral DNA polymerase. Because DHPG was rapidly converted to the triphosphate in the HSV- or CMV-infected cell (54–56), the circulating level of DHPG may not be as important as in the case of antibiotics. In July 1982, a paper by Smith, Ogilvie and colleagues (57) described the potent antiviral activities of BIOLF62, the structure of which was identical to DHPG, against HSV-1, HSV-2, VZV, and EBV, and claimed a very high selectivity index. However, the authors missed the potent activity of BIOLF62 against CMV, probably because a fluorescent antibody against the immediate early antigen was used to determine viability of the virus. BIOLF62 was prepared according to a method used previously by Ogilvie et al. for the preparation of acyclonucleosides (DHPC and DHPT), which were intermediates in the synthesis of analogues of dinucleoside monophosphate (58).

In early March 1982, the U.S. Patent Office allowed the DHPG patent. We opened a few bottles of champagne to celebrate. At the end of October 1982, the Fifth International Round Table on Nucleosides, Nucleotides and their Biological Applications was hosted by Burroughs Wellcome at Research Triangle Park, North Carolina. John Martin and I had intended to present a poster on the synthesis of 9-(4-hydroxy-2-oxo-butyl)guanine (59), which I had thought would act as a suicide inhibitor for HSV-specified thymidine kinase. This compound had shown some weak activity against HSV-1 in Vero cells, and it was also a substrate for HSV thymidine kinase, but unfortunately it did not behave as a suicide inhibitor. At the last minute we decided, because our DHPG patent had been granted, to also present all our results on DHPG, including chemistry, bioassay, bioavailability data, solubility data, and mechanism of action. The poster was intended to clarify rumors about a potent antiviral agent developed by Syntex and was, I suppose, somewhat of a surprise for the scientists at Burroughs Wellcome. At the meeting, while discussing problems of mutual interest with Dick Tolman and Kelvin Ogilvie, I learned to my great surprise that Dick and his group at Merck had independently developed DHPG (they called it 2′-NDG for 2′-nordeoxyguanosine). As often happens in research nowadays, since everyone has access to the same information, several people may conceive of the same idea independently. Albert Szent-Györgi states, "Research is to see what everyone has seen and think what no one has thought" (*60*); indeed today, with the ever-increasing amount of information, the real art is to assimilate this large body of often disparate data and to come up with that unique perspective that constitutes a breakthough. While the lawyers of each company were assessing the rights of each party, our chemists were working hard to produce DHPG in larger quantities, so that the toxicity of DHPG could be evaluated. By August 1983, we had become aware of the potent gonadal toxicity of DHPG in mice and dogs; this reinforced our intention to develop DHPG for more limited indications, such as the treatment of life-threatening CMV infection.

On May 2, 1984, I received a call from Clyde Crumpacher at the Beth Israel Hospital requesting DHPG for compassionate treatment of a young mother who had undergone bone marrow transplantation and was suffering from CMV pneumonia. Within 24 hours, Austin Brewin, the clinical representative from our project team, was on a plane to Boston with 10 g of DHPG. Unfortunately, the patient died. She had received five infusions of DHPG. Today we know that CMV pneumonia must be treated very early in order for DHPG to have a chance of inducing remission. Following the first compassionate use of DHPG further requests were made, especially for AIDS patients having CMV retinitis, which inexorably leads to blindness. The compassionate program grew exponentially and kept W. Buhles and later B. DeArmond and their associates in our Institute of Clinical Medicine busy for many months.

The extensive use of ganciclovir on a compassionate basis made double-blind, placebo-controlled studies for the treatment of CMV retinitis difficult to contemplate, because ophthalmologists who had used the drug were convinced of its efficacy and thought it would be unethical to do such a study. The members of the antiinfective drug advisory committee to the Food and Drug Administration were, however, initially concerned about maintaining the requirement for such a cornerstone study for the approval of drugs.

Nevertheless, after a careful and detailed study done at Johns Hopkins University School of Medicine by Douglas Jabs on patients with peripheral retinitis, and in conjunction with many other supportive data, ganciclovir was approved for the treatment of sight-threatening CMV retinitis in 1989, after a second advisory committee had recommended marketing approval.

Phase IV postmarketing studies are still going on. In addition, excellent results are reported in various immunocompromised transplant populations (*61, 62*). A new drug application for use in treating and preventing life-threatening CMV disease in transplant patients has recently been approved. An oral formulation is currently being developed and is in Phase II/III clinical trials. *Scrip*'s science editor John Davis wrote that in 1988 there were 53 new chemical entities, but just four breakthroughs; one of these was Syntex's antiviral drug ganciclovir. The French pharmacological review, *Prescrire*, also cited four products for their contribution to therapeutic progress; again, one of them was ganciclovir.

Ganciclovir has also become a useful reagent for molecular biologists doing gene-targeting studies via homologous recombination of DNA sequences (*63*). Rarely a day passes without us receiving a request for ganciclovir for this use from laboratories all over the world. More recently ganciclovir has been a useful adjunct for the treatment of brain tumors via gene therapy with a retroviral vector that allows intratumoral transduction of the herpes thymidine kinase enzyme (*64, 65*). Because ganciclovir is an excellent substrate for this enzyme, it is a good choice for this novel therapy.

An interference was declared by the U.S. Patent and Trademark Office between Syntex, Merck, Burroughs Wellcome, and ens Bio Logicals. After a period of discussion and negotiation among the parties, only Burroughs Wellcome and Syntex remained to contest priority of inventorship to ganciclovir. In October 1989, the Board of Patent Appeals and Interferences of the U.S. Patent and Trademark Office ruled that Syntex was entitled to the patent. This decision was upheld on appeal to the U.S. Court of Appeals for the Federal Circuit.

Epilogue

In 1984, we reported the synthesis of the carba analogue of DHPG, 9-[4-hydroxy-3-(hydroxymethyl)-1-butyl]guanine (Figure 12) (*66*). This carba derivative was not as potent as DHPG against HSV-1, HSV-2, and CMV infections, yet its activity, especially against HSV-1 and HSV-2, was not negligible. This compound had been synthesized for the first time by Pandit et al. (*67*) in 1972 but had not been tested by those authors for its antiviral activity. Had its biological activity been uncovered in 1972, perhaps a Dutch pharmaceutical company would be marketing acyclovir and ganciclovir today.

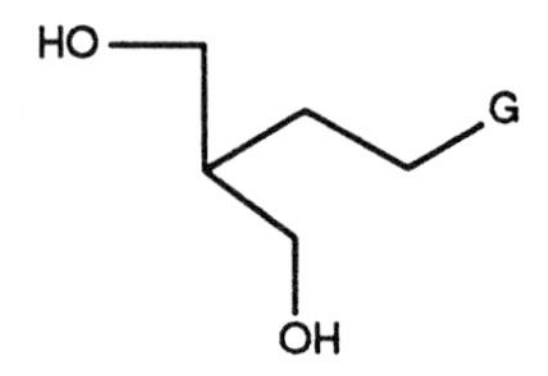

Figure 12. The carba analogue of DHPG.

In 1987, as part of a general screening of our nucleoside samples collection against HIV, the 4′-azidocytidine synthesized in 1974 appeared to have some activity, which was later shown to be mainly due to toxicity to the host cell. This little bit of data spurred us on to synthesize a series of 4′-azido-2′-deoxy nucleosides, all of which showed potent activity against HIV in vitro. One of these, ADRT, 4′-azidothymidine, had high potency and a good selectivity index and was chosen for development . . . but this is another story.

Acknowledgments

I thank J. G. Moffatt for the support and encouragement that he provided for more than 25 years. In addition to the main contributors whose names appear in the text and the references, thanks are also given to the members of the project team and the medical community who played a major role in obtaining FDA approval for ganciclovir. I also thank N. Grinder for preparing the manuscript.

Literature Cited

1. Kochetkov, N. K.; Usov, A. I. *Izv. Akad. Nauk SSSR, Ser. Khim.* **1962,** 1042.
2. Ashton, W. T.; Karkas, J. D.; Field, A. K.; Tolman, R. L. *Biochem. Biophys. Res. Commun.* **1982,** *108,* 1716.
3. Ogilvie, K. K.; Cheriyan, V. O.; Radatus, B. K.; Smith, D. O.; Galloway, K. S.; Kennell, W. L. *Can. J. Chem.* **1982,** *60,* 3005.
4. Schaeffer, H. J. In *Nucleosides, Nucleotides and Their Biological Applications*; Rideout, J. L.; Henry, D. W.; Beacham, L. M., III, Eds.; Academic: New York, 1983; pp 1–17.

5. Pfitzner, K. E.; Moffatt, J. G. *J. Am. Chem. Soc.* **1963,** 3027.
6. Landauer S. R.; Rydon, H. N. *J. Chem. Soc.* **1953,** 2224.
7. Landauer S. R.; Rydon, H. N. *Tetrahedron* **1963,** *19,* 973.
8. Verheyden, J. P. H.; Moffatt, J. G. *J. Am. Chem. Soc.* **1964,** *86,* 2093.
9. Verheyden, J. P. H.; Moffatt, J. G. *J. Org. Chem.* **1970,** *35,* 2319.
10. Verheyden, J. P. H.; Moffatt, J. G. *J. Org. Chem.* **1970,** *35,* 2868.
11. Dimitrijevich, S. D.; Verheyden, J. P. H.; Moffatt, J. G. *J. Org. Chem.* **1979,** *44,* 400.
12. Hocksema, H.; Slomp, G.; Van Tamelen, E. E. *Tetrahedron Lett.* **1974,** 1787.
13. Hough, L.; Otter, B. *Chem. Commun.* **1966,** 173.
14. Helferich, B.; Himmen, B. *Chem. Ber.* **1928,** *61,* 1825.
15. Verheyden, J. P. H.; Moffatt, J. G. *J. Am. Chem. Soc.* **1966,** *88,* 5684.
16. Verheyden, J. P. H.; Moffatt, J. G. *J. Org. Chem.* **1974,** *38,* 3573.
17. Prisbe, E. J.; Smejkal, J.; Verheyden, J. P. H.; Moffatt, J. G. *J. Org. Chem.* **1976,** *41,* 1836.
18. McCarthy, J. R.; Robins, R. K.; Robins, M. J. *J. Am. Chem. Soc.* **1968,** *90,* 4993.
19. Morton, G. O.; Lancaster, J. E.; VanLear, G. E.; Filmor, N.; Meyer, W. E. *J. Am. Chem. Soc.* **1969,** *91,* 1535.
20. Backus, E. J.; Tresner, H. D.; Campbell, T. H. *Antibiot. Chemother.* **1957,** *7,* 532.
21. Thomas, S. O.; Singleton, V. L.; Lowery, J. A.; Sharpe, R. W.; Priess, L. M.; Porter, J. N.; Mowat, J. H.; Bohonas, N. *Antibiot. Ann.* **1956–57,** 716.
22. Hall, L. D.; Manville, J. F. *Chem. Commun.* **1968,** 35.
23. Verheyden, J. P. H.; Jenkins, I. D.; Owen, G. R.; Dimitrijevich, S. D.; Richards, C. M.; Srivastava, P. C.; Le-Hong, N.; Moffatt, J. G. *Ann. NY Acad. Sci.* **1975,** *255,* 151.
24. Moffatt, J. G. In *Nucleoside Analogues*; Walker, R. T.; DeClercq, E.; Eekstein, F., Eds.; Plenum: New York, 1979; pp 71–164.
25. Jenkins, I. D.; Verheyden, J. P. H.; Moffatt, J. G. *J. Am. Chem. Soc.* **1976,** *98,* 3346.
26. Doyle, M. P.; Wierenga, W. *J. Am. Chem. Soc.* **1972,** *94,* 3896, 3901.
27. Abramovich, R. A.; Davis, B. A. *Chem. Rev.* **1964,** *64,* 149.
28. Wagner, D.; Verheyden, J. P. H.; Moffatt, J. G. *J. Org. Chem.* **1974,** *39,* 24.
29. Bornstein, J.; Skarlos, L. *Chem. Commun.* **1971,** 796.
30. Owen, G. R.; Verheyden, J. P. H.; Moffatt, J. G. *J. Org. Chem.* **1976,** *41,* 3010.

31. Robins, M. J.; McCarthy, J. R., Jr.; Robins, R. K. *J. Heterocycl. Chem.* **1967,** *4,* 313.
32. Sasaki, T. K.; Minamoto, S.; Kuroyanagi, S.; Hattori, K. *Tetrahedron Lett.* **1973,** 2731.
33. Verheyden, J. P. H.; Moffatt, J. G. *J. Am. Chem. Soc.* **1975,** *97,* 4386.
34. Lemieux, R. U.; Fraser-Reid, B. *Can. J. Chem.* **1964,** *42,* 532.
35. Richards, C. M.; Verheyden, J. P. H.; Moffatt, J. G. *Carbohydr. Res.* **1982,** *100,* 315.
36. Cook, S. L.; Secrist, J. A. III. *J. Am. Chem. Soc.* **1979,** *101,* 1554.
37. Sasaki, T.; Minamoto, K.; Hattori, K. *J. Am. Chem. Soc.* **1973,** *95,* 1350.
38. Sasaki, T.; Minamoto, K.; Asano, T.; Miyake, M. *J. Org. Chem.* **1975,** *40,* 106.
39. Ajmera, S.; Bapat, A. R.; Stephanian, E.; Danenberg, P. V. *J. Med. Chem.* **1988,** *31,* 1084.
40. Legraverend, M.; Huel, C.; Bisagni, E. *J. Chem. Res. (S)* **1990,** 102.
41. Biggadike, K.; Borthwick, A. D. *J. Chem. Soc. Chem. Commun.* **1990,** 1380.
42. Kanzaki, T.; Higashide, E.; Yamamoto, H.; Shibata, M.; Nakazawa, N.; Iwasaki, H.; Takewaka, T.; Miyake, A. *J. Antibiot.* **1962,** *15a,* 93.
43. Fox, J. J.; Kuwaka, Y.; Watanabe, K. A. *Tetrahedron Lett.* **1968,** 6029.
44. Suhadolnik, R. J. *Nucleoside Antibiotics*; Wiley-Interscience: New York, 1970; pp 170–217.
45. Damodoran, N. P.; Jones, G. H.; Moffatt, J. G. *J. Am. Chem. Soc.* **1971,** *93,* 3812.
46. Jones, G. H.; Taniguchi, M.; Tegg, D.; Moffatt, J. G. *J. Org. Chem.* **1979,** *44,* 1309.
47. Youssefyeh, R. D.; Verheyden, J. P. H.; Moffatt, J. G. *J. Org. Chem.* **1979,** *44,* 1301.
48. Schaffer, R. *J. Am. Chem. Soc.* **1959,** *81,* 5452.
49. Schaffer, R.; Isbell, H. S. *J. Am. Chem. Soc.* **1957,** *79,* 3864.
50. Swallow, D. L. *Prog. Drug. Res.* **1978,** *22,* 267.
51. Zakerinia, M.; Davary, H.; Hakimelahi, G. *Helv. Chim. Acta* **1990,** *73,* 912.
52. Martin, J. C.; Dvorak, C. A.; Smee, D. F.; Matthews, T. R.; Verheyden, J. P. H. *J. Med. Chem.* **1983,** *26,* 759.
53. Martin, J. C.; Tippie, M. A.; McGee, D. P. C.; Verheyden, J. P. H. *J. Pharm. Sci.* **1987,** *76,* 180.
54. Smee, D. F.; Boehme, R.; Chernow, M.; Binko, B. P.; Matthews, T. R. *Biochem. Pharmacol.* **1985,** *34,* 1049.
55. Biron, K. K.; Stanat, S. C.; Sorrell, J. B.; Fyfe, J. A.; Keller, P. M.; Lambe, C. U.; Nelson, D. J. *Proc. Natl. Acad. Sci. U.S.A.* **1985,** *82,* 2473.

56. Cheng, Y.-C.; Grill, S. P.; Dutschman, G. E.; Nakayama, K.; Bastow, K. F. *J. Biol. Chem.* **1983,** *258,* 12450.
57. Smith, K. O.; Galloway, K. S.; Kennell, W. L.; Ogilvie, K. K.; Radanus, B. K. *Antimicrob. Agents Chemother.* **1982,** *22,* 55.
58. Ogilvie, K. K; Gillon, M. F. *Tetrahedron Lett.* **1980,** *21,* 327.
59. Martin, J. C.; Smee, D. F.; Verheyden, J. P. H. *J. Org. Chem.* **1985,** *50,* 755.
60. Quoted by C. Honig is *Perspectives in Biology and Medicine* **1990,** *33,* 561.
61. Faulds, D.; Hael, R. C. *Drugs* **1990,** *39,* 597.
62. Schmidt, G. M.; Horak, D. A.; Niland, J. C.; Ducan, S. R.; Forman, S. J.; Zaia, J. A. *N. Engl. J. Med.* **1991,** *324,* 1057.
63. Mansour, S. L.; Thomas, K. R.; Capicchi, M. R. *Nature (London)* **1988,** *336,* 348.
64. Culver, K. W.; Ram, Z.; Wallbridge, S.; Ishii, H.; Oldfield, E. H.; Blaese, R. M. *Science (Washington, DC)* **1992,** *256,* 1550.
65. Oldfield, E. H.; Ram, Z.; Culver, K. W.; Blease, R. M.; De Vroom, H. L.; Anderson, W. F. *Human Gene Therapy* **1993,** *4,* 39.
66. Tippie, M. A.; Martin, J. C.; Smee, D. F.; Matthews, T. R.; Verheyden, J. P. H. *Nucleosides Nucleotides* **1984,** *3,* 525.
67. Pandit, U. K.; Grose, W. F. A.; Eggette, T. A. *Synthet. Commun.* **1972,** *2,* 345.

Camptothecin and Taxol

Monroe E. Wall
Research Triangle Institute

Natural compounds are enormously diverse in their structures and physical and biological properties. Most natural compounds are secondary metabolites whose functions in plants, fungi, and marine organisms are not well understood. Currently it is believed that many of these compounds act defensively against the harmful effects of toxins, carcinogens, or mutagens found in the plant (*1, 2*) or against attack by external predators (*3*).

Historical Setting

I was introduced to phytochemical research in my third year as an undergraduate at Rutgers University, in 1935, when one of my teachers, James Allison, assigned me as a laboratory exercise in biochemistry the repetition of R. Willstatter's isolation of chlorophyll (*4*). Impressed by the beautiful green and yellow colors of the plant pigments encountered during the isolation process, I became a lifelong devotee of phytochemistry and have worked in this field since 1936.

During predoctoral training I studied the role of potassium in plants (*5–11*). After this period I joined the Eastern Regional Research Laboratory (ERRL) of the U.S. Department of Agriculture (USDA), and worked there for almost 20 years. After initial studies on the fat-soluble constituents of

2523–0/93/0327$06.50/0

vegetable leaves (*12–17*), I was placed in charge of a large program from 1950 to 1959 that involved screening thousands of plants in search of steroids that might be cortisone precursors. Plant collection was conducted by botanists under the auspices of the Plant Introduction Division of the USDA. Thousands of plants were collected and identified before being shipped to the ERRL. This effort proved to be a good model for future programs, for it established that close cooperation between chemists and botanists was necessary for a successful natural products program. The survey included not only quantitative data for steroidal sapogenins but also qualitative analysis for sterols, alkaloids, tannins, and flavonoids (*18–21*). We saved thousands (but not all) of the plant alcoholic extracts, particularly of the more unusual plants.

One of the extracts so saved and stored was prepared from the leaves of *Camptotheca acuminata*, family Nyssaceae, a tree native to relatively warm areas of the southeastern provinces of China such as Szechwan, Yunnan, and Kwangsi provinces (*22*). This plant was introduced several times into the United States, and eventually a few trees were grown at the USDA Plant Introduction Garden in Chico, California, from which our sample was procured (see ref. 22 for a detailed account of the introduction of *C. acuminata* into the United States).

Discovery of Antitumor Activity in *Camptotheca Acuminata*

In addition to screening plants for various chemical constituents, I began to test some extracts for antibiotic, antitumor, and antiviral activity. In 1957 I was visited by Jonathan Hartwell of the Cancer Chemotherapy National Service Center (CCNSC), a pioneer worker in the field of plant antitumor constituents. I agreed to send him 1,000 ethanolic plant extracts for testing for antitumor activity. Almost a year later came back the astonishing result that the *Camptotheca* extracts were the only ones to have high activity in the CA-755 assay then used as a standard test system.

Discovery of Camptothecin

The discovery of high antitumor activity in the crude extract of the leaves of *C. acuminata* profoundly changed my career. Having worked in the Federal Civil Service for close to 20 years, I held a high senior position with full tenure and was assured of full support until the mandatory retirement age of 70 years. Nevertheless, work on steroids at ERRL had reached a terminus, and research priorities had shifted to areas other than natural products. I had become interested in the biodirected activities of natural products, such as the antitumor or antileukemia activities of the various plant extracts and their subsequent purified concentrates, as a tool in the isolation of active antitumor agents. Although the National Cancer Institute (NCI) was willing to support my work on the isolation of camptothecin at ERRL, it proved impossible, for the USDA felt at the time that it was not an appropriate agricultural research project. Accordingly, in July 1960 I joined the newly founded Research Triangle Institute, which had no chemistry or biology research projects of any type ongoing. I became head of the Natural Products Laboratory and eventually Vice President for Chemistry and Life Sciences.

In the spring of 1961, funds for camptothecin research were obtained from the NCI. By 1963 a sizable sample of approximately 20 kg of the wood and bark of the tree had become available to me.

Study Procedures

Use of the L1210 Assay for Fractionation

My small research group of three to five individuals began a typical biodirected isolation in which fractionation of the sample would be directed entirely by the antitumor assay. By this time it was known that crude extracts of *C. acuminata* were very active in the L1210 mouse leukemia life prolongation assay. This finding was, and still is, highly unusual. Most of the hundreds of plant extracts with which we and others subsequently worked were never sufficiently active in the L1210 assay to warrant fractionations in this system. However, this was the case for camptothecin and aroused very high interest at the NCI. In general, after conducting a particular phase of the isolation, it would often take us 3 months or more before the results were received. The long time horizon was primarily due to the fact that the L1210 assay is based on life prolongation. In the case of highly active extracts, the study would have to be carried out for 30 days or more. Although it was a slow process, we persisted, working at the same time on other plant bioassay directed programs, of which one, taxol, is discussed subsequently.

Fractionation of **Camptotheca acuminata**

Chart 1 shows the results obtained from the extraction of almost 20 kg of dry plant material consisting of the wood plus wood bark of the tree. This sample was obtained from the Plant Introduction Station in Chico, California, or in some cases from trees growing nearby in California. Our procedure involved a continuous hot extraction with heptane. The residual plant material was extracted with hot 95% ethanol, and, after concentration, the aqueous ethanolic residue was extracted with chloroform (Chart 1). The fractions shown in Chart 1 were studied in the L1210 assay; only the chloroform fractions were active.

Craig Countercurrent Partition

Several procedures involving chromatography, particularly on alumina, were tested on a small scale. They were all unsuccessful for reasons unknown to us at the time, although we now know that camptothecin is adsorbed very tightly onto alumina, and that alumina is not an appropriate chromatographic agent for camptothecin. We then tested the Craig countercurrent partition procedure. This methodology was designed and perfected by L. C. Craig, who received the Nobel Prize for his work (*23, 24*). Various types of countercurrent equipment were available in the early 1960s. Most of this equipment is now obsolete, as more convenient and compact equipment is now available. The Craig partition technique involved liquid–liquid partition and avoided harsh treatments that could cause changes in a substance of unknown constitution.

Most of the chloroform phase shown in Chart 1 after concentration was subjected to an 11-stage preparative Craig countercurrent distribution (Table 1). In this simple use of the Craig methodology, the partition was carried out in large separatory funnels using a chloroform–carbon tetrachloride–methanol–water partition system with the ratios shown in Table 1. All of the fractions were tested in both the in vivo L1210 mouse life prolongation assay and an in vitro cytotoxicity assay utilizing inhibition of growth of KB cells. There was reasonable correlation between the results of the two assays. Tubes 2–6 were judged to contain the most active material. It should be kept in mind that in the in vivo life prolongation assay, it is the combination of the lowest dose with the largest T/C activity that is the basis for selecting active fractions (T/C is defined as the number of days survived by treated (T) mice divided by the survival time of control animals (C) times 100). In the cytotoxicity assay it is simply the lowest ED_{50} dose that will inhibit the growth of the KB cells (ED_{50} is defined as the dose in micrograms

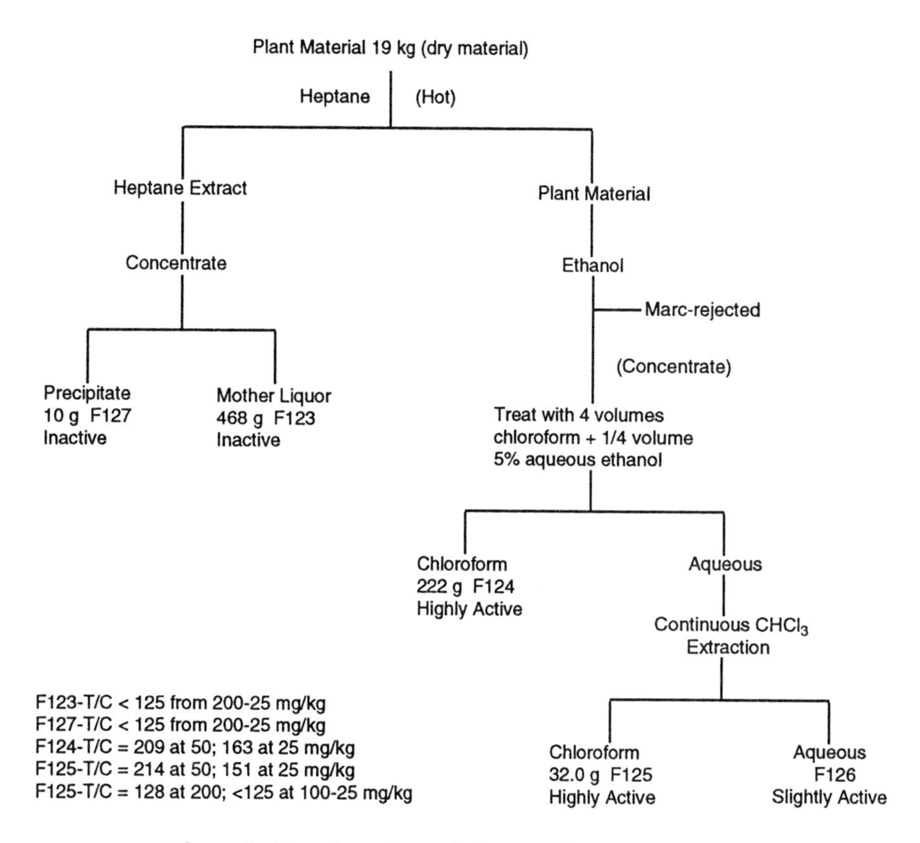

Chart 1. Fractionation of Camptotheca acuminata.

per milliliter that produces 50% inhibition of growth of cells). We had thus removed about 80% of the total weight while concentrating the most active fractions in 20% of the original weight of the chloroform extracts. By comparison with a pure sample obtained later but again subjected to the standard L1210 assay (Table 1, K-128), it was estimated that the most active fractions contained 1–2% of the active compound. It should be noted that KB activity was reasonably correlated with the L1210 assay result.

Table 1. Craig Countercurrent Distribution of Chloroform Fraction from Chart 1

NSC No.	*Tube No.*	*Wt. Fraction (grams)*	*Dose (mg/kg)*[2]	*T/C*	*9KB*
F098	[1]	—	125	153	<1.0
F099	1	21.5	250	179	5.3
F0100	2	8.7	62.5	201	0.8
F0101	3	7.2	62.5	201	0.5
F0102	4	6.9	31.2	206	0.09
F0103	5	7.6	31.2	156	0.3
F0104	6	7.2	31.2	198	0.5
F0105	7	9.3	31.2	160	<1.0
F0106	8	11.5	250	172	0.5
F0107	9	16.1	250	148	4.3
F0108	10	19.2	250	96	33.0
F0109	11	62.7	250	100	>100.0
K128 (camptothecin)	—	—	0.5	163	0.7

[1] Starting material.
[2] Dose at which miximum T/C in LE-1210 was observed.

Isolation of Pure Camptothecin

When fractions 2–6 were combined and the solvent was partially concentrated, a yellow precipitate was formed and collected by filtration. This material was subsequently further purified by chromatography on a silica gel column and crystallization. The pure compound is very active in the L1210 life prolongation assay. Dosages as low as 0.5 mg/kg appreciably prolonged life. A dosage of 4 mg/kg was the maximum possible prior to occurrence of toxicity.

Some of the major physical properties of camptothecin are as follows. The compound is a high-melting substance with a molecular weight of 348.111 daltons as determined by high-resolution mass spectrometry, corresponding to the formula $C_{20}H_{16}N_2O_4$. It fluoresces intensely blue under ultraviolet light and is optically active $[\alpha]_D^{25}$, +31.3°. The compound gives negative reactions on the phenol ($FeCl_3$) and indole tests and negative reactions on the Dragondorf and Mayer tests. No crystalline salts could be obtained with a variety of acids. The compound cannot be methylated with diazomethane or dimethyl sulfate under a variety of conditions. The compound does not react with bicarbonate or carbonate, but it can be quantitatively reconverted to the sodium salt with sodium hydroxide at room temperature. On acidification, the sodium salt regenerates camptothecin. Other spectral properties are given in detail in ref. 25.

Structure and Biological Activity of Camptothecin

Camptothecin can be readily converted to an acetate and a chloroacetate. The chloroacetate was converted to the corresponding iodoacetate by treatment with sodium iodide–acetone. The iodoacetate crystallized in orthorhombic crystals suitable for X-ray analysis, which was carried out by A. T. McPhail and G. A. Sim, then in the Department of Chemistry, University of Illinois. The structure of camptothecin and the corresponding sodium salt are shown in Figure 1 (*25*). This structure was completely in accord with ultraviolet and infrared spectral data, NMR spectral data, and mass spectral data (*25*).

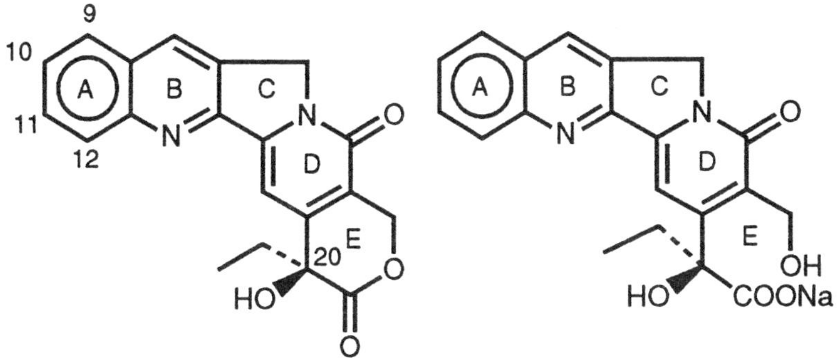

Camptothecin (1)   Camptothecin Sodium (2)

Figure 1. Structures of camptothecin and camptothecin sodium.

The structure of camptothecin is unique. Camptothecin has been shown to be related to the indole alkaloids, and the six-membered ring B and a five-membered ring C are formed by a ring expansion-ring contraction sequence of reactions. The pentacyclic ring structure is highly unsaturated. Some of the unique structural features involve the presence in ring E of an α-hydroxy lactone system and, in ring D, an unsaturated conjugated pyridone moiety. On treatment with alkali, the compound readily opens, forming an open lactone sodium salt, as shown in Figure 1 (*25*). On acidification, the extremely water-soluble sodium salt is readily relactonized. The parent compound, however, is extremely water-insoluble, and indeed is insoluble in virtually all organic compounds except dimethyl sulfoxide, in which it exhibits moderate solubility.

Biological Activity of Camptothecin and Its Analogues

The isolation and structure proof for camptothecin and some of the analogues obtained at an early stage, 10-hydroxy and 10-methoxy camptothecin (*26*), enabled extensive studies to be conducted on camptothecin and a few of its analogues. Camptothecin was remarkably active in prolonging the life of mice treated with L1210 leukemia cells, showing activity at dosages between 0.5 and 4.0 mg/kg in this assay. It showed activity of a similar order in the P388 leukemia life prolongation assay. The compound was also very active in the inhibition of solid tumors that were being studied at this early stage, including the Walker WM tumor, which was completely inhibited by camptothecin. 10-Methoxy camptothecin was found to be somewhat less active than camptothecin, whereas 10-hydroxy camptothecin was the most active compound in the series and was more active than camptothecin in both L1210 and P388 leukemia life prolongation assays (*26–28*). Unfortunately, 10-hydroxy camptothecin is found in only very small quantities in nature, amounting to about 10% of the camptothecin content. All of these previous compounds were extremely insoluble in water, but the lactone could be opened under mild circumstances with sodium hydroxide or sodium methoxide, and this sodium salt was very soluble in water. Not until much later was it definitively shown that this compound is only one-tenth as active as camptothecin in the P388 assay (*27*).

Early Clinical Trials

Encouraged by the broad scope of antitumor activity of camptothecin in animal studies, the NCI decided to go to clinical trial with the sodium salt (**2**, Figure 1). In contrast to camptothecin, the sodium salt was water-soluble and hence was easily formulated for intravenous administration. In a Phase I trial, Gottlieb and Luce reported partial responses in five of 18 patients (*29*). The responses occurred primarily in patients with gastrointestinal tumors and were of short duration. Toxicity involving mainly dose-limiting hematological depression was noted, along with some vomiting and diarrhea. In another Phase I trial conducted by Muggia et al. (*30*), only two partial responses occurred in ten evaluable patients. Because of the somewhat encouraging results obtained in the Phase I study of Gottlieb and Luce, a Phase II study was undertaken in 61 patients with adenocarcinomas of the gastrointestinal tract, but only two patients showed objective partial responses (*31*). The drug has also been under study as a sodium salt in the People's Republic of China. Up to 1,000 patients were enrolled in the trial, and effective results were reported in patients with gastric cancers, intestinal cancers, head and neck tumors, and bladder carcinomas (*32*). These results were more promising than those in the U.S. trial but may have been due to the fact that in the United States, only patients that had been treated with many other drugs previously and had become resistant were given camptothecin.

Structure–Activity Relationship Studies

Although our work on camptothecin nominally ceased by 1970, our interest in camptothecin and its analogues did not wane, and because some of this work led to clinical studies, it is briefly described here.

In 1969 we reported the isolation of 10-hydroxy- and 10-methoxy camptothecin as minor components of *C. acuminata.* Later we found that 10-hydroxy camptothecin was much more active than camptothecin in a number of assays, and this finding stimulated synthetic efforts by both SmithKline Beecham and a Japanese pharmaceutical company to prepare water-soluble 10-hydroxy camptothecin analogues. This work, however, was based on our earlier work showing the activity of 10-hydroxy camptothecin (*26, 27*). 10-Methoxy camptothecin is somewhat less active than camptothecin. We initiated structure–activity studies of camptothecin (Chart 2) because we felt that much more information would be needed before other analogues could be made, and because we wanted to compare the sodium salt with the parent compound. In these studies the sodium salt had considerable activity, although it was less potent than camptothecin (*27*). However, in later studies, particularly in the P388 leukemia model, the sodium salt was found to have only one-tenth the activity of camptothecin. Oxidation of the ring B nitrogen with perchloric acid or with chloroperbenzoic acid led to considerable loss of activity. Camptothecin reacted readily with amines, such as ethylamine, resulting in ring opening and lower activity. Of great interest was the observation that hydroxylation in ring A, at least at the 10-position, was compatible with activity, and indeed this compound, 10-hydroxy camptothecin, has greater activity than camptothecin. A major reduction in antineoplastic activity was noted as a result of reactions involving the hydroxyl or lactone moiety in ring E (Figure 1). After acetylation of camptothecin, the resulting acetate is virtually inactive. Other reactions also point to the absolute requirement for the α-hydroxy group. For example, after replacement of this group by chlorine, both the resultant chloro analogue and the corresponding reduction product, desoxycamptothecin, are inactive in L1210 leukemia (*27*). Reduction of the lactone under mild

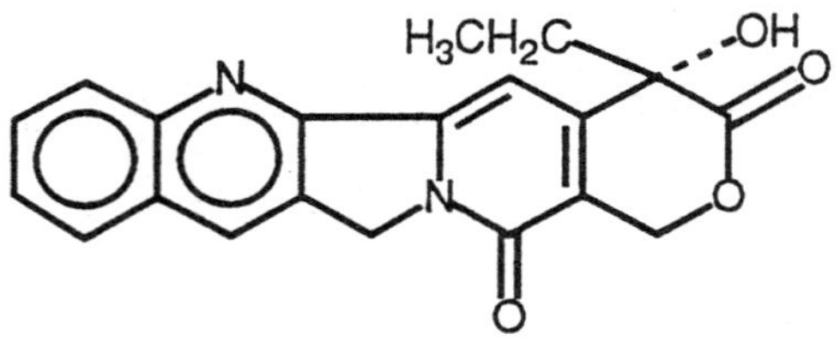

Camptothecin (1)
L1210, T/C 220 at 2 mg/kg
WM, T/C 0 at 16 mg/kg
9 at 7 mg/kg
18 at 5 mg/kg

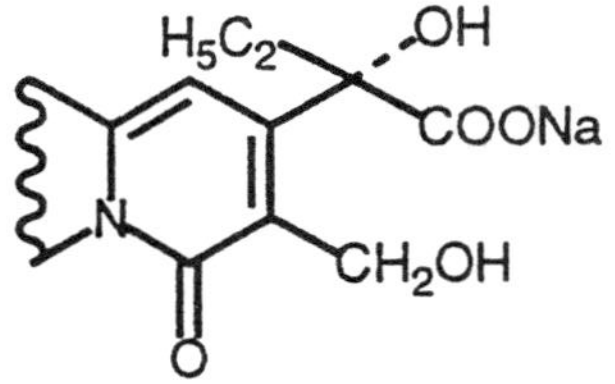

Camptothecin Sodium (2)
L1210, T/C 209 at 3.0 mg/kg

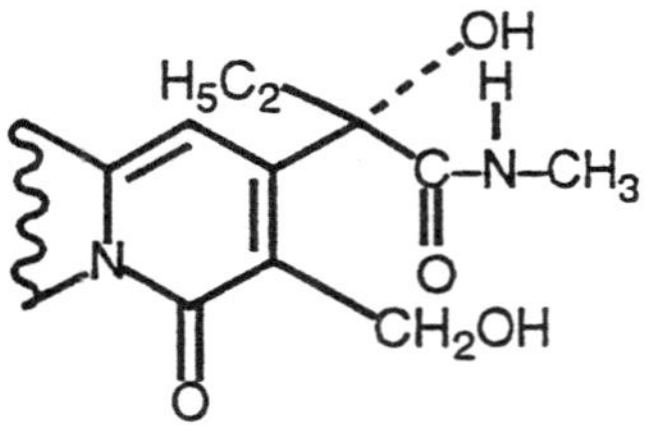

L1210, T/C 172 at 3.5 mg/kg

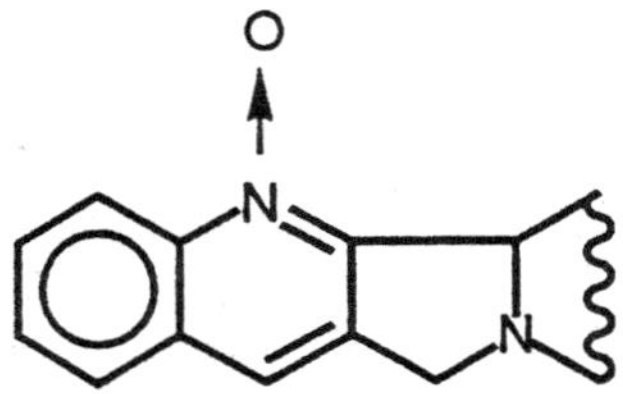

L1210, T/C 144 at 2.0 mg/kg

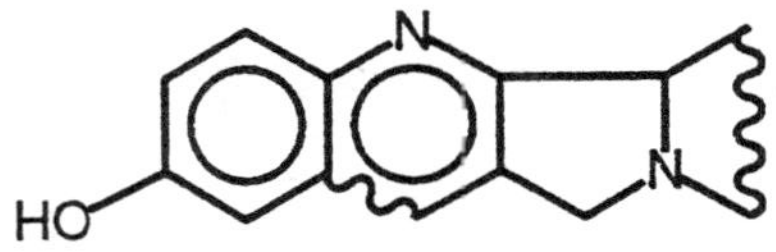

10-Hydroxycamptothecin (3)
L1210, T/C 230 at 0.5 mg/kg

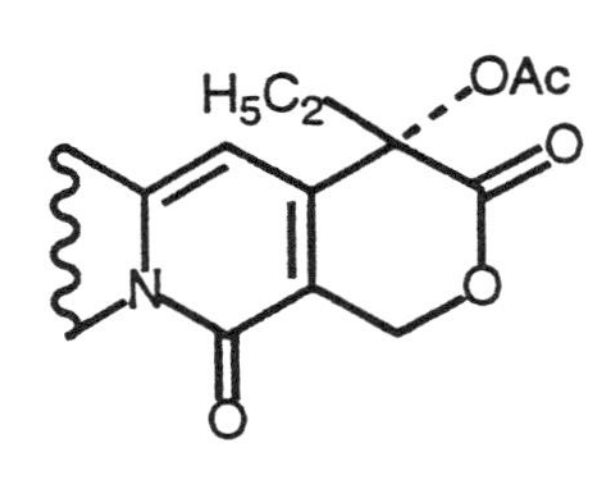

Slightly Active

Camptothecin Acetate (8)

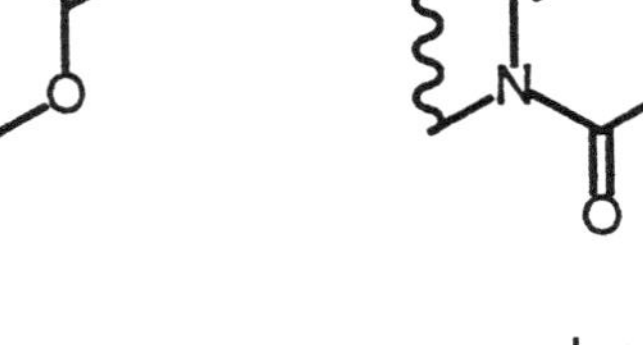

a, R = H (10)
b, R = Cl (9)
Inactive

Inactive (11)

Chart 2. Structure–activity relationships in the camptothecin series.

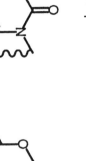

conditions to give the lactol also results in complete loss of activity (*27*). It is thus quite clear that the activity of the sodium salt is due entirely to the ability of this compound to recyclize to camptothecin, at least to a limited extent, under physiological conditions.

After the initial trials, clinical interest in camptothecin languished from about 1972 to 1988. Nevertheless, several advances during this period were of interest to the NCI. In 1980 we greatly improved the synthesis of camptothecin and prepared a number of analogues. The initial compounds, while active, were not significantly better than camptothecin (*28*). They were also racemic. Subsequent syntheses led to methods for the synthesis of 9-, 10-, 11-, and 12-amino camptothecin (*33–35*). The 9- and 10-amino compounds were very active and generated considerable interest, whereas the 11- and 12-amino analogues were essentially inactive. Then, in 1987, we found a way to resolve the racemic camptothecin compounds and prepare the pure 20(*S*) and 20(*R*) isomers (*36*). Since then we have made only the naturally occurring 20(*S*) form, and we proved immediately that the 20(*R*) form of camptothecin was inactive. Concomitantly with our improved synthetic abilities there came a biological finding of noteworthy importance, the discovery that camptothecin inhibits topoisomerase I activity (*37*).

Many of the camptothecin analogues that were synthesized by our new procedures at Research Triangle Institute (*33–35*) were tested for their topoisomerase I inhibition. The inhibition of topoisomerase I activity closely parallels in vivo mouse leukemia assays, as shown in Table 2 (*38, 39*). Finally, we were able to separate the racemic synthon by which we synthesized various camptothecin analogues into the corresponding 20(*S*) and 20(*R*) analogues and were able to demonstrate that the 20(*R*) form is inactive, both in the topoisomerase inhibition assay and in the in vivo assays (Table 2) (*38*).

In a study conducted cooperatively with the Stehlin Institute, New York University Medical School, Johns Hopkins University, and the Research Triangle Institute, both the 9-amino and the 10,11-methylenedioxy camptothecin compounds had great potency in inhibiting human colon cancer xenografts in nude mice (*40*). A large number of reference compounds commonly used in cancer chemotherapy were totally ineffective.

Recent Preclinical and Clinical Trials

The results of in vitro and in vivo studies on new camptothecin analogues prompted renewed interest in the clinical possibilities for a number of the camptothecin analogues synthesized in our laboratory and elsewhere. The NCI is moving 9-amino-20(*S*) camptothecin rapidly to the clinic; similarly,

Table 2. Comparison of In Vivo and In Vitro Activity of Camptothecin and Analogs

Camptothecin Derivative	*% Inhibition of Relaxation*[1]	*%DNA Scission*		*Antitumor Activity*[2] *(T/C)×100*		*9KB*[3]	*9PS*[3]
		Super-coiled[1]	*Linear*[1]	*L1210*	*P388*		
S-	52	48	41	197 (8)	197 (4)	10^{-2}	10^{-2}
R-	20	12	9	<125	—	10^{-1}	10^{0}
RS-	30	25	15	—	222 (8)	10^{-1}	—
20-deoxy-*RS*-	0	0	1	<125	—	—	
10-OH-*S*-	73	71	40	348 (20)	297 (3)	10^{-2}	
11-OH-*RS*-	74	35	31	357 (60)	—	—	
10-OCH_3-*RS*-	—	—	—	167 (2)	—	10^{-2}	
10,11-$(CH_3O)_2$-*RS*-	0	0	2	<125	<125	10^{-1}	
10,11-OCH_2O-*RS*-	62	82	40	325 (2)	—	10^{-2}	
10-NO_2-*RS*-	19	23	22	219 (16)	—	10^{-1}	
11-NO_2-*RS*-	0	11	11	147 (80)	—	10^{0}	
12-NO_2-*S*-	0	10	3	151 (40)	151 (40)	—	
9-NO_2-*S*-	40	50	22	348 (10)	—	10^{-2}	
9-NH_2-*S*-	91	35	31	348 (2.5)	—	10^{-2}	
10-NH_2-*RS*-	16	16	18	325 (8)	—	10^{-2}	
11-NH_2-*RS*-	33	7	14	147 (40)	—	10^{-1}	
12-NH_2-*S*-	0	0	3	<125	<125	—	
21-lactam-*S*-	3	0	2	178 (40)	—	10^{-1}	
tricyclic-*RS*-	0	0	—	<125	—	—	

[1]Effect of 10 μM of test compound, as percent of control (cf. reference 38).
[2]Mouse leukemias L1210 and P388, inoculated intraperitoneally; drug administered intraperitoneally on days 1 and 5. In parentheses is the dose (mg/kg) giving the greatest reported T/C value. Antitumor data are from references 28 and 33–36.
[3]Inhibition of cell growth in 9KB or 9PS as defined by Geran et al. (54) expressed as ED_{50} (μg/mL).

the Glaxo Pharmaceutical Company U.S.A. is conducting studies with an analogue of 9-amino camptothecin, which is very active and is being moved as quickly as possible toward clinical trial. Already in clinical trial is an analogue of 10-hydroxy camptothecin called topotecan, produced by SmithKline Beecham. This compound is considerably less active than the active compounds discussed previously. However, it has the advantage of water solubility, and it is now in Phase II trial. Another analogue of 10-hydroxy camptothecin, a long-chain water-soluble compound (CPT-11), is in clinical trial in Japan. It seems evident that the discovery of camptothecin has led to the development of more potent synthetic analogues that are being rapidly developed for clinical use. After a rapid rise and fall in clinical uses for camptothecin in the early 1970s, two decades later the analogues of camptothecin are receiving expedited study. There are many problems, particularly for the water-insoluble analogues, but these problems are being solved quickly.

Taxol

Since 1988, the remarkable clinical efficacy of taxol, resulting in numerous partial and complete remissions of advanced ovarian cancer and, more recently, reported efficacy in breast, lung, and prostate cancer, has aroused great interest in this antitumor compound, which was discovered in my laboratory at Research Triangle Institute many years ago (*41, 42*). As with many other investigations of this type, a combination of serendipity and hard work led to the discovery of this very active antitumor agent.

Initial Procurement and Isolation

A screening program for antitumor agents in the Plant Kingdom was initiated in 1960 under Jonathan L. Hartwell. In this program, plant samples collected at random were supplied by the USDA under an interagency agreement with the NCI. In August 1962, USDA botanist Arthur S. Barclay and three college-student field assistants collected 650 plant samples in California, Washington, and Oregon, including bark, twigs, leaves, and fruit of *Taxus brevifolia* in Washington state (*43*).

T. brevifolia is a slow-growing tree that is primarily found in the coastal areas of the three West Coast states. It had never received any chemical investigation until it was assigned to my laboratory by Hartwell. The assignment of the plant was not accidental. Some of these samples had been shown to have 9KB cytotoxicity. At that time there were only three groups working under contract to NCI: Jack Cole's group, at the University of Arizona; S. Morris Kupchan's group, at the University of Wisconsin; and my laboratory at Research Triangle Institute. At the time the other groups were not particularly interested in plants with 9KB activity. I had noted an excellent correlation in my camptothecin studies between activity in the in vivo L1210 assay and 9KB cytotoxicity. Accordingly, I had requested Hartwell to assign me as many 9KB actives as possible, and from this request arose the assignment to my laboratory of *T. brevifolia.* A number of other plants highly active in the 9KB assay were also assigned to me, and several highly active novel compounds were found in these cases as well, including colubrinol, a maytansine analogue (*44*), carminomycin, related to daunomycin (*45*), and an active quasinoid (*46*).

Initial samples of plant material had arrived in my laboratory by 1964. The method finally adopted after several unsuccessful trials is shown in Chart 3. It involved our standard ethanol extraction procedure, partition of

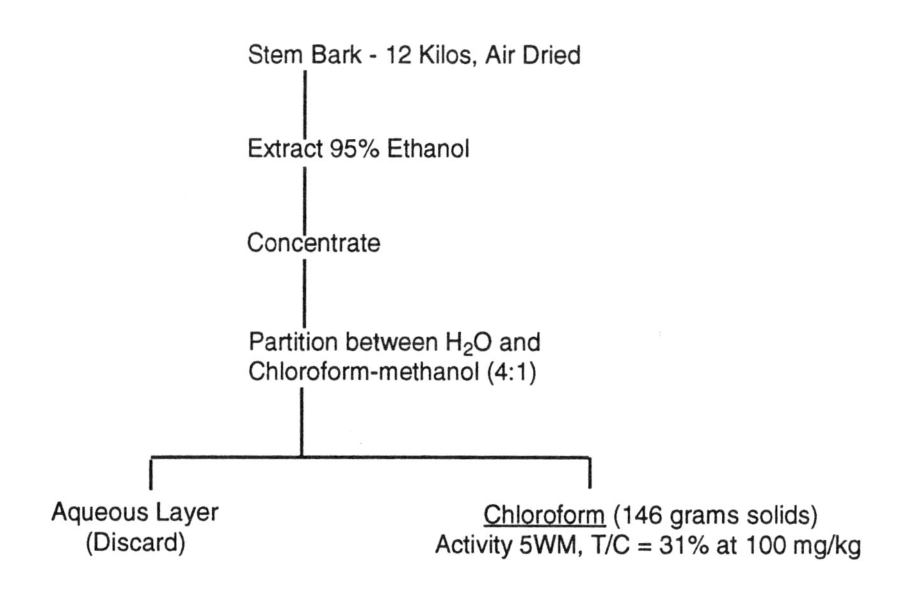

Chart 3. Isolation of taxol from Taxus brevifolia.

the ethanolic residue between water and chloroform, and a large number of Craig countercurrent distribution treatments, the last of which involved a 400-tube Craig countercurrent distribution. In this manner, approximately 0.5 g of taxol was isolated from 12 kg of air-dried stem and bark, for a yield of about 0.004%. All the various steps were monitored by an in vivo bioassay that involved inhibition of the Walker WM solid tumor. As is shown in Chart 3, increased purification is accompanied by lower T/C values and dosages. The isolation procedure therefore was laborious, but performed in a manner in which no losses due to treatment and no changes in the unknown chemical constitution of the eventual product occurred because of the mild countercurrent distribution methodology. Much simpler procedures have subsequently been developed at Research Triangle Institute and elsewhere.

Biological Activity of Crude and Purified Taxol

I assigned the name taxol to this compound before we knew its complete structure, but it was evident that it did contain some hydroxyl groups. In addition to isolating the pure material, we subjected the crude extracts to a large number of tests. In early work we found that the crude extracts were active not only in the Walker tumor inhibition assay, but also had modest activity in an L1210 leukemia model and particularly high activity in the 1534 (P4) leukemia assay. The latter assay, a life prolongation assay in mice, had been used some years before by scientists at Eli Lilly to isolate the vinca alkaloids, which greatly prolonged life in this system. The same was noted for taxol, with T/C values in the P4 system in excess of 300, even with crude extracts. The activity of pure taxol in a number of in vivo rodent assays is shown in Table 3. Particularly high activity was shown in the B-16 melanoma assay. Years later it was one criterion by which taxol was moved to clinical trial (*47*).

At least a year prior to our isolating the pure material, in a letter dated April 15, 1966, to Jonathan Hartwell, I requested that the extracts we sent "receive a special priority with the biological screeners as I regard it as one of the most important samples we have had in a long time." By May 1966, in RTI Progress Report No. 18 ("The Screening of Fractionated Plants for Antitumor Inhibitory Substances"), I stated, "At present, a major effort by our group is being placed on this plant (*Taxus brevifolia*)." By November 1966 we were able to isolate a purified fraction, and we presented some physical constants in RTI Progress Report No. 20. The actual isolation was completed by June 1966 (Chart 3). Indeed, by this time we were able to compare crude chloroform extracts from various samples of *T. brevifolia* collected in Alaska, California, Washington, Idaho, Oregon, and Montana

Table 3. Biological Activity of Taxol Cytotoxicity

System Tested	*Administration*	*Activity (% T/C)*
i.p. P388 Leukemia	intraperitoneal	+(164)
i.p. B16 Melanoma	intraperitoneal	++(283)
i.p. L1210 Leukemia	intraperitoneal	+(139)
S.R.C. CX-1 Colon Xenograft	subcutaneous	++(3)
S.R.C. LX-1 Lung Xenograft	subcutaneous	+(8)
S.R.C. MX-1 Mammary Xenograft	subcutaneous	++(−77)

NOTE: KB (human carcinoma of the nasopharynx) was $ED_{50} = 3.5 \times 10^{-5}$ µg/mL.

and present the data on their cytotoxicity and WM inhibition. In 1967 I presented the first report in which taxol was mentioned to the American Chemical Society (*41*).

The Structure Determination of Taxol

As soon as we had isolated taxol in pure form, a great deal of work on the structure of the compound was carried out by available spectroscopic methods. Although methods for ultraviolet, infrared, and mass spectrometry were at a reasonably advanced stage in the late 1960s, NMR technology was relatively primitive compared to the very sophisticated instrumentation now available. The physical properties of taxol are as follows. Taxol crystallized from methanol as needles and melted at 213–216 methanol solution; the ultraviolet spectrum in methanol had λ_{max} 227 nm with extinction values (ε) 29,800 at 227 nm and 1700 at 273 nm; the infrared spectrum showed characteristic bands at 3300–3500 (hydroxyl and amine), 1730 (ester), 1710 (ketone), and 1650 (amide) cm^{-1}.

It was evident by this time that taxol probably contained a taxane skeleton. A number of taxane derivatives had been previously reported in the literature (*48–50*). It was evident that taxol was much more complex, for its molecular weight from high-resolution mass spectrometry was $C_{47}H_{51}NO_{14}$, corresponding to a molecular weight of 853 daltons. The evidence then indicated that taxol consisted of a taxane nucleus to which an ester was attached, as preliminary experiments had indicated that an ester moiety was easily cleaved from the rest of the molecule. Attempts were made to make crystalline halogenated derivatives of taxol. However, none had properties suitable for X-ray analysis. Taxol was therefore subjected to a mild base-catalyzed methanolysis at 0 °C, which yielded a nitrogen-containing α-hydroxymethyl ester, $C_{17}H_{17}NO_4$; a tetrol, $C_{29}H_{36}O_{10}$; and methyl acetate (for details, see ref. 42). The methyl ester obtained by the mild methanolysis procedure was converted to a parabromobenzoate ester and characterized by X-ray analysis as $C_{24}H_{20}BrNO_5$. The tetrol produced by methanolysis was converted to a bisiodoacetate, $C_{33}H_{38}O_{12}I_2$, which again underwent X-ray analysis (*42*).

Because the ester could have come originally from hydroxyl groups either at C_7 or at C_{13} (Figure 2), it was necessary to establish by a few chemical steps on which of these hydroxyl moieties the ester had originally been located. The esterified position was found to be the allylic C_{13} hydroxyl moiety. Consequently, taxol has the structure shown in Figure 2.

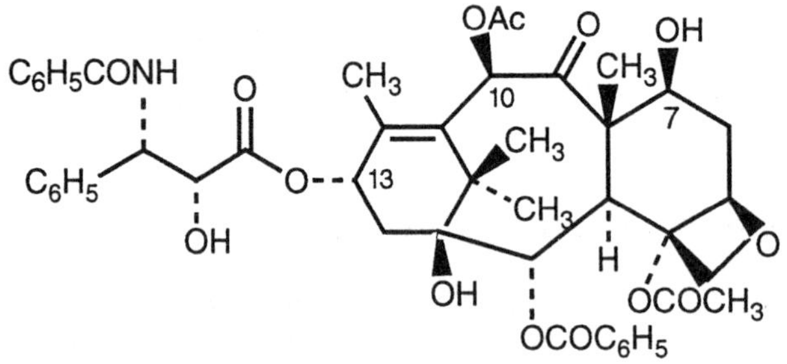

Figure 2. Structure of taxol.

There were no natural taxane compounds shown to have activity prior to the time of our isolation of taxol. There are many interesting features of the molecule, particularly the four-membered oxide ring 4, which is not found in any of the other natural taxanes. The ester moiety itself is of interest, for it contains two phenyl groups, one of which is attached as part of an amide function. The large number of asymmetric carbon atoms renders total synthesis of taxol most difficult.

For antitumor activity, it is essential that the entire taxane molecule be present. We have shown that the ester and the tetraol formed by low-temperature cleavage of taxol are each essentially inactive.

Subsequent Developments

Our initial discovery of taxol essentially ended with the publication of our paper on the structure (*42*). I made repeated efforts to interest NCI administrators in obtaining larger amounts of the bark and wood of *T. brevifolia* so that advanced animal studies, toxicology studies, and eventually clinical trials could ensue. The response was that the compound was present in too limited a quantity in the natural product, that the extraction and isolation procedures were difficult, and that the supply of the tree was limited. These are essentially the same problems NCI and its contractors face currently, although the extraction and isolation of taxol offer much less difficulty as a consequence of our original work. Hence, after some years of fruitless efforts on this matter, I was forced to turn my attention to other research areas. Fortunately, two major developments occurred. Some years later taxol was found to have very high activity in the B-16 melanoma assay (*47*). In a second development, studies by Susan Horwitz at the Albert Einstein Medical Center in New York showed dramatically that the mode of activity of taxol was due to its binding with tubulin in a novel manner. Tubulin is a protein that is involved in mitosis. Horwitz and her collaborators found that taxol in concentrations completely inhibitory to cell division had no significant effects on DNA, RNA, or protein synthesis (*51*). In contrast to other antimitotic agents such as colchicine, vincristine, or podophyllotoxin, taxol promotes assembly of calf brain microtubules, and the microtubules so produced are resistant to depolymerization by cold or $CaCl_2$ (*52*). This was a unique mode of action for mitotic inhibitors. On the basis of the B-16 melanoma findings and the mechanism of activity of taxol, the NCI instituted a very large effort to collect sufficient bark and wood, extract the compounds, and make them available to clinical investigators (*53*).

The current situation in regard to taxol is promising. Clinical investigators have already found striking evidence that taxol brings about partial and in some cases complete remission of advanced ovarian cancer. Articles that have appeared recently in the lay press state, apparently correctly, that taxol is also active against breast and lung cancer and may have activity against prostate cancer. In the chemistry area, partial syntheses have isolated baccatin (essentially that portion of taxol lacking the ester group) in larger quantities than taxol in another species, *Taxus baccata.* Both French and American investigators have successfully used this substance, potentially available in larger quantities than taxol. After synthesis of the side chain, it can be linked to baccatin, thus forming taxol. In addition to these very promising efforts, reports are appearing almost every day on efforts to use either tissue culture or molecular biology techniques to clone the gene responsible for the production of taxol and insert it into an appropriate carrier.

Work on the camptothecin and taxol series has taken almost 25 years to come to a fruitful outcome. Both may prove useful in a large number of cancer patients because of their unique mode of action: each inhibits either an enzyme or a protein that is absolutely required for DNA formation or cell division.

Acknowledgments

The aid of M. C. Wani, my colleague for 25 years, has been invaluable in both the camptothecin and the taxol investigations. Further thanks are due to Andrew McPhail of Duke University, who performed the fundamental X-ray analysis in both the camptothecin and taxol series. The National Cancer Institute provided long-term contract and grant support in both of these activities.

Literature Cited

1. Mitscher, L. A.; Drake, S.; Gollapudi, S. R.; Harris, J. A.; Shankel, D. M. In *Antimutagenesis and Anticarcinogenesis Mechanisms;* Shankel, D. M.; Hartman, P. A.; Cotta, T.; Hollaender, A., Eds.; Plenum: New York, 1986; p 153.
2. Williams, D. H.; Stone, M. J.; Hauck, P. R.; Rahman, S. R. *J. Nat. Prod.* **1989,** *52,* 1189.
3. Woodbury, A. M.; Wall, M. E.; Willaman, J. J. *J. Econ. Bot.* **1961,** *15,* 79.
4. Willstatter, R.; Stoll, A. *Investigations on Chlorophyll;* Schertz; Merz, Trans.; Lancaster, England, 1928.
5. Tiedjens, V. A.; Wall, M. E. *Proc. Am. Soc. Hort. Sci.* **1938,** *36,* 740.
6. Wall, M. E. *Soil Sci.* **1939,** *47,* 143.
7. Wall, M. E. *Soil Sci.* **1940,** *49,* 315.
8. Wall, M. E. *Soil Sci.* **1940,** *49,* 393.
9. Wall, M. E.; Tiedjens, V. A. *Soil Sci.* **1940,** *49,* 221.
10. Wall, M. E. *Chron. Bot.* **1940,** *6,* 133.
11. Wall, M. E. *Plant Physiol.* **1940,** *15,* 537.
12. Kelley, E. G.; Wall, M. E. *Annual Report of the Vegetable Association;* 1942.
13. Wall M. E.; Kelley, E. G. *Ind. Eng. Chem. Anal. Ed.* **1943,** *15,* 18.
14. Wall, M. E.; Kelley, E. G. *Ind. Eng. Chem. Anal. Ed.* **1946,** *18,* 198.
15. Wall, M. E.; Kelley, E. G. *Anal. Chem.* **1947,** 19, 677.
16. Wall, M. E. *Ind. Eng. Chem.* **1949,** *41,* 1465.
17. Wall, M. E. "Preparation of Chlorophyll Derivatives for Industrial and Pharmaceutical Use"; U.S. Department of Agriculture circular, AIC 299, March 1951.
18. Wall, M. E.; Krider, M. M.; Krewson, C. F.; Eddy, C. R.; Willaman, J. J.; Correll, D. S.; Gentry, H. S. *J. Am. Pharm. Assoc.* **1954,** *43,* 1.

19. Wall, M. E.; Eddy, C. R.; Willaman, J. J.; Correll, D. S.; Schubertand, B. G.; Gentry, H. S. *J. Am. Pharm. Assoc.* **1954,** *43,* 503.
20. Wall, M. E.; Fenske, C. S.; Kenney, H. E.; Willaman, J. J.; Correll, J. S.; Schubert, B. G.; Gentry, H. S. *J. Am. Pharm. Assoc.* **1957,** *46,* 653.
21. Wall, M. E.; Garvin, J. W.; Willaman, J. J.; Jones, Q.; Schubert, B. G. *J. Pharm. Sci.* **1961,** *50,* 1001.
22. Perdue, R. E., Jr.; Smith, R. L.; Wall, M. E.; Hartwell, J. L.; Abbott, B. J. "*Camptotheca acuminata* Decaisne (Nyssaceae): Source of Camptothecin, an Antileukemic Alkaloid"; U.S. Department of Agriculture, Agricultural Research Service, Technical Bulletin No. 1415, April 1970.
23. Craig, L. C. *J. Biol. Chem.* **1944,** *155,* 519.
24. Craig, L. C.; Post, O. *Anal. Chem.* **1949,** *21,* 500.
25. Wall, M. E.; Wani, M. C.; Cook, C. E.; Palmer, K. H.; McPhail, H. T.; Sim, G. A. *J. Am. Chem. Soc.* **1966,** *88,* 3888.
26. Wani, M. C.; Wall, M. E. *J. Org. Chem.* **1969,** *34,* 1364.
27. Wall, M. E. *Abstracts of Papers*, International Symposium on Biochemistry and Physiology of the Alkaloids; Mothes, K.; Schreiber, K.; Schutte, H. R., Eds.; Academie-Verlag: Berlin, 1969; pp 77–87.
28. Wani, M. C.; Ronman, P. E.; Lindley, J. T.; Wall, M. E. *J. Med. Chem.* **1980 ,** *23,* 554–560.
29. Gottlieb, J. A.; Luce, J. K. *Cancer Chemother. Rep.* **1972,** 103.
30. Muggia, F. M.; Creaven, P. J.; Hanson, H. H.; Cohen, M. C.; Selawry, O. S. *Cancer Chemother. Rep.* **1972,** *56,* 515.
31. Moertel, C. G.; Schutt, A. J.; Reitemeier, R. G.; Hahn, R. G. *Cancer Chemother. Rep.* **1972,** *56,* 95.
32. Xu B. *Abstracts of Papers,* U.S./China Pharmacology Symposium; Burns, J. J.; Tsuchiatani, G. J., Eds.; Washington, D.C.: National Academy of Sciences, 1980; p 156.
33. Wall, M. E.; Wani, M. C.; Natschke, S. M.; Nicholas, A. W. *J. Med. Chem.* **1986,** *29,* 1553.
34. Wani, M. C.; Nicholas, A. W.; Wall, M. E. *J. Med. Chem.* **1986,** *29,* 2358.
35. Wani, M. C.; Nicholas, A. W.; Manikumar, G.; Wall, M. E. *J. Med. Chem.* **1987,** *30,* 1774.
36. Wani, M. C.; Nicholas, A. W.; Wall, M. E. *J. Med. Chem.* **1987,** *30,* 2317.
37. Hsiang, Y.-H.; Hertzberg, R.; Hecht, S.; Liu, L. *J. Biol. Chem.* **1985,** *260,* 14873.
38. Jaxel, C.; Kohn, K. W.; Wani, M. C.; Wall, M. E.; Pommier, Y. *Cancer Res.* **1989,** *49,* 1465.

39. Hsiang, Y.-H.; Liu, L. F.; Wall, M. E.; Wani, M. C.; Kirschenbaum, S.; Silber, R.; Potmesil, M. *Cancer Res.* **1989,** *49,* 4385.
40. Giovanella, B. C.; Wall, M. E.; Wani, M. C.; Nicholas, A. W.; Liu, L. F.; Silber, R.; Potmesil, M. *Science (Washington, DC)* **1989,** *246,* 1046.
41. Wall, M. E.; Wani, M. C. *Abstracts of Papers*, 153rd National Meeting of the American Chemical Society, Miami Beach, FL, 1967; No. M-006.
42. Wani, M. C.; Taylor, H. L.; Wall, M. E.; Coggin, P.; McPhail, A. T. *J. Am. Chem. Soc.* **1971,** *93,* 2325.
43. "Washington Insight"; Persinos, E., Ed.; Sept. 15, 1990.
44. Wani, M. C.; Taylor, H. L.; Wall, M. E. *Chem. Commun.* **1973,** 390.
45. Wani, M. C.; Taylor, H. L.; Wall, M. E.; McPhail, A. T.; Onan, K. D. *J. Am. Chem. Soc.* **1975,** *97,* 5955.
46. Wani, M. C.; Taylor, H. L.; Thompson, J. B.; Wall, M. E. *Lloydia* **1978,** *41,* 578.
47. Suffness, M., National Cancer Institute personal communication, 1989.
48. Woods, M. C.; Nakanishi, K.; Bhacca, N. S. *Tetrahedron* **1966,** *22,* 243.
49. Harrison, I. W.; Scrowston, R. M.; Lythgoe, B. *J. Chem. Soc.* **1966,** 1933.
50. Della, D. P.; de Marcano, C.; Halso, T. G. *Chem. Commun.* **1970,** 1382.
51. Parness, J.; Kingston, D. G. I.; Powell, R. G.; Harracksingh C.; Horwitz, S. B. *Biochem. Biophys. Res. Commun.* **1982,** *105,* 182.
52. Schiff, P. B.; Fant, J.; Hortwitz, S. B. *Nature (London)* **1979,** *277,* 665.
53. Workshop on Taxol and *Taxus*: Future Perspectives. Sponsored by the National Cancer Institute, Bethesda, MD, June 26, 1990.
54. Gevan, R. I.; Greenberg, N. H.; MacDonald, M. M.; Schumacher, A. M.; Abbott, B. J. *Cancer Chemother. Rep.* **1972,** *3,* 1–63.

Etoposide

Albert von Wartburg and Hartmann Stähelin
Sandoz AG

The synthesis of etoposide and teniposide represented the culmination of almost two decades of close research collaboration between medicinal chemists and pharmacologists at Sandoz AG, in Basel, Switzerland. The stepwise conversion of a long-known natural product into a valuable agent in cancer chemotherapy is marked by several interesting chemical and pharmacological problems and finally by the discovery of a new mechanism of action of a group of novel synthetic compounds (*1, 2*).

The investigation began in the early 1950s, when *Podophyllum* species aroused our interest among the indigenous and exotic plants evaluated as potential sources for the preparation of anticancer drugs. In 1942, Kaplan had reported that podophyllin, an alcoholic resinous extract from the rhizomes of *Podophyllum* species, exerted a curative effect in tumorous growths (condylomata acuminata) in humans (*3*). Subsequently King and Sullivan found the mechanism of the antiproliferative effect of podophyllin to be similar to that of colchicine (*4*). In several laboratories these publications prompted extensive chemical and biological research activities on podophyllin.

Based on our experience in the chemistry of natural products we discussed the possibility that *Podophyllum* species might contain cytostatic active glycosides. Our investigation showed that this was indeed the case. From the isolated glucosides the development of new compounds went via

2523–0/93/0349$09.00/0

benzylidene glucosides and the first clinical preparations SP-G and SP-I (an aglucone), to the discovery of 1-epi glucosides, and finally to etoposide.

The path to etoposide was not a straight one but rather a meander, and paved with hopes and enthusiasm alternating with disappointments and disillusions. The final success must be attributed to a combination of research, luck, serendipity, and perseverance, especially in critical phases. As a result, however, nearly 600 derivatives of podophyllotoxin have been produced and biologically tested.

The large detours that were necessary to arrive at the final drug have interest in their own right. We started with natural *Podophyllum* compounds, then proceeded to semisynthetic and fully synthetic derivatives. Only after producing and testing hundreds of derivatives, some (including the aglucones) structurally quite distant from the genuine products, did we arrive at the epipodophyllotoxins teniposide and etoposide, which were again chemically quite close to the starting compounds and differ from natural products only by the content of an aldehyde residue.

In the present account of the development of etoposide, chemical aspects prevail; a more extensive review emphasizing the biological side has been published (*1*).

Early Lignans from *Podophyllum* Species

Podophyllum emodi Wall. (family Berberidaceae), growing in the Himalayan region, and the American *P. peltatum* L. (mandrake, May apple) are medicinal plants with a long history of use. The dried rhizomes, known as podophyllum, were used by the natives of both continents as cathartics, anthelmintics, emetics, and cholagogues. Percolation of the roots with alcohol yields a sirupy extract that, when poured into water, forms a resinous precipitate. The dried and powdered preparation, named podophyllin (resina podophylli), was for many years included as a cathartic in the pharmacopoeia of the United States and several European countries. A comprehensive review of the history, biological effects, and chemical composition of podophyllin has been published by Kelly and Hartwell (*5*).

A first chemical investigation carried out by Podwyssotzki (*6*) and reported in 1880 led to the isolation of podophyllotoxin (Figure 1), the main compound in podophyllin. A second crystallized compound, named picropodophyllin (= picropodophyllotoxin, Figure 1) was later recognized as an artifact, produced from podophyllotoxin by an alkali-catalyzed rearrangement. The correct structure and relative configuration of podophyllotoxin were established by Hartwell and Schrecker (*7*) and subsequently confirmed by Gensler and Gatsonis (*8, 9*) in the first total synthesis. The absolute configuration determined by Schrecker and Hartwell (*10*) using a series of stereochemical correlation reactions was later corroborated in our laboratories by X-ray analysis of 2′-bromopodophyllotoxin (*11*).

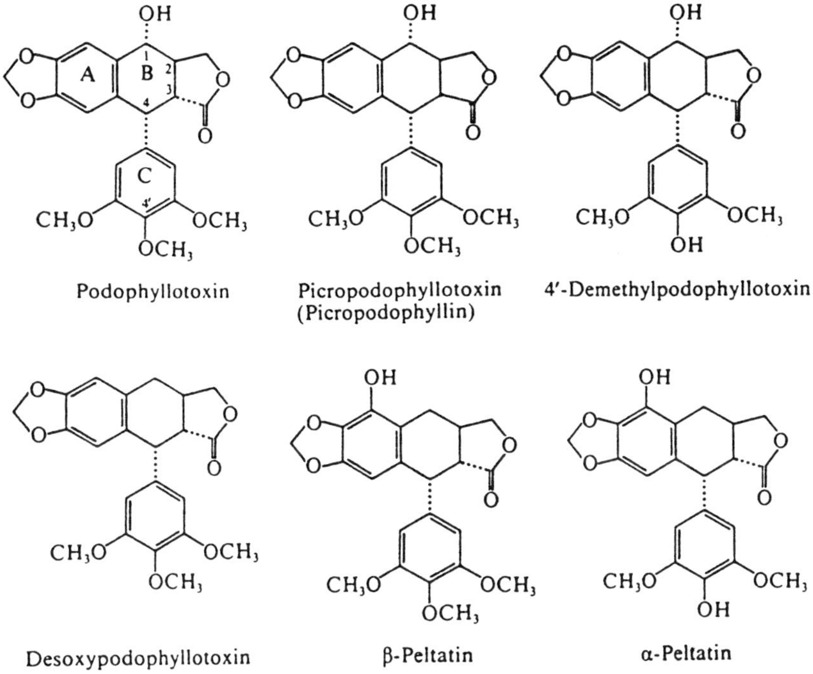

Podophyllotoxin

Picropodophyllotoxin (Picropodophyllin)

4′-Demethylpodophyllotoxin

Desoxypodophyllotoxin

β-Peltatin

α-Peltatin

Figure 1. Podophyllum *lignans known in the early 1950s.*

The new research activities provoked by the discovery of the antineoplastic effects of podophyllin (*3, 4*) resulted in the isolation of the following minor components: 4′-demethylpodophyllotoxin, desoxypodophyllotoxin, and α- and β-peltatin (Figure 1) (*12*). All these compounds are structurally closely related and belong to the class of lignans, which comprises natural products containing the 2,3-dibenzylbutane skeleton.

Another cytostatic-active lignan, named sikkimotoxin (*13*), was obtained from a new *Podophyllum* species, described as *P. sikkimensis* R. Chatterjee and S. K. Mukerjee. Sikkimotoxin differs from podophyllotoxin only in the replacement of the methylenedioxy group by two methoxyls (*14*). However, the proposed structure (or the purity of the natural product) seems doubtful.

The physical data on 6,7-*O*-demethylene-6,7-*O*-dimethylpodophyllotoxin (= sikkimotoxin) synthesized in our laboratories by Emil Schreier deviated significantly from the reported values (*15, 16*). Unfortunately, a direct comparison was not possible. An excellent account of the chemistry of *Podophyllum* and the state of the art in the mid-1950s has been compiled by Hartwell and Schrecker (*12*).

Although podophyllin is still used for the local treatment of condylomata acuminata, attempts to treat malignant tumors in humans by the systemic administration of podophyllotoxin or the peltatins were disappointing, and clinical use was given up.

Genuine Lignan Glucosides from *Podophyllum* Species

In the early 1950s we began investigating *Podophyllum* species with the somewhat vague idea that antimitotic-active lignans could occur in the plant as glycosides that might be lost or degraded in the course of the isolation procedure in use at that time. We also hoped that lignan glycosides might be water-soluble and possibly exert enhanced biological activity. This expectation had been generated by similar phenomena in the field of cardiac glycosides, an area where Sandoz chemists and pharmacologists had years of experience. While our studies were under way, the isolation of picropodophyllotoxin-β-D-glucoside was reported (*17*). However, this compound is nearly inactive biologically, poorly water-soluble, and probably an artifact.

For the separation of genuine glycosides it was important to extract fresh or carefully dried rhizomes of *P. emodi*. Moreover, we applied special isolation procedures, developed in the course of our investigations on *Digitalis* and *Strophanthus* glycosides, to inhibit enzymatic degradation reactions. In this way we succeeded in obtaining a mixture of glucosides that could be separated by partition chromatography into podophyllotoxin-β-D-glucopyranoside (*18, 19*) and 4′-demethylpodophyllotoxin-β-D-glucopyranoside (Figure 2) (*20*). Both glucosides were also found in *P. peltatum*, which in addition contained the β-D-glucopyranosides of α- and β-peltatin (Figure 2) (*21–23*).

As expected, the new lignan glucosides were fairly soluble in water and easily split into D-glucose and their respective aglucones by emulsin and other enzyme preparations containing β-glucosidases. In animal experiments the *Podophyllum* glucosides showed considerably less toxicity than their parent aglucones (Table 1) (*24*). Thus, our hypothesis concerning better

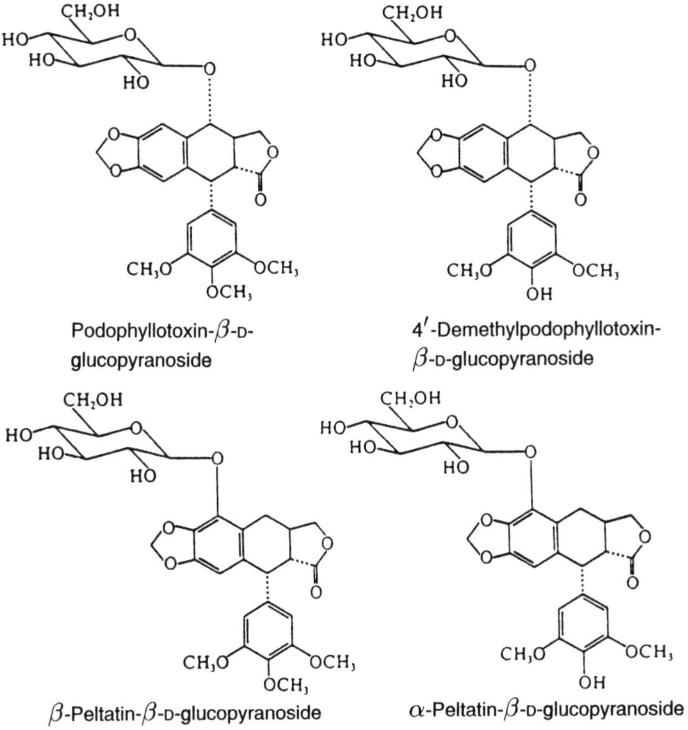

Podophyllotoxin-β-D-glucopyranoside

4′-Demethylpodophyllotoxin-β-D-glucopyranoside

β-Peltatin-β-D-glucopyranoside

α-Peltatin-β-D-glucopyranoside

Figure 2. Major Podophyllum *glucosides of the Indian and American species.*

water solubility and lower toxicity proved correct. However, our enthusiasm quickly dissipated when it was found that the therapeutic potency and efficacy of the glucosides were also substantially diminished. The concentrations required to arrest cell division in metaphase of mitosis in tissue cultures were higher by two to four orders of magnitude for the glucosides (Table 1), and the results in mouse tumors were not encouraging either (*24, 25*). Furthermore, the glucosides showed a short duration of action in vivo. Finally, we had to consider the possibility of unpredictable enzymatic cleavage to more toxic aglucones in the intestinal tract after oral administration of the glucosides.

Nevertheless, we continued to hope that *Podophyllum* glucosides could eventually be converted by chemical modification into useful antineoplastic

Table 1. Cytostatic and Toxic Effects of Podophyllotoxin (P), Its Glucoside (PG), the Benzylidene Derivative of the Glucoside (PBG), and SP-G and SP-I

Compound	*IC_{50} for P-815 Cells (μg/mL)*	*Sarcoma Growth Inhibition (%)*	*L-1210 ILS (%)*	*LD_{50} in Mice (mg/kg)*
P	0.005	29	35	35
PG	6	40	7	297
PBG	3	n.s.	5	240
SP-G	0.5	47	65	214
SP-I	0.5	46	17	283

NOTE: IC_{50} is the concentration inhibiting by 50% the proliferation of P-815 mastocytoma cells in vitro. Sarcoma growth inhibition (by weight) was determined for sarcomas 37 and 180 by 8 days of treatment with maximal tolerated daily doses. L-1210 ILS is the increase in life span of mice inoculated subcutaneously with 10^6 L-1210 cells by daily treatment in doses resulting in maximal ILS. LD_{50} is the 50% lethal dose on single parenteral administration. n.s. denotes results not significant.

agents. We therefore embarked on an extensive derivation program that included glucosides as well as aglucones.

Derivatives of *Podophyllum* Glucosides

In a first attempt to obtain more information on structure–activity relationships, we synthesized several higher alkyl homologues of the genuine glucosides. The glucosides of 4′-demethylpodophyllotoxin and α-peltatin, both containing a free phenolic hydroxyl group, reacted easily with diazoalkanes to yield the corresponding 4′-O-alkyl ethers (*25*). Using ^{14}C-labeled diazomethane we could prepare radioactive ^{14}C-podophyllotoxin glucoside, which proved valuable for pharmacokinetic studies (*26*). The 4′-O-ethyl, -propyl, -butyl, and -isoamyl homologues of podophyllotoxin glucoside displayed in fibroblast cultures a cytostatic effect of the same order of magnitude as that of the genuine glucoside. In Ehrlich's murine ascites tumor the alkyl homologues were found to have a better oncostatic activity than the respective natural 4′-methoxy derivative, but they showed pronounced nonspecific cytotoxicity for normal cells. In short, no therapeutic advantage over the genuine glucosides could be gained with these 4′-O-alkyl homologues (*24, 25*).

In a parallel series of experiments we prepared acyl derivatives of the glucosides in order to block enzymatic cleavage and to enhance absorption.

Derivatives with an esterified glucose were only slightly soluble in water and, at least in vitro, were resistant to glucosidases. Their antimitotic activity in fibroblast cultures and in Ehrlich's ascites tumors in mice turned out to be rather inferior to that of the free glucosides. Furthermore, a decrease in potency was observed with increasing chain length of the fatty acid residue (*24, 25*).

The uninspiring pharmacological results obtained with peracylated derivatives showed clearly that total blocking of the sugar hydroxyl groups did not improve the therapeutic value of these products. A possible chemical alternative consisted of a selective partial substitution in the glucose unit. This idea could be realized by condensing podophyllotoxin glucoside with benzaldehyde in the presence of Lewis acids to yield the cyclic acetal podophyllotoxin benzylidene glucoside (R = phenyl in Figure 3) (*25*). As expected, the reaction involved only the hydroxyl groups at C-4 and C-6 of the hexose unit. The aldehyde residue occupies the energetically favorable equatorial position, as indicated by a characteristic shift of the axial proton in the NMR spectrum (*27*). An isomer with an axial bonded aldehyde residue is also formed, but only in minimal amounts.

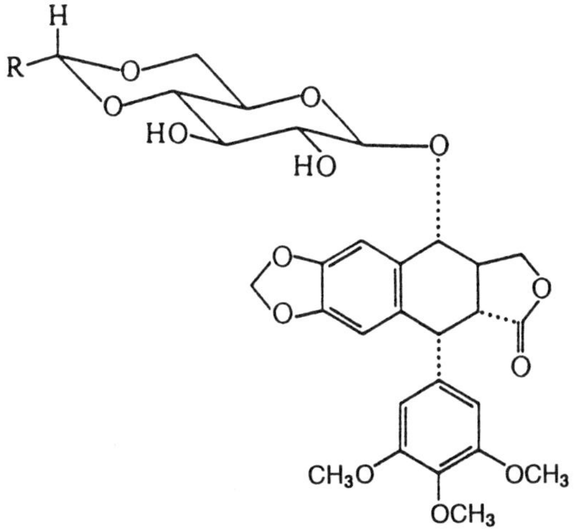

Figure 3. Cyclic acetals of podophyllotoxin glucoside.

The acid-catalyzed condensation reaction turned out to be a versatile method and, with the use of various aldehydes, allowed the production of a large series of cyclic acetals with glucosides of both the podophyllotoxin and the peltatin type. Aldehyde condensation products are only slightly soluble in water and are resistant to enzymatic degradation by β-glucosidases. On the other hand, the intact glucoside can easily be regenerated by cleavage of the aldehyde residue with diluted acids (*25*).

The specific antimitotic activity in vitro of some cyclic acetals corresponds approximately to that of podophyllotoxin glucoside (Table 1), but they display remarkably few toxic side effects. The extensively tested benzylidene (R = phenyl) and anisylidene (R = *p*-methoxyphenyl) derivatives (Figure 3) were very well absorbed by the intestinal tract, and their half-life in the body was markedly increased over that of the genuine glucoside. In chick embryo fibroblast cultures, both compounds, like the free glucosides, clearly demonstrated spindle-poison activity, with large increases in the mitotic index (*1, 25*).

Certain advantageous properties of podophyllotoxin benzylidene glucoside, in particular chemical stability, good absorption by the gastrointestinal tract, and positive effects in a few selected cancer patients (*28*), encouraged us to look more deeply into this type of derivative (*29*). For large-scale production of the benzylidene derivative, a more convenient and cheaper procedure had to be developed. By using, instead of the pure podophyllotoxin glucoside, the extracted mixture containing all glycosidic natural *Podophyllum* compounds for the condensation reaction with benzyldehyde, we obtained a less pure preparation that consisted of about 80% podophyllotoxin benzylidene glucoside, about 10% 4′-demethylpodophyllotoxin benzylidene glucoside, small amounts of free glucosides, and some incompletely identified by-products. This preparation was given the code designation SP-G 827, or just SP-G. The poor solubility of SP-G in aqueous media presented some galenical problems, and powerful solvents such as polysorbate 80 (Tween 80) and dimethyl sulfoxide had to be used to obtain clear solutions with optimal and reproducible activity in vitro and in vivo (including oral administration). The inhibition of cell multiplication in vitro and of tumor growth in vivo displayed by SP-G is superior to the cytostatic effects of the pure main component, podophyllotoxin benzylidene glucoside (Table 1). A summary of the pharmacological and clinical results with SP-G is given subsequently in the section on Proresid.

Nitrogen-Containing Podophyllotoxin Derivatives

Among the simple acyl derivatives of podophyllotoxin, its ester with carbamic acid is one of the most active compounds in vitro; it also displayed considerable tumor inhibition in the mouse sarcomas 37 and 180. The chemically quite stable podophyllotoxin carbamate (Figure 4) also showed a favorable ratio between cytostatic potency in vitro and toxicity in animals (*1*). Inhibition of tumor growth is equally substantial in the solid rat tumors tested (Walker carcinosarcoma and Yoshida sarcoma). Another interesting feature is the high binding to human serum proteins (*30*). In 1967 podophyllotoxin carbamate was tested in humans under the code name VB 74. The results, in a limited number of patients, were not sufficiently encouraging, and the compound was dropped.

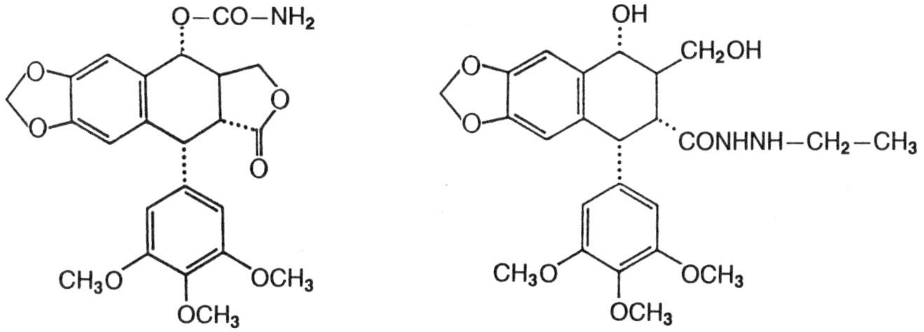

Figure 4. Podophyllotoxin carbamate (VB 74) (left), podophyllinic acid ethyl hydrazide (SP-1) (right).

Another series of nitrogen-containing podophyllotoxin derivatives is represented by amines containing the lignan ring system. Compounds of this type may be regarded as structural intermediates between the antimitotic-active natural products podophyllotoxin and colchicine. In an attempt to synthesize amines of this kind, Rutschmann and Renz at Sandoz achieved a stereoselective cleavage of the lactone ring by reacting podophyllotoxin with hydrazine in buffered solution (*31*). The main reaction product was identified as podophyllinic acid hydrazide, displaying the intact trans configuration of the substituents at C-2 and C-3. The isomeric 2,3-*cis*-picropodophyllinic acid hydrazide, described earlier by Borsche and Niemann (*32*), was formed as a by-product, generated by base-catalyzed epimerization. Hydrazine cleavage turned out to be a general method that could be applied to glucosides as well as to peltatins and other lignans (*25*).

Subsequently more than 200 *Podophyllum* compounds with an open lactone ring were synthesized, of the 2,3-trans as well as of the 2,3-cis (picro) series. All derivatives were tested in cell cultures, and some were also tested in vivo. Unsubstituted hydrazides of the lignan type and their acyl derivatives, as well as acyl hydrazones (obtained by reaction of hydrazides with carbonyl reagents), displayed, in general, low to medium cytostatic activity. The differences between 2,3-trans and 2,3-cis forms were only minor in most cases. An increased antiproliferative effect was found with several substituted hydrazides, especially with the dodecyl hydrazide and the ethyl hydrazide of podophyllinic acid (Figure 4) (*1*). The ethyl hydrazide (code name SP-I, later called Proresid intravenous) was more active in solid mouse tumors and less problematic in respect to stability and galenical aspects; the compound could be dissolved in aqueous media with the help of alcohol, thus allowing the preparation of ampules for intravenous injection and infusion. After appropriate toxicological assays, SP-I was tested clinically, often in combination with SP-G (which was later called Proresid oral), and finally commercialized in 1963, together with SP-G.

A critical examination of the efficacy of numerous *Podophyllum* compounds with an open lactone ring showed a not wholly attractive picture. Although considerable effort had been invested, no major therapeutic progress could be achieved beyond SP-I. Our eagerness to pursue this research line, therefore, gradually faded.

SP-G, SP-I (Proresid)

In the latter half of the 1950s, the preliminary biological results obtained with SP-G (see the earlier section "Derivatives of *Podophyllum* Glycosides") looked quite promising. This preparation was obviously superior to the pure podophyllotoxin benzylidene glucoside and did not show the main side effect of the few cytostatic compounds then in use, namely, bone marrow depression when the compound was given in doses that inhibited tumor growth in experimental animals. It was therefore tested, together with SP-I, more extensively (for details see ref. 29) in order to be evaluated in humans.

In chick embryo fibroblast cultures, a full cytostatic effect—that is, no anaphases or telophases of mitosis visible—was obtained with a concentration of SP-G of 1 μg/mL, and about the same concentration completely prevented proliferation of P-815 mouse mastocytoma cells in vitro (Table 1). The mechanism of action of the mixture SP-G, as deduced from results in fibroblast cultures, is the same as that of the pure benzylidene glucoside: it is a spindle poison that arrests cells in metaphase of mitosis (as does the prototype of spindle poisons, colchicine, and as do podophyllotoxin and the vinca alkaloids), thus producing an accumulation of c- or ball metaphases with clumped chromosomes. This effect is due to the inhibition of the assembly of tubulin into the microtubules that form the mitotic spindle. At much higher concentrations (20 μg/mL), SP-G begins to slow down cell entry into mitosis (*29*), probably owing to its content of antileukemia factor (see the section "The Antileukemia Factor of SP-G"). The effect of SP-I in cell and tissue cultures is quantitatively and qualitatively very similar to that of SP-G.

In solid experimental mouse and rat sarcomas, both preparations, SP-G and SP-I, exhibited a clear-cut inhibitory effect. With regard to prolongation of survival time in mice inoculated subcutaneously with the lymphatic leukemia L-1210 cell line, however, only SP-G showed good activity—an increase in life span of over 100% in some experiments—whereas SP-I was but marginally active. Both drugs had comparatively few depressing effects on the bone marrow (*29*).

Since 1961 (*33*), the clinical effects of SP-G and SP-I have been described in several hundred publications, and after appropriate toxicological evaluation, both preparations were introduced into the market in 1963; they later received the commercial designation Proresid. A considerable percentage of the treated patients showed measurable tumor remission and, particularly, improvement in their general condition. However, long-term results with both forms of Proresid were not as good as those achieved with some of the newer and more aggressive anticancer drugs being developed at that time. The clinical use of Proresid therefore gradually declined (*1*).

Neopodophyllotoxin and Podophyllinic Acid

The positive pharmacological results with podophyllinic acid ethylhydrazide (SP-I) clearly indicated that the five-membered lactone ring is not an indispensable structural element for the antimitotic effect of *Podophyllum* lignans, as had been earlier assumed. Among the derivatives with an open lactone ring, one compound was still lacking, namely, podophyllinic acid, the prototype of this series. We considered the unsubstituted 2,3-*trans*-hydroxy acid, unknown at that time, as an informative substance with regard to structure–activity relationships, and its synthesis as a chemical challenge. All attempts to date in several laboratories to produce the (2,3-*trans*)podophyllinic acid by alkaline cleavage of the lactone group of podophyllotoxin had failed; epimerization of the highly strained lactone ring to the more stable 2,3-cis configuration and hence formation of (2,3-*cis*)picropodophyllinic acid proved to be the kinetically predominant reaction (Figure 5). For the same reason conversion of podophyllinic acid hydrazides or podophyllinic acid methyl ester into the parent hydroxy acid could not be realized (*31*).

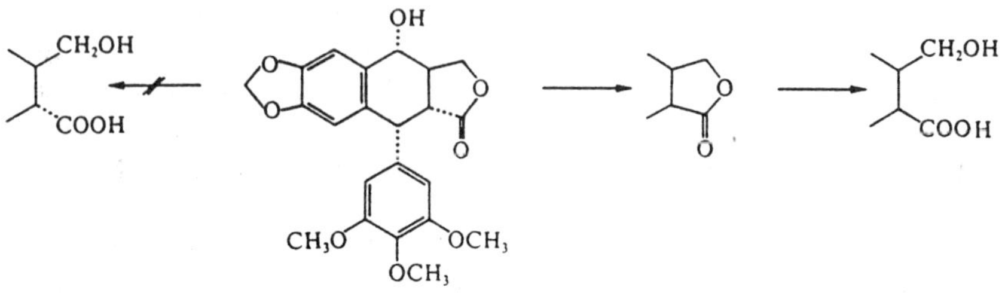

Podophyllinic acid　Podophyllotoxin　Picropodophyllotoxin　Picropodophyllinic acid

Figure 5. Reaction of podophyllotoxin with base yielding picropodophyllinic acid by epimerization.

Our new approach emerged from prior experiments with desoxypodophyllotoxin. We could convert this lignan by a transesterification reaction in methanol with perchloric acid into desoxypodophyllinic acid methyl ester (*34*). Application of this reaction procedure to podophyllotoxin in methanol using a Lewis acid ($ZnCl_2$) as catalyst furnished a mixture of unchanged podophyllotoxin, podophyllinic acid methyl ester, and a new 1,3-lactone named neopodophyllotoxin (Figure 6). The acid-catalyzed methanolysis proceeded in an equilibrium reaction with preservation of the vital configuration at C-2 and C-3. Neopodophyllotoxin turned out to be a very useful intermediate. Due to steric hindrance, base-catalyzed epimerization of neopodophyllotoxin did not occur; hence ring opening with 2n NaOH led smoothly to the desired (2,3-*trans*)podophyllinic acid (Figure 6) (*35, 36*). The structure and configuration of the free hydroxy acid were proved by chemical correlation reactions, spectroscopic evidence, and comparison of molecular rotations.

To our great disappointment, podophyllinic acid (and neopodophyllotoxin) exhibited a low cytostatic potency (*1*). Its methyl ester is almost 100 times more active and showed only low toxicity. However, the methyl ester did not increase the survival time of mice inoculated with leukemia L-1210 cells and exhibited a weak and variable inhibitory effect on the growth of mouse sarcoma 37 cells (*1*).

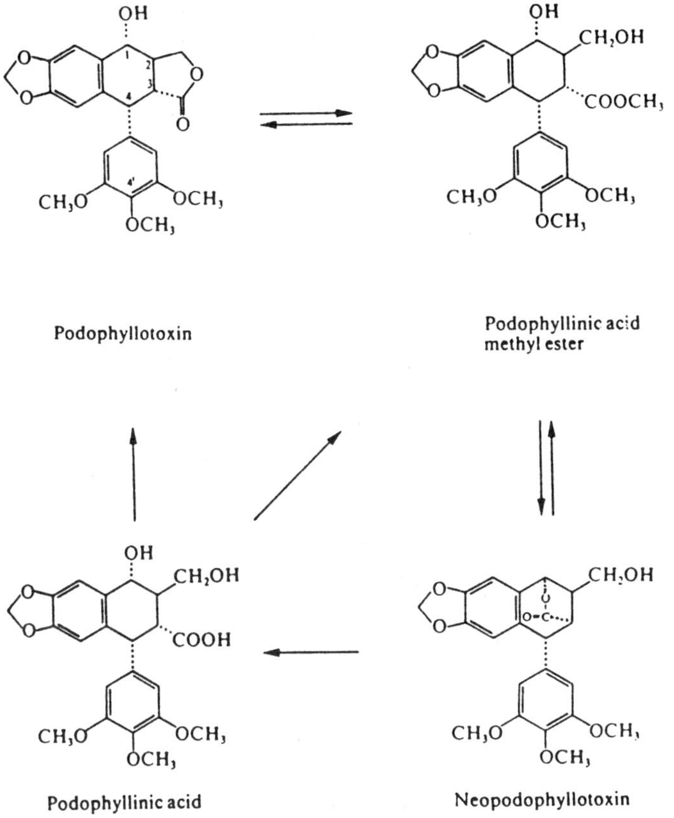

*Figure 6. Synthesis of (2,3-*trans*)podophyllinic acid via neopodophyllotoxin.*

Interlude

With the isolation of the genuine *Podophyllum* glucosides, the synthesis of their aldehyde condensation products, and the ongoing clinical evaluation of SP-G and SP-I, we believed we had realized a breakthrough in the development of useful anticancer drugs of the lignan type. The biological results obtained with these compounds confirmed our hypothesis that chemical modification of the podophyllotoxin molecule could eventually lead to improved chemotherapeutics for the treatment of malignancies. With this conviction we pursued our chemical program on a broad basis. Our main lines of research involved new derivatives of desoxypodophyllotoxin, compounds with an open lactone ring, derivatives of the apopicropodophyllins, and substances of the podophyllol type, as well as miscellaneous lignan derivatives (*1*). To study the influence of methoxy groups on biological activity, a series of demethylated and desmethoxy compounds were synthesized. To reveal the role of the methylene dioxy group for the antimitotic effect, Emil Schreier achieved the synthesis of 6,7-O-demethylene and 6,7-O-demethylene-6,7-O-dimethyl analogues of podophyllotoxin, including sikkimotoxin (*15, 16*) (see the section "Early Lignans from *Podophyllum* Species"). In the course of our research activities, several hundred derivatives of the lignan type were produced and biologically tested. Many of these compounds exhibited high cytostatic potency in vitro. Some derivatives displayed significant inhibition of experimental tumors but failed on in-depth evaluation. A summary of the chemistry and pharmacology of the most important preparations has been published (*1*).

By the early 1960s we had reached a very labile phase in our work with no improved follow-up preparation in the pipeline and the rational design of new potential leads becoming more and more elusive. In addition, many hypotheses on structure–activity relationships turned out to be unreliable and led us into blind alleys. At this critical point a new impulse emerged from careful pharmacological analysis of SP-G.

The Antileukemia Factor of SP-G

In early 1962, analysis of SP-G lots by means of a novel cell culture assay (*37*) revealed that the cytostatic potency of SP-G could not be accounted for by the activities of the known components of the mixture. In addition, we observed that SP-G led to significant increases in survival time in mice inoculated with the lymphocytic leukemia L-1210 cell line, an effect not seen with the previously isolated benzylidene glucosides. Because initially there was some reluctance to accept that a significant part of the effects of SP-G could not be explained by the activity of its known single components, a major effort was undertaken by Sandoz biologists to determine whether the known constituents of SP-G would potentiate each other's effects. Potentiation, however, did not seem to occur, and the search for the "missing components," presumably highly active but present in very small quantities, was taken up with more confidence. Besides SP-G, crude, unmodified extracts of *Podophyllum* species were also included in these analyses.

Systematic and meticulous fractionation of SP-G, guided by thin-layer chromatography and biological assays, soon yielded a new component that

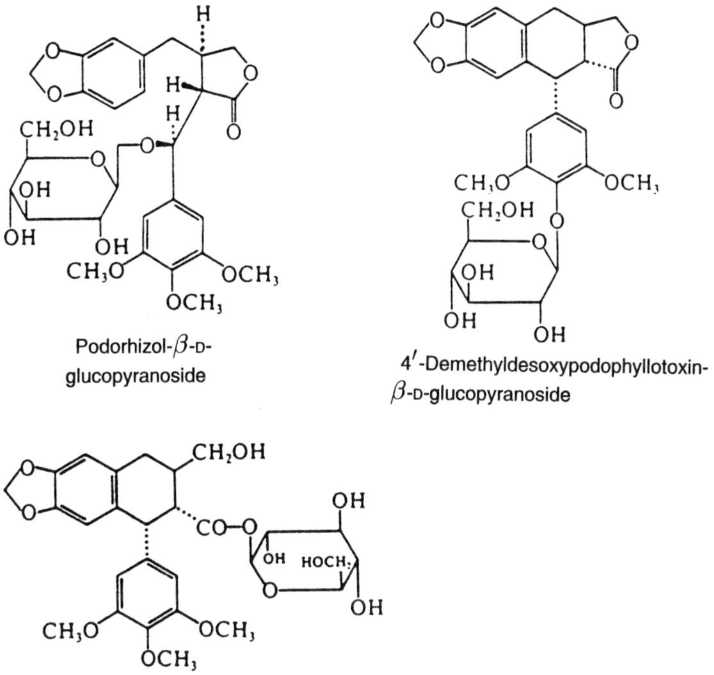

Podorhizol-β-D-glucopyranoside

4′-Demethyldesoxypodophyllotoxin-β-D-glucopyranoside

Desoxypodophyllinic acid-1β-D-glucopyranosyl ester

Figure 7. New glucose-containing natural lignans from Podophyllum *species.*

was identified as the anomer of podophyllotoxin benzylidene glucoside and that displayed the axial bond of the aldehyde residue. Subsequently two additional novel components were isolated, representing the benzylidene derivatives of podorhizol glucoside and 4′-demethyldesoxypodophyllotoxin glucoside. The parent lignans have also been encountered as genuine glucosides (named lignan F and H, respectively) of *P. emodi* (Figure 7) (*38, 39*). All these products inhibited cell multiplication in vitro, a result especially remarkable for podorhizol glucoside, a derivative of a dibenzylbutyrolactone (*38*). However, their antiproliferative activity was too low to account for the potency of SP-G. Eventually, a fourth new substance was separated that exerted a powerful cytostatic effect in cell cultures but, to our disappointment, lacked activity against leukemia L-1210. Its chemical structure was established as the benzylidene derivative of desoxypodophyllinic acid glucopyranose ester. The parent free glucose ester (named lignan J) was also found in both *Podophyllum* species as a natural product (Figure 7) (*34*). Lignan J is easily split into glucose and a secondary product, identified as desoxypodophyllotoxin. Cleavage of the ester bond and ring closure to desoxypodophyllotoxin proceed under very mild conditions (e.g., with diluted acids, on silica gel columns, and even in cell culture medium). The formation of the highly active desoxypodophyllotoxin by hydrolysis explains the powerful antiproliferative effect of lignan J.

The unusual and striking structural features of the genuine lignans F, H, and J (Figure 7) suggested that they might play a role in the course of the *Podophyllum* lignan biosynthesis.

In addition to lignan compounds, we encountered several yellow pigments such as the flavonoid astragalin (kaempferol-3-glucoside), which occurs in both *Podophyllum* species (*40*). These companions are devoid of any cytostatic activity.

Finally, in late 1964, after more than 2 years of endeavors to trace down the antileukemia factor of SP-G, our perseverance was rewarded. We isolated a further novel compound (named benzylidene lignan P) that not only showed a powerful inhibition of cell proliferation in vitro but also considerably prolonged survival time in L-1210 leukemic mice at low doses. Benzylidene lignan P, which seemed to be the long-sought antileukemia factor of SP-G, exhibited several surprising and unique features: an interesting chemical structure (closely related to compounds we had already synthesized and tested), high antiproliferative potency, and a new mechanism of action. Chemical investigations established its structure as a benzylidene derivative of 4′-demethylepipodophyllotoxin-β-D-glucopyranoside (Figure 8) (C. Keller, M. Kuhn, A. von Wartburg, H. Stähelin, unpublished data). Important structural elements of the new lignan consist in the free phenolic OH group at C-4′ and, more striking, in the 1-epi configuration (that is, the OH group bearing the glucose group occupies the inverse configuration from that in podophyllotoxin).

The very close chemical relationship of the antileukemia factor to other constituents of SP-G, the scarce content, and the time-consuming biological assays explain the difficulties encountered in the course of the isolation procedure. Today the separation proceeds much faster with the use of high-performance liquid chromatography. The parent free glucoside of the antileukemia factor, 4′-demethylepipodophyllotoxin glucoside (see Figure 10), was also traceable in crude extracts of *P. emodi*; however, only

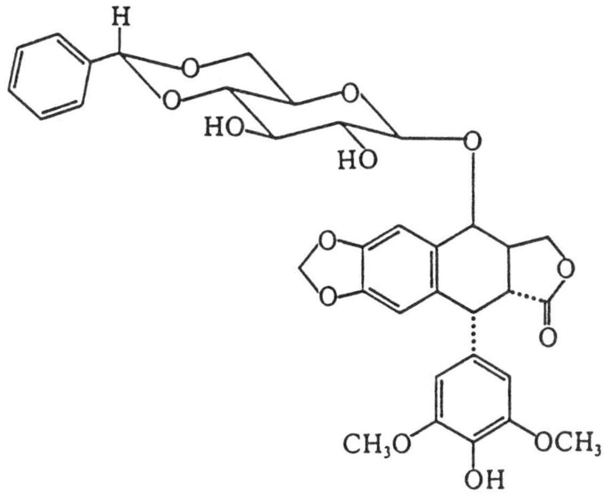

Figure 8. 4′-Demethylepipodophyllotoxin benzylidene glucoside, the antileukemia factor of SP-G.

minimal amounts could be isolated (C. Keller, M. Kuhn, A. von Wartburg, unpublished data).

Pharmacological assays using cultures of chick embryo fibroblasts immediately showed that benzylidene lignan P had a different mechanism of action. Although the high cytostatic potency and good anti-L-1210 activity had been expected, based on the biological results with SP-G compared with those of its components, the new mechanism was surprising. A look through the microscope at stained fibroblast cultures (see the section "Biological Test Systems and Their Role") treated with benzylidene lignan P immediately revealed that we were dealing with a compound that inhibited cell proliferation in another way than the *Podophyllum* drugs known so far. There was no accumulation of arrested metaphases with clumped chromosomes, the picture characteristic for spindle poisons and easy to recognize. On the contrary, the number of mitoses was reduced to almost zero after a 6-h treatment with benzylidene lignan P. Most of the cells present (the resting cells) looked normal. No prophases or early metaphases were visible, just a few cells in the process of disintegration. Because of this unusual aspect, there were some doubts among the biologists as to whether benzylidene lignan P really was a compound of the *Podophyllum* class, but the chemists soon identified it as a lignan and elucidated the complete structure.

This highly interesting and surprising discovery regarding the mode of action (which was then investigated in greater depth for VM and VP) and the observed antitumor effect prompted further investigations in a joint pharmacological and chemical effort. The poor availability of the antileukemia factor and its parent glucoside from natural sources was a barrier that had to be surmounted before we could proceed to more extensive biological testing and the preparation of new derivatives of the 1-epi series. Therefore, our next task was to elaborate a synthesis of 4′-demethylepipodophyllotoxin glucoside adaptable to large-scale production.

Synthesis of 4′-Demethylepipodophyllotoxin Glucoside

The elaboration of a sterospecific synthesis of 4′-demethylepipodophyllotoxin glucoside (lignan P) was a fascinating chemical challenge, owing to the stereochemical peculiarities and the sensitivity of the aglucone to acids and bases. Numerous preliminary experiments were necessary before the final synthetic route could be established. In the course of our studies, additional results were obtained, such as the first total synthesis of podophyllotoxin-β-D-glucopyranoside (*41*) and epipodophyllotoxin-β-D-glucopyranoside (*42*).

The synthesis of 4′-demethylepipodophyllotoxin glucoside presented three major problems: (1) the preparation of large amounts of 4′-demethylepipodophyllotoxin; (2) the development of a new stereospecific glycosidation method leading to β-glycosides; and (3) the removal of protecting groups, because of the acid and base sensitivity of the products.

The previously unknown 4′-demethylepipodophyllotoxin was easily obtained by epimerization of 4′-demethylpodophyllotoxin via the 1-chloride (Figure 9). Unfortunately, the content of the latter lignan in plant extracts is rather poor. An improved method consisted of a selective ether cleavage at C-4′ of the readily accessible podophyllotoxin, using HBr at low temperature. Subsequent hydrolysis of the intermediate 1-bromo derivative led smoothly to 4′-demethylepipodophyllotoxin (Figure 9) (*43*).

The next step, the development of a useful glycosidation reaction, was far more intricate because classical methods were not applicable in the 1-epi series. Thus, reaction of epipodophyllotoxin with tetra-O-acetyl-α-glucopyranosyl bromide and a suitable catalyst (HgO, $Hg(CN)_2$, or ZnO), a procedure that proved successful in the synthesis of podophyllotoxin glucoside (*41*), gave only poor yields of the expected tetraacetyl epiglucoside and furnished mainly degradation products (*42*). This was also true when

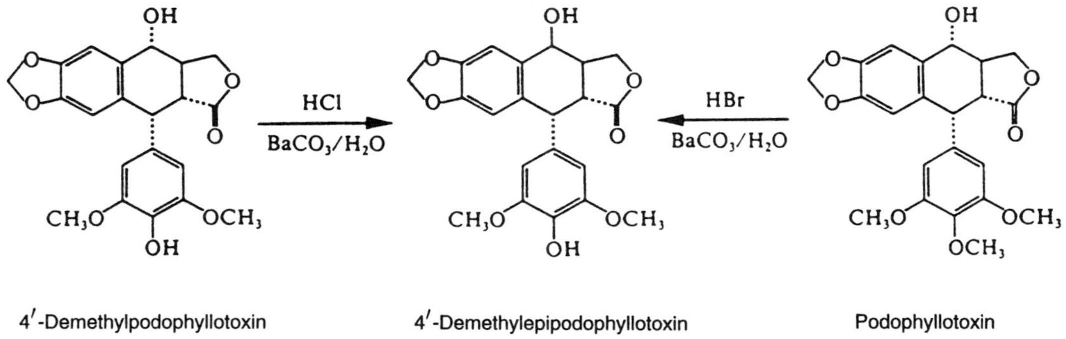

4'-Demethylpodophyllotoxin

HCl, $BaCO_3/H_2O$

4'-Demethylepipodophyllotoxin

HBr, $BaCO_3/H_2O$

Podophyllotoxin

Figure 9. Preparation of 4'-demethylepipodophyllotoxin.

other reactive sugar reagents were applied, such as tri-*O*-acetyl-1,2-ethylorthoacetyl-α-D-glucopyranose or penta-*O*-acetyl-β-D-glucopyranose. The failure of the known glycosidation methods can be explained by the specific stereochemical situation at C-1 in the epi series. Protonation or complex formation with the catalyst converts the 1-epi hydroxyl into a highly reactive leaving group, thus opening ways to side reactions. Based on these findings, we treated epipodophyllotoxin with 2,3,4,6-tetra-*O*-acetyl-β-D-glucose at low temperature in the presence of BF_3-etherate. The resulting main product was characterized as the desired tetraacetate of epipodophyllotoxin β-D-glucoside (*42*). Podophyllotoxin was converted under the same conditions to the identical acetylated 1-epi glucoside. Evidently, inversion of the configuration at C-1 had occurred. Presumably the reaction proceeds through a common carbonium ion at C-1 of the aglucone generated by BF_3, with subsequent substitution by the pyranose from the less hindered side to yield exclusively the 1-epi derivative. Another consequence of the proposed mechanism concerns the stereochemistry of the glycosidic linkage, which is determined by the beta configuration of the glycosidating pyranose. The new method is not restricted to glucose; other hexoses, for example β-D-galactose, are also suitable sugar reagents (*42*).

The outlined glycosidation procedure was then applied to 4′-demethylpodophyllotoxin and its 1-epi isomer (Figure 10). Both aglucones, protected as 4′-benzyloxycarbonyl derivatives, yielded an identical product characterized as tetra-*O*-acetyl-4′-benzyl-oxycarbonyl-4′-demethylepipodophyllotoxin-β-D-glucopyranoside (*42, 44*).

The last problem to be solved was removal of the protecting groups. Removal was accomplished by zinc acetate catalyzed methanolysis of the tetraacetyl glucosides. By means of this transesterification reaction, podophyllotoxin glucoside and its 1-epimer could be synthesized from their corresponding acetyl derivatives (*41, 42*). In the case of the 4′-demethyl compound, methanolysis was followed by hydrogenolysis of the benzyloxycarbonyl group at C-4′ using H_2/Pd to yield the desired 4′-demethylepipodophyllotoxin-β-D-glucopyranoside (lignan P) (Figure 10). The structure of the synthetic compound, especially the vital 1-epi configuration and the β-glycosidic linkage of the glucose unit, was confirmed chemically and spectroscopically (*42, 44*).

Compared to podophyllotoxin glucoside, the glucosides of the 1-epi series did not display striking biological peculiarities with regard to either the quality or the potency of their antimitotic activity. This finding is in marked contrast to the biological effects exerted by their condensation products with carbonyl compounds (*1*).

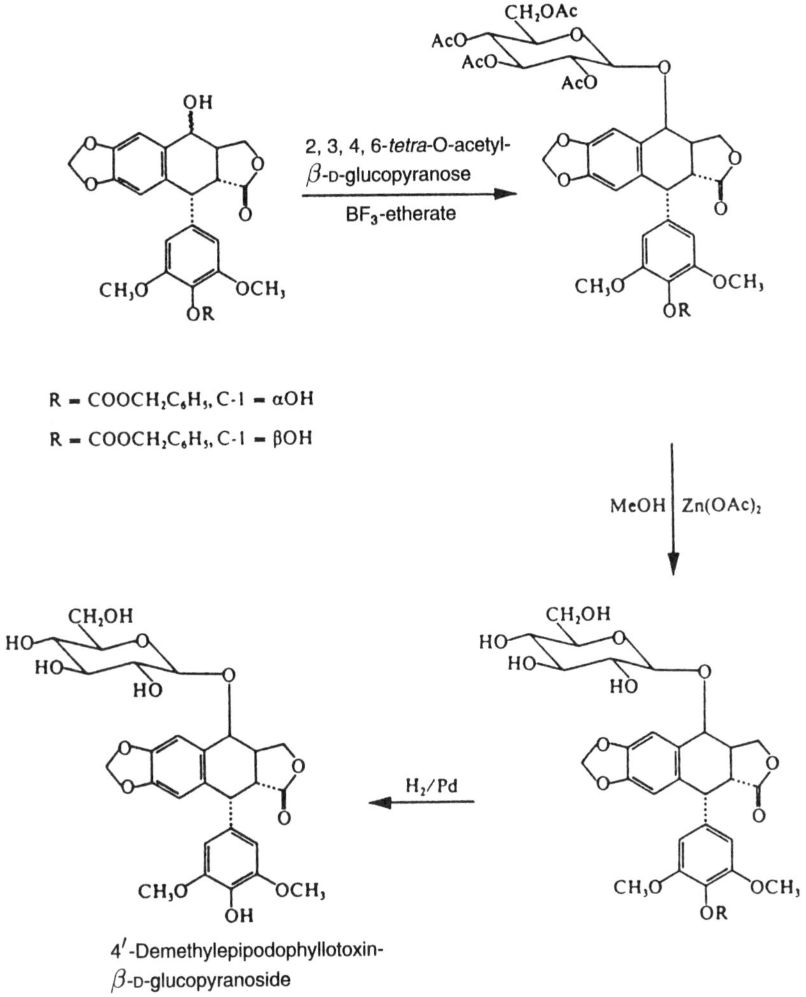

Figure 10. Synthesis of 4′-demethylepipodophyllotoxin-β-D-glucopyranoside.

Cyclic Acetals of 4′-Demethylepipodophyllotoxin Glucoside

Synthetic access to *Podophyllum* glucosides of the 1-epi series now allowed the preparation of numerous derivatives. Because of the outstanding pharmacological effects exhibited by benzylidene lignan P, the antileukemia factor of SP-G, we concentrated on condensation products with aldehydes and ketones. The synthesis of these products was generally accomplished by acid-catalyzed reaction of 1-epi glucosides with appropriate carbonyl compounds or their respective acetals and ketals. The transacetalization reaction was especially indicated for simple aliphatic carbonyl reagents (*27*). The formation of the two isomers, differing from one another in the configuration at the newly introduced asymmetric C-atom and the almost exclusive domination of the isomer with an equatorial bond of the aldehyde residue, was discussed earlier.

The increase in biological activity achieved by aldehyde condensation to epipodophyllotoxin glucoside is less dramatic than that achieved by acetal or ketal formation with 4′-demethylepipodophyllotoxin glucoside (analogues of benzylidene lignan P). The latter compounds not only exhibit a higher cytostatic potency in vitro but also produce a significant increase in survival time in the leukemia L-1210 assay (Table 2). The biological activity, especially the effect in mouse leukemia L-1210, depends on the aldehyde type, and even more on the nature of the sugar moiety. Cyclic acetals of 4′-demethylepipodophyllotoxin galactoside did not give comparable results (*1*, *27*).

Among the more than 60 analogues produced, the condensation products of 4′-demethylepipodophyllotoxin glucoside with 2-thiophene aldehyde (teniposide) and with acetaldehyde (etoposide) (Figure 11) were selected for extensive pharmacological evaluation and clinical trials. The selection was based to a minor degree on potency in vitro, but mainly on efficacy in vivo (particularly leukemia L-1210, where etoposide is one of the most effective drugs; it produces large increases in survival time and cures a large percentage of animals, depending on the treatment schedule, the size and site of the tumor inoculum, the route of drug administration, and other factors). Another important reason for selecting etoposide was its effectiveness after oral administration, which made it particularly suitable as a complement to teniposide (which is almost inactive when given orally).

The structural requirements for the new mechanism have been reported elsewhere (*45*). Starting from podophyllotoxin, four alterations are essential: demethylation in the 4′ position, 1-epi configuration, the β-linked D-glucose moiety in position 1, and the aldehyde residue at the glucose. Details of these

Table 2. Biological Activity of 4′-Demethylpodophyllotoxin Benzylidene Glucoside (DPBG), Epipodophyllotoxin Benzylidene Glucoside (EPBG), 4′-Demethylepipodophyllotoxin Glucoside (DEPG), 4′-Demethylepipodophyllotoxin Benzylidene Glucoside (DEPBG), VM 26, and VP 16

Compound	*IC_{50} for P-815 Cells (µM)*	*Relative Mitotic Index*	*L-1210 ILS (%)*
DPBG	1.2	3.7	29
EPBG	0.50	10	60
DEPG	7.8	26	34
DEPBG	0.010	0.1	91
VM 26	0.0076	0.070	121
VP 16	0.078	0	167

NOTE: VM is teniposide, VP is etoposide. IC_{50} is the concentration inhibiting by 50% the proliferation of P-815 mastocytoma cells in vitro. Relative mitotic index indicates the factor by which the number of mitoses in fibroblast cultures is increased over control cultures after 6 h of incubation with a fully active drug concentration. L-1210 ILS is the increase in life span observed for mice inoculated subcutaneously with 10^6 L-1210 cells by daily treatment in doses resulting in maximal ILS.
The treatment schedule used here for leukemia L-1210 is not optimal for DEPBG, VM, and VP; much higher ILS and many cures are obtained with twice-weekly treatment.

structure-activity relationships have been discussed (*1*) (see also Tables 1 and 2).

Further Development of Teniposide and Etoposide

The biological effects of teniposide (VM) and etoposide (VP) were found to be qualitatively about the same as those of benzylidene lignan P, particularly as to mechanism of action, but quantitatively (i.e., in regard to potency) etoposide was less active in vitro than teniposide by a factor of about 10 (Table 2). The cytostatic (in vitro) and oncostatic (in vivo) effects of both drugs as determined in our laboratories have been described (*46–48*). These results have been corroborated and extended by many other investigators (see ref. 1). The results of toxicological studies were unremarkable insofar as the (histological and other) alterations found after prolonged treatment could all be explained by the cytostatic activity of the drugs.

Further studies on the effects of the epipodophyllotoxins (as the group of aldehyde derivatives of 4′-demethylepipodophyllotoxin glucoside, including

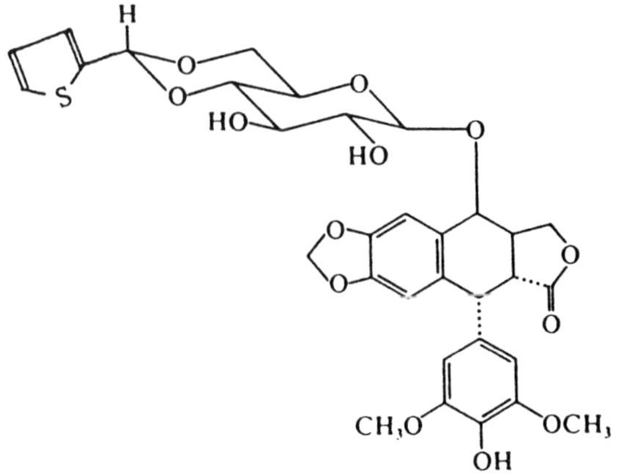

Teniposide

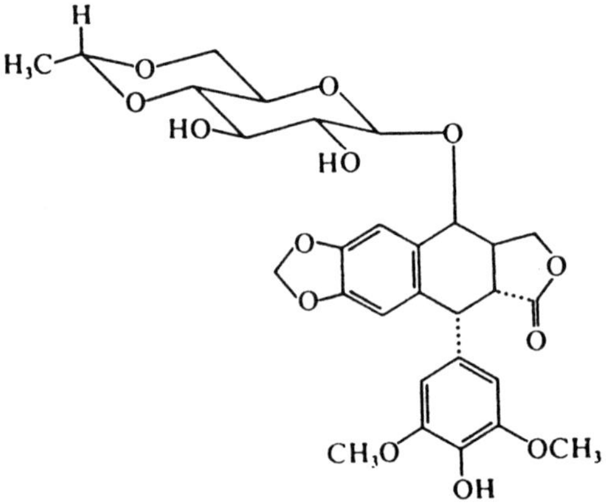

Etoposide

Figure 11. Teniposide (Vumon, Vehem), VM 26, NSC 122829, Sandoz code no. 15-426 (top); etoposide (Vepesid), VP 16, NSC 141540, Sandoz code no. 16-213 (bottom).

teniposide and etoposide, is often called) were performed mainly for teniposide and etoposide. An interdependence of the two prominent biological effects of the aldehyde derivatives of lignan P—anti-L-1210 activity and a new mechanism of action—had to be presumed. This presumption proved correct, insofar as other podophyllotoxin derivatives exhibiting preprophase activity, as the new mechanism was originally called (*47*), also produced considerable increases in life span in animals inoculated with the leukemia L-1210 cell line (see refs. 1 and 2).

From the beginning, observations in fibroblast cultures showed that benzylidene lignan P and its congeners, VM 26 and VP 16, still retained some spindle-poison activity (*47*). That effect becomes visible, however, only at much higher concentrations than the preprophase activity—that is, at concentrations at which podophyllotoxin glucosides are active as spindle poisons (*49*). Podophyllotoxin derivatives that exhibit preprophase activity and spindle-poison effect at about the same concentrations are, at least in leukemia L-1210, therapeutically less useful because the spindle-poison activity contributes very little to the anti-L-1210 effect but apparently a lot to general toxicity (see ref. 1).

Establishing the new mode of action in fibroblast cultures at the cellular level was then supplemented by further studies involving the reversibility and time course of the effect, the incorporation of biochemical precursors, and other parameters. It became apparent that the epipodophyllotoxins act in late S or in G_2 phase (or both, depending on the concentration) of the cell cycle. Fibroblast cultures had already revealed a low degree of reversibility (*47*), and subsequent experiments with P-815 mastocytoma cells showed that a 1-h exposure of proliferating cells to the epipodophyllotoxins VM and VP was, depending on the drug concentration, already sufficient to permanently prevent all of them from multiplying further (*50, 51*). Thymidine incorporation into P-815 mastocytoma cells is inhibited by VM and VP without an immediate corresponding depression of DNA synthesis (*52*). A comparison with other anticancer agents revealed a surprising similarity to some cellular and biochemical effects of X-rays (*53*).

A strange phenomenon observed in rodents when they were injected intraperitoneally with VM or VP was the development of a late but usually lethal chronic peritonitis (*54*). The mechanisms involved in this delayed toxicity are unclear, and similar observations in humans have not yet been reported.

Although these studies in our biological laboratories characterized the effects of teniposide and etoposide and showed their difference from other cytostatic compounds in use at that time, a first step to a real understanding of the biochemical basis for the mechanism of action came only with the investigations of Horwitz's group (55, 56). They showed a DNA-breaking effect of etoposide and teniposide at concentrations close to those inhibiting cell proliferation. Some years later, Long and Minocha (57) and others found that DNA breaking by etoposide correlates with interference by the drug with the activity of topoisomerase type II. Although it is not yet clear whether all effects of the epipodophyllotoxins relevant for cancer therapy are based on derangement of topoisomerase activity, this mechanism seems to be the most important one (see refs. 1 and 2).

Clinical testing of VM 26 started less than 2 years after synthesis, first in Switzerland, and soon thereafter in the United States, France, and Denmark. Evaluation of VP in cancer patients followed somewhat later, but this compound soon generated more interest than VM, because early clinical results seemed superior and because the compound could also be given orally. More recent clinical results again suggest similar merits of VM and VP (58). Activities in many types of cancer were found with the epipodophyllotoxins. Among the most important cancer types are small-cell lung cancer, testicular cancer, lymphomas, leukemias, and brain tumors. Advances in etoposide therapy and pharmacokinetics were discussed at a 1991 symposium (59).

In the early 1970s, experimental work with *Podophyllum* compounds in our laboratories was decreasing because we were awaiting clinical data on VM and VP and because it appeared difficult to find podophyllotoxin derivatives with significantly better efficacy. Furthermore, because of a reorientation of research goals in the pharmaceuticals department of Sandoz, interest in cancer chemotherapy decreased, and VM (which had already been commercialized in a few countries) and VP were licensed out to Bristol-Myers in 1978. This U.S. company successfully continued the clinical development of both drugs and in 1983 introduced etoposide as VePesid in the U.S. market.

Biological Test Systems and Their Role

In the collaboration between chemists and pharmacologists that eventually led to teniposide and etoposide, the development, use, and adaptation of rapid, reliable, and reproducible biological test systems played a paramount role, since their results permitted the deduction and evaluation of working hypotheses on structure–activity relationships. The availability of suitable biological assays was also crucial in the detection of the antileukemia factor of SP-G and of the new mechanism of action of the epipodophyllotoxins.

In the beginning of our work, the antiproliferative effect of *Podophyllum* substances and other potentially cytostatic compounds was assayed in vivo mainly by using Ehrlich's ascites tumor in mice, and in vitro by tissue culture, in the fibroblast test (*24, 25*). The latter test, a modification of an assay communicated to us by M. Allgöwer and L. Hulliger (then at the surgical University Clinic in Basel), entails explanting pieces of blood vessel walls of 14-day-old chick embryos onto cover slips with the help of coagulated chicken plasma. The monolayer formed by fibroblasts migrating out of the explant exhibits numerous mitoses and is very suitable for observing the effect of drugs on dividing and resting cells. An important aspect of the method, which is more demanding in manual skill than more popular techniques, is that mitotic cells do not detach from the cover glass as they easily do in monolayer cultures of most cell lines; mitotic counts are therefore more reliable with this type of tissue culture (for a later modification of the method, see ref. *29*).

Another in vitro assay, the P-815 mastocytoma test, proved particularly useful in establishing that the components of SP-G known up to 1961 could, quantitatively, not explain the potency of the whole preparation. This test, developed by us a few years earlier (*37*), is quick and simple to perform and gives accurate, reproducible results. This is probably due to the high proliferation rate of the P-815 murine cells (they have perhaps the shortest doubling time—about 8 h under favorable conditions—ever reported for mammalian cells in vitro) and to the fact that the cells do not adhere to the glass or plastic of the culture vessel, which makes transfer and counting much easier than in other cytostatic assay systems in use at that time.

A third crucial test was the murine lymphocytic leukemia L-1210 assay. This transplantable tumor not only was considered in the 1960s as among the animal models for testing anticancer drugs with the highest predictive value for human malignancies (*60*), but, like the mastocytoma test, it has good reproducibility, that is, a low variability of results within and between experiments as to survival time of inoculated control or treated animals. We used to inoculate L-1210 subcutaneously (into the hind leg) and to administer the drugs intraperitoneally, orally, or, quite often, intravenously. Intravenous injection was possible because one of our technical co-workers had great manual dexterity, which enabled him to inject the drugs intravenously into mice daily for a period of 2 weeks. This mode of testing seemed to us to provide a more valid indication of anticancer activity of a drug than the more popular method—but more remote from the clinical situation—of intraperitoneal tumor inoculation and intraperitoneal treatment. Our method of using the L-1210 assay contributed to our ability to distinguish the effects of SP-G from those of its components: when tumors inoculated intraperitoneally are treated by intraperitoneal drug injections, the proportion of false-positive results is likely to be higher because the tumor cells are exposed to unpharmacologically high drug concentrations, and subtle differences between drugs are less likely to be recognizable.

The analysis of SP-G by means of the mastocytoma test initiated the search for additional compounds in the mixture, and the L-1210 assay was the basis for the investigations leading to the discovery of the antileukemia factor. Our hope that the unknown compounds would not only be of high cytostatic potency but also of improved therapeutic value was finally realized with the isolation of benzylidene lignan P and the synthesis of its congeners, teniposide and etoposide.

The importance of immunology in medicine was increasingly evident in the 1960s, and we added an immunology laboratory to our biological cancer chemotherapy group where the appropriate tests for measuring the immune response were set up. Among the *Podophyllum* drugs, SP-G (H. Stähelin, unpublished data, 1964; *61*), VM, and particularly VP (S. Lazary, Sandoz Ltd., unpublished data) were found to be quite efficacious immunosuppressants in the hemagglutination assay, which later was instrumental in discovering the immunosuppressive activity of cyclosporin (Sandimmun) (*62, 63*).

Serendipity and Coincidence

Serendipity, or finding something interesting while searching for something else, played an important role in many discoveries (*64, 65*). A significant element of serendipity was involved in arriving at etoposide. Originally, condensation of aldehydes to podophyllotoxin glucoside was performed in order to stabilize the molecule and to improve its absorption from the gastrointestinal tract. This goal was achieved, but without significantly changing the cytostatic potency or mechanism of action. When the same was done with a crude plant extract (resulting in SP-G), it produced a completely unexpected effect: it converted an unknown natural compound in the mixture into a derivative with an up to 1,000-fold higher potency (see Table 2), a different mechanism of action, and a much greater therapeutic value.

Also somewhat serendipitous was the use of the P-815 mastocytoma test for evaluating cytostatic potency of compounds, an assay method that played an essential role in the development of etoposide. We had acquired this cell line for a different purpose and then found that it was much more suitable for testing anticancer drugs (*62*).

Another type of chance, coincidence, should be mentioned briefly in connection with etoposide. There are numerous duplicities or coincidences between etoposide and cyclosporin: (1) Both were discovered (*66, 67*) and developed (*68–71*) by the same chemical and biological groups, headed by the present authors. (2) They were approved by the U.S. Food and Drug Administration on the same day, although submitted by different companies and for different indications. (3) Both act on an intranuclear isomerase. (4) Polysorbate played a role in the galenical investigations of both drugs. These and other coincidences are described in ref. 2. It is left to the reader to make assumptions about the heuristic significance of these numerous duplicities.

Outlook

What can we expect from future investigations on *Podophyllum* compounds? There is certainly room for additional chemical work on glucosides of the 1-epi series and for the synthesis of structurally new derivatives of the lignan type; this work may eventually lead to second-generation drugs of greater clinical utility. Additional, systematic efforts are necessary to gain more precise information on structure–activity relationships in order to allow a more rational design of new active compounds. Although the structural elements required for the new mechanism (as opposed to spindle-poison activity) were listed in 1972 (*45*), present knowledge may allow better-founded hypotheses as to structure–activity relationships (see also ref. 1). Attempts in this direction are going on in a number of laboratories. Clinical evaluation and application of etoposide and teniposide will no doubt progress, as will studies on their mechanism of action, especially with regard to their interference with topoisomerase activity and the radiomimetic cellular effects.

Acknowledgments

Over the more than 20 years of our investigation of *Podophyllum*, many of our colleagues and co-workers actively participated in the development of etoposide and teniposide, in particular the chemists C. Keller-Juslén, E. Angliker, and M. Kuhn, and, on the pharmacological side, H. Emmenegger, P. Gradwohl, A. Trippmacher, and M. Hernandez-Straessle.

Literature Cited

1. Stähelin, H.; von Wartburg, A. *Prog. Drug Res.* **1989,** *33,* 169.
2. Stähelin, H.; von Wartburg, A. *Cancer Res.* **1990,** *51,* 5.
3. Kaplan, I. W. *New Orleans Med. Surg. J.* **1942,** *94,* 388.
4. King, L. S.; Sullivan, M. *Science (Washington, DC)* **1946,** *104,* 244.
5. Kelly, M. G.; Hartwell, J. L. *J.N.C.I.* **1954,** *14,* 967.
6. Podwyssotzki, V. *Arch. Exp. Pathol. Pharmakol.* **1880,** *13,* 29.
7. Hartwell, J. L.; Schrecker, A. W. *J. Am. Chem. Soc.* **1951,** *73,* 2909.
8. Gensler, W. J.; Gatsonis, C. D. *J. Am. Chem. Soc.* **1962,** *84,* 1748.
9. Gensler, W. J.; Gatsonis, C. D. *J. Org. Chem.* **1966,** *31,* 4004.
10. Schrecker, A. W.; Hartwell, J. L. *J. Org. Chem.* **1956,** *21,* 381.
11. Petcher, T. J.; Weber, H. P.; Kuhn, M.; von Wartburg, A. *J. Chem. Soc., Perkin Trans. 2* **1973,** 288.
12. Hartwell, J. L.; Schrecker, A. W. *Fortschr. Chem. Org. Naturst.* **1958,** *15,* 83.
13. Chatterjee, R.; Datta, D. K. *Indian J. Physiol. All. Sci.* **1950,** *4,* 7.
14. Chatterjee, R.; Chakravarti, S. C. *J. Am. Pharmacol. Assoc.* **1952,** *41,* 415.
15. Schreier, E. *Helv. Chim. Acta* **1963,** *46,* 75.
16. Schreier, E. *Helv. Chim. Acta* **1964,** *47,* 1529.
17. Nadkarni, M. V.; Hartwell, J. L.; Maury, P. B.; Leiter, J. *J. Am. Chem. Soc.* **1953,** *75,* 1308.
18. Stoll, A.; Renz, J.; von Wartburg, A. *J. Am. Chem. Soc.* **1954,** *76,* 3103.
19. Stoll, A.; Renz, J.; von Wartburg, A. *Helv. Chim. Acta* **1954,** *37,* 1747.
20. Stoll, A.; von Wartburg, A.; Angliker, E.; Renz, J. *J. Am. Chem. Soc.* **1954,** *76,* 5004.
21. Stoll, A.; von Wartburg, A.; Angliker, E.; Renz, J. *J. Am. Chem. Soc.* **1954,** *76,* 6413.
22. Stoll, A.; von Wartburg, A.; Renz, J. *J. Am. Chem. Soc.* **1955,** *77,* 1710.
23. von Wartburg, A.; Angliker, E.; Renz, J. *Helv. Chim. Acta* **1957,** *40,* 1331.
24. Cerletti, A.; Emmenegger, H.; Stähelin, H. *Actual. Pharmacol. (Paris)* **1959,** *12,* 103.
25. Emmenegger, H.; Stähelin, H.; Rutschmann, J.; Renz, J.; von Wartburg, A. *Drug Res.* **1961,** *11,* 327, 459.
26. Stoll, A.; Rutschmann, J.; von Wartburg, A.; Renz, J. *Helv. Chim. Acta* **1956,** *39,* 993.
27. Keller-Juslén, C.; Kuhn, M.; von Wartburg, A.; Stähelin, H. *J. Med. Chem.* **1971,** *14,* 936.

28. Weder, A. *Schweiz. Med. Wochenschr.* **1958,** *88,* 625.
29. Stähelin, H.; Cerletti, A. *Schweiz. Med. Wochenschr.* **1964,** *94,* 1490.
30. Stähelin, H. *Abstracts of Papers,* 9th International Cancer Congress, Tokyo, 1966; Vol. I, p 342.
31. Rutschmann, J.; Renz, J. *Helv. Chim. Acta* **1959,** *42,* 890.
32. Borsche, W.; Niemann, J. *Liebigs Ann. Chem.* **1932,** *499,* 59.
33. Hubacher, O. *Schweiz. Med. Wochenschr.* **1961,** *91,* 1316
34. Kuhn, M.; von Wartburg, A. *Helv. Chim. Acta* **1963,** *46,* 2127.
35. Kuhn, M.; von Wartburg, A. *Experientia* **1963,** *19,* 391.
36. Renz, J.; Kuhn, M.; von Wartburg, A. *Liebigs Ann. Chem.* **1965,** *681,* 207.
37. Stähelin, H. *Med. Exp.* **1962,** *7,* 92.
38. Kuhn, M.; von Wartburg, A. *Helv. Chim. Acta* **1967,** *50,* 1546.
39. von Wartburg, A.; Kuhn, M.; Lichti, H. *Helv. Chim. Acta* **1964,** *47,* 1203.
40. von Wartburg, A.; Kuhn, M. *Experientia* **1965,** *21,* 67.
41. Kuhn, M.; von Wartburg, A. *Helv. Chim. Acta* **1968,** *51,* 163.
42. Kuhn, M.; von Wartburg, A. *Helv. Chim. Acta* **1968,** *51,* 1631.
43. Kuhn, M.; Keller-Juslén, C.; von Wartburg, A. *Helv. Chim. Acta* **1969,** *52,* 944.
44. Kuhn, M.; von Wartburg, A. *Helv. Chim. Acta* **1969,** *52,* 948.
45. Stähelin, H. *Planta Med.* **1972,** *22,* 336.
46. Stähelin, H. *Proc. Am. Assoc. Cancer Res.* **1969,** *10,* 86.
47. Stähelin, H. *Eur. J. Cancer* **1970,** *6,* 303.
48. Stähelin, H. *Eur. J. Cancer* **1973,** *9,* 215.
49. Stähelin, H.; Poschmann, G. *Oncology* **1978,** *35,* 217.
50. Stähelin, H. In *Progress in Chemotherapy*; Daikos, G. K., Ed.; Hellenic Society of Chemotherapy: Athens, 1974; Vol. 3, p 819.
51. Stähelin, H. In *Workshop on Clinical Usefulness of Cell Kinetic Information for Tumor Chemotherapy*; van Putten, L. M., Ed.; REP-TNO: Rijswijk, The Netherlands, 1974; p 21.
52. Grieder, A.; Maurer, R.; Stähelin, H. *Cancer Res.* **1974,** *34,* 1788.
53. Grieder, A.; Maurer, R.; Stähelin, H. *Cancer Res.* **1977,** *37,* 2998.
54. Stähelin, H. *Eur. J. Cancer* **1976,** *12,* 925.
55. Loike, J. D.; Horwitz, S. B.; Grollman, A. P. *Pharmacologist* **1974,** *16,* 209.
56. Loike, J. D.; Horwitz, S. B. *Biochemistry* **1976,** *15,* 5443.
57. Long, B. H.; Minocha, A. *Proc. Am. Assoc. Cancer Res.* **1983,** *24,* 1271.
58. Hansen, H. H., Ed. *Vepesid and Vumon: A Status Report,* 1988; report on a meeting in Charlottenlund, Denmark.

59. Aisner, J., et al. *Cancer* **1991,** *67* (Suppl. 1), 215.
60. Goldin, A.; Serpick, A. A.; Mantel, N. *Cancer Chemother. Rep.* **1966,** *50,* 173.
61. Lazary, S.; Stähelin, H. In *Fifth International Congress of Chemotherapy*; Spitzy, K. H., Ed.; Verlag Wiener Medizinische Akademie: Vienna, 1967; Vol. 3, p 317.
62. Stähelin, H. *Prog. Allergy* **1986,** *38,* 19.
63. Stähelin, H. In *Immunology 1930–1980: Essays on the History of Immunology*; Mazumdar, P. M. H., Ed.; Wall and Thompson: Toronto, 1989; p 185.
64. Humphrey, J. H. *Annu. Rev. Immunol.* **1984,** *2,* 1.
65. Roberts, R. M. *Serendipity: Accidental Discoveries in Science*; Wiley: New York, 1989.
66. Stähelin, H. *Sandoz Internal Screening Report,* Jan. 31, 1972.
67. Rüegger, A.; Kuhn, M.; Lichti, H.; Loosli, H. R.; Huguenin, R.; Quiquerez, C.; von Wartburg, A. *Helv. Chim. Acta* **1976,** *59,* 1075.
68. Borel, J.F.; Rüegger, A.; Stähelin, H. *Experientia* **1976,** *32,* 777.
69. von Wartburg, A.; Traber, R. *Prog. Allergy* **1986,** *38,* 28.
70. Borel, J. F.; Feurer, C.; Gubler, H. U.; Stähelin, H. *Agents Actions* **1976,** *6,* 468.
71. von Wartburg, A.; Traber R. *Prog. Med. Chem.* **1988,** *25,* 1.

Amsacrine

William A. Denny
University of Auckland

Historical Background

Origins

Amsacrine had its origins in work carried out in the Cancer Research Laboratory (CRL), Auckland, New Zealand, under the direction of Bruce Cain in the late 1960s. Responding to theories that tumor cells possessed a more negative surface charge than normal cells (*1*), and prompted by reports of the good antileukemic activity in vivo of the series of bisamidines (e.g., **1**) known collectively as the phthalanilides or "Wander compounds" (originating from the Swiss firm Dr. A. Wander SA) (*2, 3*), Cain undertook the synthesis of a large series of bisquaternary ammonium heterocycles (summarized in ref. 4). Early members of this class (e.g., **2**) showed high antileukemic activity in vivo, but were later (*5*) shown to possess a chronic toxicity that resulted in late deaths in animals cured of leukemia.

During this work, Cain became convinced that the critical property required of all these compounds was a geometry and charge separation that allowed them to fit neatly into the minor groove of double-helical DNA. Parallels were drawn between these compounds and the antitrypanosomal agent homidium (ethidium; **3**), the charged phenanthridinium chromophore of which had been shown to bind to DNA in a different manner, by

2523–0/93/0381$07.00/0

intercalating between the base pairs. It was suggested that the more active members of the series (e.g., quinolinium compounds such as **4**) bound to DNA initially in the minor groove, a step that preceded final intercalation of the chromophore (*6*). The first assumption was later proved correct by an NMR study of the complex between **4** and an oligonucleotide (*7*), as well as by more indirect methods, but no evidence has been found for intercalation by these compounds (*8*). A series of acridinium analogues of **4** (e.g., **5**) were then prepared and found to be highly active (*9*). Later work showed that these compounds, with the larger chromophore, did intercalate DNA (*8*).

The acridinium series was unique in that several unquaternized derivatives (e.g., **6**) also proved to have antileukemic activity in vivo (*9*). A literature survey done at that time showed that 9-(4-dimethylaminoanilino)acridine (**7**; NSC-13002) had been shown (*10*) to have activity against both L-1210 leukemia and Walker 256 carcinosarcoma in mice, but no follow-up work had been reported. A series of progressively simpler analogues of **6** were then prepared, culminating in 9-(4-aminoanilino)acridine (**8**) and the corresponding methanesulfonamide (**9**) (*11*). Exploration of the 1′-position showed that electron-withdrawing substituents abolished activity, and that weakly electron-withdrawing groups with H-bond donor capability were favored. In this latter category, the superiority of the methanesulfonamide substituent in terms of water solubility, stability, and biological activity was so marked (Table 1) that **9** became the "parent" molecule in which the effects of further aniline substitution were explored (*12*). Variation of substituents in the 2′- and 3′-positions produced a pattern of activity consistent with the hypothesis that high electron density was required at the 6′-position (*12*). The most important practical finding (*13*) was that a 3′-methoxy group greatly increased potency, with **10** being sevenfold more dose-potent in vivo than **9** (Table 1). Biochemical studies showed that amsacrine and several related analogues were DNA-intercalating agents, unwinding and rewinding closed circular supercoiled DNA (*14*). Following detailed animal testing, the U.S. National Cancer Institute (NCI) in 1974 made the decision to take **10** to clinical trial.

The abbreviation AMSA was devised for this subclass of the 9-anilinoacridines by taking letters from the formal chemical name, 9-acridinylaminomethanesulfonanilides (*15*). Thus the unsubstituted compound 9-(4′-acridinylamino)methanesulfonanilide (**9**) became known within the CRL as AMSA, the 3′-OMe derivative 9-(4′-acridinylamino)-methanesulfon-*m*-anisidide (**10**) as *m*-AMSA, and the isomeric 2′-OMe compound (**11**) as *o*-AMSA. Unfortunately, some confusion later arose in the literature, with other workers using the names AMSA and *m*-AMSA

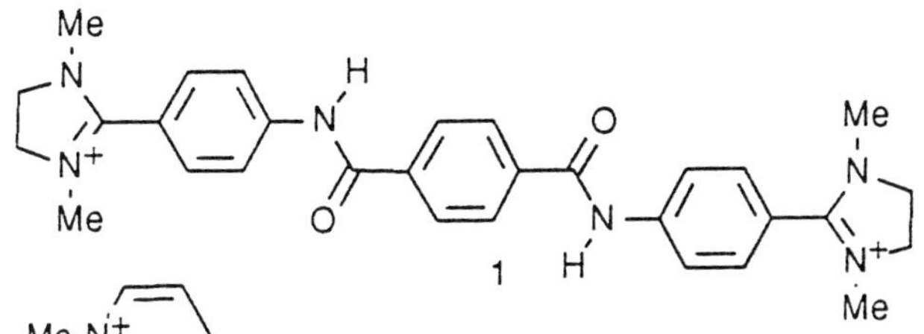

1

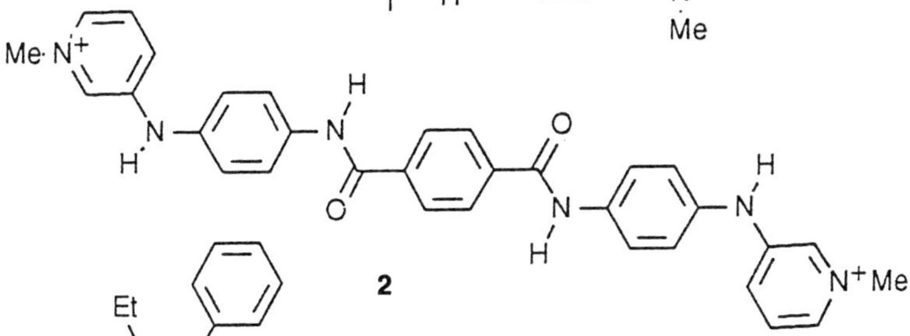

2

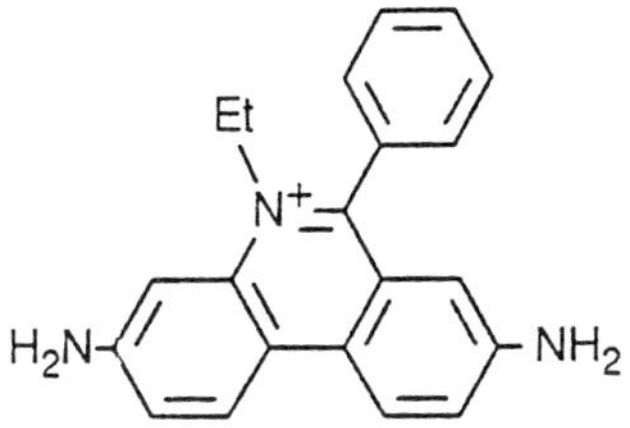

3 : ethidium

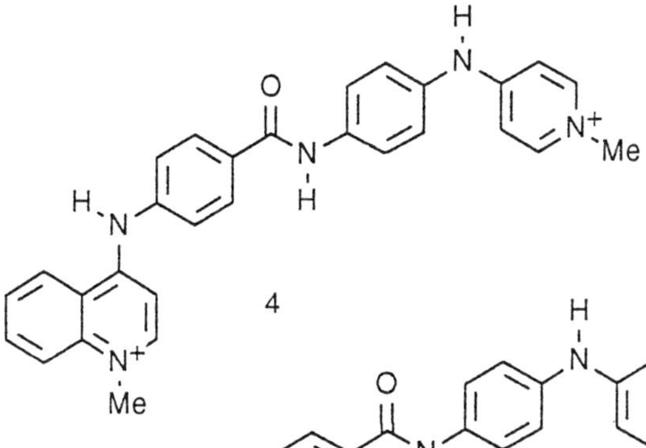

4

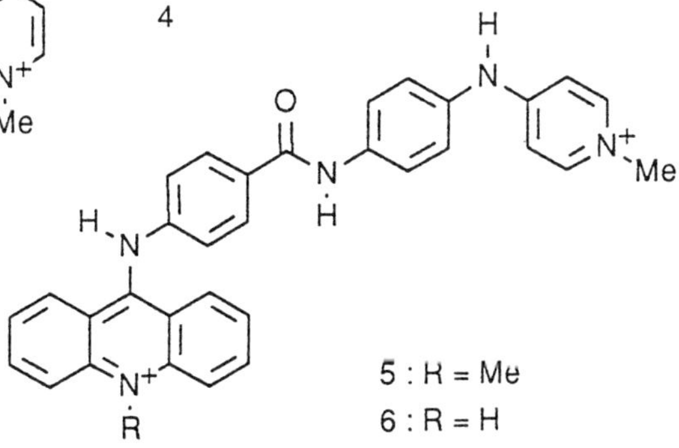

5 : R = Me
6 : R = H

Table 1. Comparison of the Activity of Selected Aniline-Substituted 9-Anilinoacridines Against L-1210 Leukemia

Aniline Substituent(s) R	*IC_{50}[a]*	*OD[b]*	*%ILS[c]*
H	2,600	200	NA[d]
1′-$CONH_2$	6,000	>500	NA
1′-Cl	12,000	100	NA
1′-NH_2	350	33	72
1′-NHMe	180	40	51
1′-NHCOMe	100	20	53
1′-$NHSO_2Me$ (AMSA: **9**)	35	45	107
1′-$NHSO_2Me$, 3′-F	800	>500	NA
1′-$NHSO_2Me$, 3′-NH_2	1,200	45	106
1′-$NHSO_2Me$, 3′-Me	120	97	106
1′-$NHSO_2Me$, 3′-OMe (amsacrine: **10**)	35	6.7	114
1′-$NHSO_2Me$, 2′-OMe (*o*-AMSA: **11**)	550	>500	NA

[a] IC_{50} is the concentration of drug (nanomolarity) required to reduce the growth of L-1210 cells in culture to 50% of control cultures after 3 days.
[b] OD is the optimal dose of drug (in milligrams per kilogram per day) administered intraperitoneally on days 1, 5, and 9 after tumor inoculation.
[c] %ILS is the percentage increase in life span of treated animals (at the optimal dose) compared with tumor-bearing control animals.
[d] No activity (ILS > 20%) at all nontoxic doses.
SOURCE: Data are taken from ref. 52.

interchangeably, and matters became even more confusing when *m*-AMSA received the generic name amsacrine.

Synthesis

The synthesis of the 9-anilinoacridines is straightforward (Scheme 1), involving acid-catalyzed coupling of 9-chloroacridine (**12**) with the appropri-

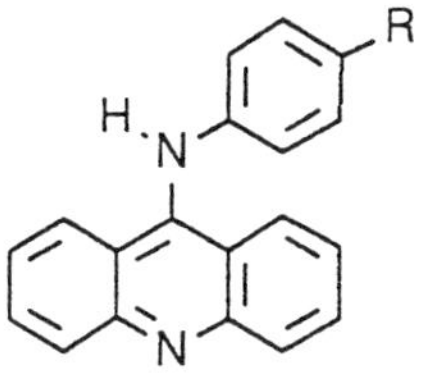

7 : R = NMe_2
8 : R = NH_2
9 : R = $NHSO_2Me$

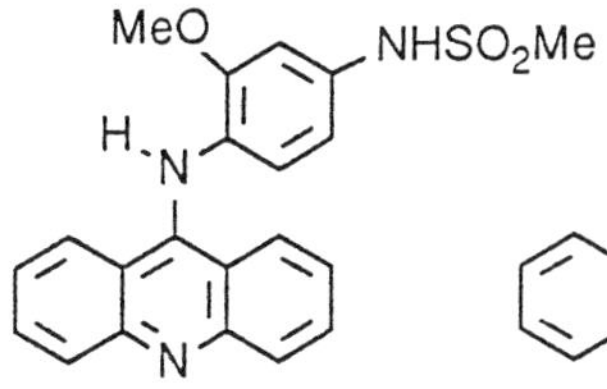

10 : amsacrine
(*m*-AMSA)

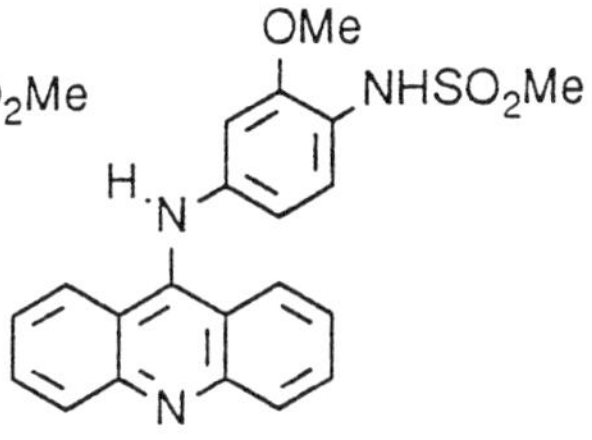

11 : *o*-AMSA

ate aniline derivative. For example, amsacrine is prepared by mixing equimolar solutions of *N*-(4-amino-3-methoxyphenyl)methanesulfonamide (**13**) (also known as 4-aminomethanesulfon-*m*-anisidide) and 9-chloroacridine in anhydrous methanol. Adding a catalytic amount of dry HCl and heating the solution briefly to boiling results in immediate crystallization of the deep red hydrochloride salt of amsacrine (**10**) in 90–95% yield. The only problem that had to be solved for large-scale preparation, following the NCI's decision to take amsacrine to clinical trial, was a suitable synthesis of the side chain. The method used in the initial synthesis of amsacrine (*13*)—nitration of 3-methoxyacetanilide, followed by a separation of the isomers, hydrolysis, and mesylation of 3-methoxy-4-nitroaniline—was not suitable for kilogram-scale preparations.

The initial large-scale route developed in the CRL (*13*) was from the commercially available 2-methoxy-4-nitroaniline (**14**), by protection via the butamide (**15**), Fe^{+}/HCl iron reduction to the amine (**16**), formation of the sulfonamide (**17**) with methanesulfonyl chloride, and hydrolysis (Scheme 1). The butamide was used to provide a product of suitable lipophilicity for workup of the iron reduction step, but had two drawbacks. The major drawback, a technical one, was that the butamides **15** and **16** were noncrystalline and thus difficult to purify. The aesthetic disadvantage to using large quantities of *n*-butyric anhydride will be apparent to anyone who has had occasion to use this malodorous reagent. Even scheduling such work for Friday afternoons and going away for the weekend did not help very much! Nevertheless, the method was patented (*16*) and used for the commercial production of amsacrine. Later improvements in the synthesis (*17*) avoided the use of butyric anhydride by beginning with the corresponding crystalline acetanilide derivative (**18**) and employing catalytic reduction of the nitro group.

The hydrochloride salt of amsacrine (**10**) prepared directly from this synthesis was the form initially studied by NCI (as NSC-141549), but had

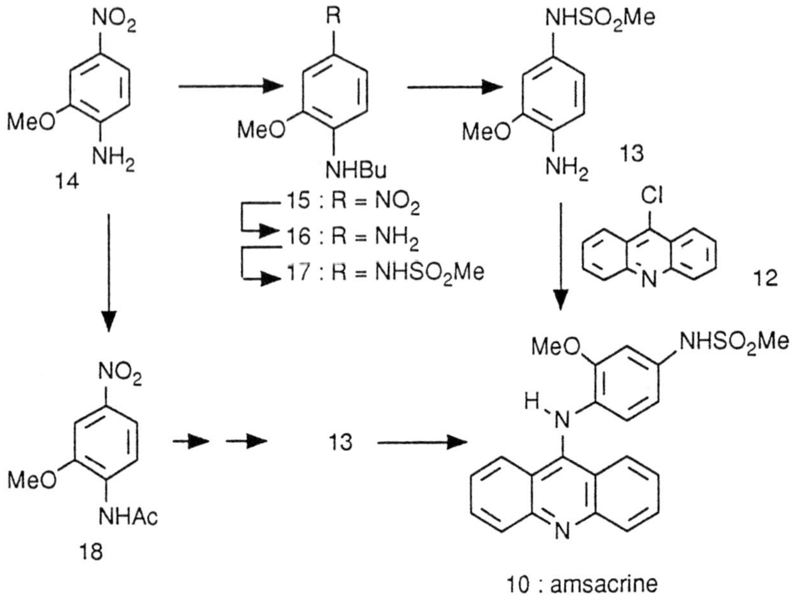

Scheme 1

poor water solubility. Even the methanesulfonate salt (NSC-156303) was not sufficiently water-soluble, and the first clinical trials of amsacrine were carried out using a two-ampule formulation. A solution of the anhydrous free base in dimethylacetamide in one ampule was mixed with a solution of dilute lactic acid (provided in the second ampule) to provide an injectable solution of the lactate salt (designated NSC-249992).

Clinical Development

Phase I trials began in 1976 under the auspices of the NCI, and the encouraging results seen against both leukemias and lymphomas were quickly confirmed in Phase II trials (*18*), with a potentially useful aspect of the clinical activity being an apparent lack of cross-resistance with Adriamycin (doxorubicin hydrochloride) (*19*). Amsacrine thus became the first totally synthetic drug of the DNA-intercalating type to show clinical efficacy, joining a number of other natural products including the anthracyclines daunorubicin and doxorubicin, actinomycin D, and 9-methoxyellipticine (*20*). The success of the Phase I trials resulted in numerous inquiries from pharmaceutical companies, and in 1979 an agreement was signed between the CRL and the Parke-Davis Division of the Warner-Lambert Company, under which they undertook to develop amsacrine toward marketing. An event of particular significance for the CRL came in 1984, when amsacrine was registered (under the trade name Amsidyl) for use in New Zealand, becoming the first pharmaceutical developed in that country to achieve such status. The main clinical use of amsacrine now is in the treatment of adult leukemia, where it is used in combination with antimetabolites (*21*), although clinical studies on other applications continue (*22*).

Properties and Mechanism of Action

Structure

Simple minimum-energy calculations (*23*) for amsacrine predict the anilino ring to lie almost at right angles to the acridine to relieve nonbonded interactions. However, crystallographic studies (*24, 25*) show much less rotation, owing to the considerable degree of conjugation of the C14-N15 bond, which is significantly shorter (1.35–1.36 Å) than normal (cf. the C19-C22 bond, 1.43 Å). This conjugation is also demonstrated by the high degree of transmission of electronic effects from the anilino to the acridine ring, as summarized by equation 1 for the σ effects of anilino ring substituents on acridine pK_a (*26*):

$$pK_a = -2.03(\pm 0.08)\sigma + 7.27 \qquad (1)$$

for $n = 17$, where n is the number of compounds in the equation; $r = 0.984$, where r is the correlation coefficient; and $s = 0.172$, where s is the standard error of the regression.

DNA Binding

Amsacrine binds to DNA by reversible, enthalpy-driven (*27*) intercalation of the acridine chromophore, with an association constant of $1.8 \times 105\ M^{-1}$ for calf thymus DNA in 0.01 M salt (*28*). By analogy with the crystal structure determined for 9-aminoacridine binding to a dinucleotide (*29*), amsacrine was postulated to bind with the anilino ring lodged in the minor groove, with the 1′-substituent pointing tangentially away from the helix, and the possibility of it thus interacting with another (protein) macromolecule to form a ternary complex was noted (*28*). This conformation was supported by later energy calculations (*30*). Intercalation is the favored binding mode for molecules with polycyclic aromatic chromophores, including the antitumor antibiotics actinomycin, doxorubicin, and ellipticine. The driving force for the interaction is stacking interactions between the drug and the base pairs, and also the increase in entropy by the release of structured water from both the DNA minor groove and the ligand (*31*). Because this binding mode distorts DNA structure, and because these compounds selectively kill S-phase cells (*32*), the mechanism by which amsacrine (and other DNA intercalators) expressed their cytotoxicity and anticancer activity was considered to be inhibition of nucleic acid synthesis, possibly by prevention of template or polymerase binding (*33*).

Thiol Reactivity

Amsacrine is not as chemically inert as might be thought from its structure. Early observations (*34*) in the CRL showed its reactivity toward thiols, resulting in displacement of the side chain to give unstable 9-thio species (**19**), and finally acridone (**20**) (Scheme 2). The reaction of 2-mercaptoethanol with amsacrine derivatives has been studied in detail (*34, 35*) and is general-acid-catalyzed, with the rate-determining step the expulsion of the anilino leaving group. However, although the potency of amsacrine derivatives was inversely proportional to their relative rates of thiolysis (*36*), this had little bearing on their antitumor activity.

Redox Properties

The redox properties of amsacrine were first demonstrated by studies of its biliary metabolism in rats, where more than 50% of the dose was excreted as the 6′-glutathione conjugate (**22**), resulting from two-electron oxidation of amsacrine to the quinonedimine (AQDI; **21**) followed by 1,4-addition of GSH (Scheme 3) (*37*). We later showed this oxidation to be a two-step process, proceeding through a one-electron oxidation product (the radical anion) which was detectable by pulse radiolysis (*38*). The two-electron product, the quinonediimine (**21**), is isolable and can be produced electrochemically by cyclic voltammetry, where a redox potential of 280 mV can be measured (*39, 40*); chemically (*37*) by activated MnO_2; biologically by liver microsome preparations (*41*); and even by leaving dilute aqueous solutions exposed to the air. However, it is very reactive, undergoing quantitative conversion to amsacrine with mild reducing agents such as ascorbic acid, and

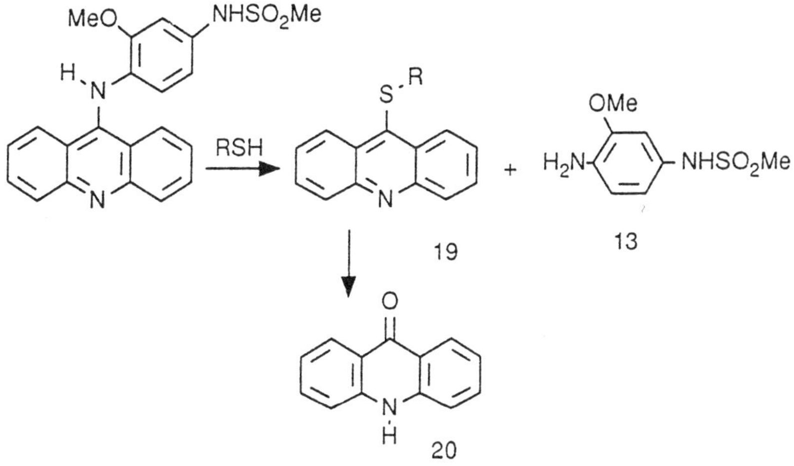

Scheme 2

1,4-additions with nucleophiles such as amines and thiols (*37, 42*). It has been suggested to be the active species responsible for the cytotoxicity of amsacrine (*41*), but recent studies (*43*) showing a slow rate of formation of AQDI in AA8 cells indicate that this is unlikely.

Metabolism and Pharmacokinetics

As noted in the preceding discussion, the primary route of elimination of amsacrine in rodents is via biliary metabolism, and both the 6′- and 5′-GSH conjugates (**22** and **23**, respectively) have been identified (*37, 43*), together with the 9-GSH adduct (**24**) (*44*), possibly formed by direct displacement of the side chain (Scheme 3). Much of the pharmacokinetic work on amsacrine has been carried out by James Paxton and colleagues in the Department of Pharmacology and Clinical Pharmacology at the University of Auckland School of Medicine. Studies in rabbits (*45*) indicated high plasma–protein binding and nonlinear elimination kinetics, and a review of pharmacokinetic data obtained in several animal species suggested relationships that could be used to predict suitable human doses (*46*). In humans (*47*), amsacrine elimination was rapid and biexponential, with half-lives of 0.8 and 5.3 h.

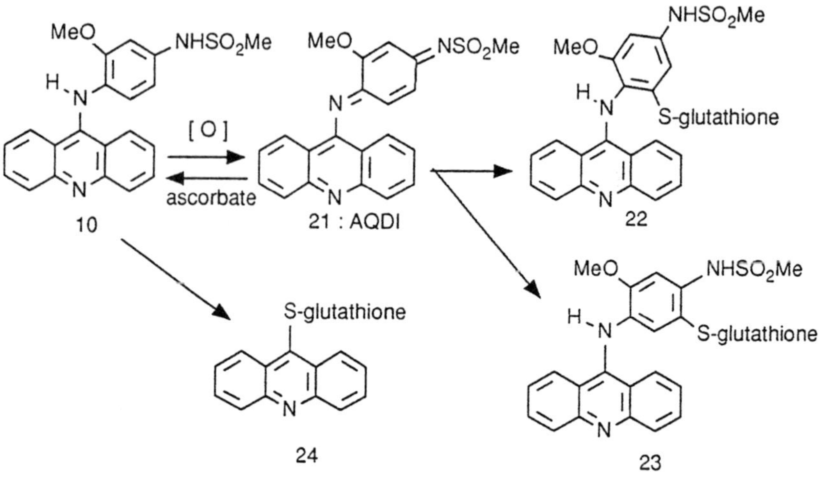

Scheme 3

Inhibition of Topoisomerase II

A major step forward was the demonstration that amsacrine and other DNA intercalators induced protein-linked DNA breaks, resulting from interference with the action of the DNA-processing enzyme topoisomerase II (*48*). These drugs form ternary complexes with the enzyme and DNA, as proposed earlier (*28*), altering the position of equilibrium and trapping a reaction intermediate termed the "cleavable complex" (*49, 50*). Although now shown to be general for DNA-intercalating agents, this mechanism was first described for amsacrine (*49*), which has also become a valuable biological reagent for studying this phenomenon.

The expression of amsacrine's cytotoxicity via formation of a ternary drug-DNA-protein complex provided an explanation for the failure of DNA-binding studies alone to fully predict the activities of amsacrine analogues. Thus, amsacrine can be seen as having two distinct functional domains: the "protein-binding" anilino side chain and the "DNA-binding" acridine chromophore. For acridine-substituted analogues, cytotoxicity varies with DNA-binding efficiency (*51*), but changes in the anilino ring can also evoke large alterations in cytotoxicity, essentially independent of DNA-binding properties (*52*). More recent work shows that the cytotoxic activities of amsacrine derivatives correlate more closely with their ability to stabilize formation of the topoisomerase II cleavable complex than with their DNA-binding ability (53). This mechanism of action also accounts for the fact that these compounds show marked selectivity for cycling cells (32), since topoisomerase II is highly cell-cycle dependent (54).

Structure–Activity Relationships for Amsacrine Derivatives

Empirical Studies

The selection of amsacrine for clinical evaluation was a great stimulus for further development of the 9-anilinoacridine class of compounds in the CRL. As noted previously, at that time the mode of action of such DNA intercalators was considered to be via inhibition of nucleic acid synthesis, and the important property was considered to be some aspect of DNA binding. This concept led to a great deal of work, by ourselves and others (summarized briefly in ref. 20), to develop more tightly binding agents. In the amsacrine series, we explored analogues with additional cationic side chains (55), and many compounds in which the anilino ring or the acridine chromophore (or both) was modified. Association constants for DNA binding could be varied manyfold by making changes in the acridine ring system, but whereas tighter binding did generally correlate with improved potency, the resulting compounds did not necessarily show higher or broader antitumor activity in experimental systems. However, changes in the aniline ring, while having relatively little effect on DNA binding, also had profound effects on biological activity.

Modeling in Terms of Measured Physicochemical Properties

Extensive quantitative structure–activity relationship (QSAR) studies were carried out (summarized in ref. 56) relating the in vivo antileukemic potency of both 9-anilinoacridines in general and amsacrine analogues in particular to several measured properties, including lipophilicity, stability to cleavage by thiols (36), DNA binding (51), and in vitro cytotoxic potency (57). However, the resulting equations never encompassed more than a small fraction of the 9-anilinoacridines available.

Modeling in Terms of Computed Parameters

An opportunity to attempt a more extensive QSAR study using the large amount of data available on the in vivo antileukemic activity of the 9-anilinoacridines came during a period of sabbatical leave in 1980, which I spent in Corwin Hansch's laboratory (58). Because of the large number of compounds involved, it was necessary to use only calculated parameter values as independent variables. Global lipophilicity ($\Sigma\pi$) values were calculated by the fragment constant method (59), and global electronic effects ($\Sigma\sigma$) were summed for their effects on pK_a, but steric effects were parameterized at each position. A number of indicator variables were also used to parameterize specific subclasses. The best-fit equation covered 509 active compounds, with a correlation coefficient of 0.88. There was only a small lipophilicity effect, but a large negative correlation with $\Sigma\sigma$. The most interesting aspect of the work was the pattern of steric effects, which were entirely consistent with the biologically important receptor being DNA and the acridine chromophore intercalating longitudinally into a restricted site (58). Thus, no steric tolerance was evident at the 1-position of the acridine ring, very little at the 2-position, and considerable at the 3-position, with essentially no steric restrictions existing at the 4-position. In the anilino ring there were steric restrictions at the 2′- and the 3′-positions but not at the 1′-position. As well as condensing a large amount of data into manageable form, this study highlighted areas of chemistry of the 9-anilinoacridines that had not been fully explored, and that looked worthwhile.

CI-921

Selection

The success of amsacrine as an antileukemia drug led to a search in the CRL, from among the numerous analogues available, for a "second-generation" compound with a broader spectrum of action. At the time, high DNA binding was still considered to be an important parameter, but it was also thought that the high aqueous pK_a of amsacrine (8.02, implying 87% ionization at physiological pH) played an important part in limiting its in vivo distribution. Therefore, the initial search was for compounds with a particular set of physicochemical properties, including improved aqueous solubility, high DNA binding, and lower pK_a. The structure–activity relationships observed for the side chain suggested that the existing structure was close to optimal, and changes were therefore sought mainly in the acridine chromophore. Previous work with carboxamide derivatives (*60*) had shown the suitability of these substituents, and the large QSAR study mentioned earlier had focused attention on the 4- and 5-positions as being the most suitable for modification. Among the compounds studied in detail were several 4- and 3-carboxamides, including the 4-methyl-5-methylcarboxamide (**25**), which was the most active of a series prepared and evaluated in a search for orally active analogues of amsacrine (*61*).

These compounds were evaluated for their in vivo solid tumor activity using the Lewis lung carcinoma as the primary screen (Table 2) (*62*), because at the time this was considered to be the mouse tumor least responsive to intercalators in the main NCI panel (*63*). The most active compound against both the Lewis lung carcinoma and the cell line panel was the 4,5-disubstituted derivative (**25**) just mentioned. This compound also had the best combination of physicochemical properties compared to amsacrine (*64*), including higher aqueous solubility and DNA binding, together with a significantly lower pK_a and higher lipophilicity as measured chromatographically (Table 3). Following further comparative studies of **25** and amsacrine against a range of human solid tumor cell lines in culture (*65*) and a wide range of murine solid tumors in vivo (Table 4) (*66*), the former compound (CI-921; NSC-343499; 9-[(2-methoxy-4-methylsulfonylamino)phenylamino]-*N*,5-dimethyl-4-acridinecarboxamide isethionate) was selected for clinical trials.

Synthesis

When CI-921 was selected for preclinical evaluation in 1983, three different syntheses of the key precursor acridone (**28**) were evaluated by the CRL (Scheme 4) (*65*). The shortest route (method A) is by decarboxylation of the 9-carboxyacridine (**27**), which can be obtained directly by reaction of 7-methylisatin (**26**) with 2-chlorobenzoic acid. However, the yields were only moderate, and the isatin itself had to be prepared. A higher yielding route (method B) resulted from other work in the CRL (*67*) seeking a general route to substituted acridone-4-carboxylic acids. Reaction of 2-iodoisophthalic acid (**29**) with 2-methylaniline gave **30**, which could be cyclodehydrated to give the desired acridone (**28**) in 65% overall yield. However, the necessity to synthesize 2-iodoisophthalic acid made this route commercially unattractive as well.

The most cost-efficient preparation, and the one finally employed in the large-scale synthesis at Warner-Lambert/Parke-Davis (*17*), used the classic Jourdan-Ullmann coupling of 2-chlorobenzoic acid with 3-methylanthranil-

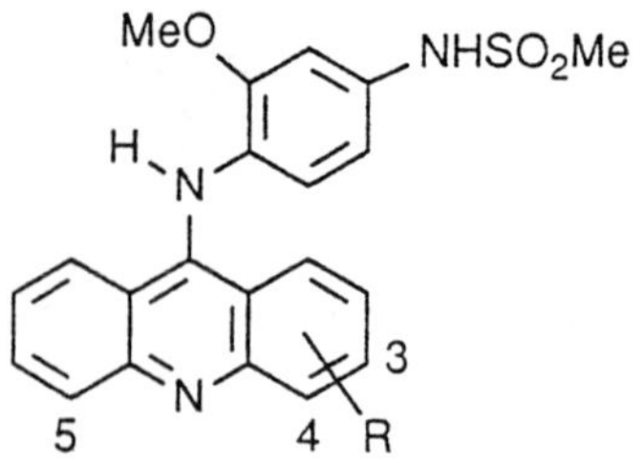

Table 2. Cytotoxicity of Selected Acridine-Substituted Amsacrine Derivatives Against P388 Leukemia and the Lewis Lung Carcinoma

Acridine Substituent(s) R	P388 Leukemia[a]			Lewis Lung Carcinoma[b]		
	OD[c]	%ILS[d]	Survivors[e]	OD	%ILS	Survivors
H (amsacrine: **10**)	13.3	78	0/6	13.3	42	0/6
3-Me	10	120	1/6	13.3	54	0/6
3-OMe	8.9	196	4/6	8.9	35	0/6
3-Cl	20	232	0/6	30	100	0/6
4-Me	20	172	3/6	20	130	0/6
4-OMe	30	139	5/6	45	147	2/6
4-CONHMe	30	127	0/6	37.5	126	4/6
3-Me, 5-CONHMe	30	171	1/6	30	128	0/6
3-OMe, 5-CONHMe	30	182	4/6	30	89	0/6
4-Me, 5-CONHMe (CI-921: **25**)	20	200	3/6	30	192	4/6
4-OMe, 5-CONHMe	20	139	1/6	30	74	0/6

[a] Intraperitoneal inoculation of 10^6 P388/W leukemia cells.
[b] Intravenous inoculation of 10^5 Lewis lung carcinoma cells.
[c] OD is the optimal dose of drug (in milligrams per kilogram per day) administered intraperitoneally on days 1, 5, and 9 after tumor inoculation.
[d] %ILS is the percentage increase in life span of treated animals (at the optimal dose) compared with tumor-bearing control animals.
[e] Average number of long-term survivors/group size.
SOURCE: Data are taken from ref. 56.

ic acid (**31**) (both commercially available). This coupling proceeded virtually quantitatively to **32**, resulting in an overall 80% yield of the desired acridone. Methodology previously developed by the CRL (*68*) then gave 9-chloro-5-methylacridine-4-carbonyl chloride (**33**), which can be treated sequentially with aqueous methylamine to give **34**, followed by treatment with 4-aminomethanesulfon-*m*-anisidide to give the hydrochloride salt of CI-921 (**25**). For clinical trials, this compound was formulated as the isethionate salt (NSC-343499) in a single-vial package (*69*).

Table 3. Comparison of Physicochemical Properties of Amsacrine (10) and CI-921 (25)

Properties Evaluated	*Amsacrine*	*CI-921*
pK_a	8.02	6.99
$E_{1/2}$ (mV)[a]	280	240
K (M^{-1})[b]	1.8×10^5	2.1×10^6
Off-rate (s^{-1})[c]	5.5	108
log P[d]	0.60	1.10
Aqueous solubility (mg/mL)[e]	0.12	0.72

[a] Half-wave potential, measured by cyclic voltammetry (*39*).
[b] For binding to calf thymus DNA, 0.01 ionic strength, 20 °C (*64*).
[c] Slowest off-rate, as measured by stopped-flow spectrophotometry (*80*).
[d] For the cations, computed from chromatographic data (*64*).
[e] For the hydrochloride salts in water at 20 °C (*75*).

Clinical Trials

Compared with amsacrine, CI-921 shows much higher levels of protein binding, resulting in very small free-drug fractions in both animal and human studies (*69–71*). However, although the plasma half-life was correspondingly longer in mice, this was not the case in humans (*69*). The results of Phase I (*69*) and limited Phase II (*72*) clinical trials suggest that CI-921 has activity in non-small-cell cancer and possibly breast cancer, and further studies are proceeding.

Future Developments

Concomitant with the development of CI-921, the CRL continued screening other amsacrine variants for in vivo solid tumor activity. Replacement of the 3′-OMe group with the more powerful electron donors NHMe and NMe_2 gave compounds **35** and **36**, whose acridine-substituted analogues provided two series of compounds with excellent activity against the Lewis lung carcinoma (*73–75*). However, although the best of these appeared equivalent to CI-921 in their profiles of solid tumor activity, no compounds with clear advantages could be identified.

The focus of experimental work with the 9-anilinoacridines has now moved to compounds with activity against multi-drug-resistant (MDR) cell lines. The most widely studied and clinically important form of MDR, particularly evident with (but not restricted to) DNA-binding agents, arises from increased production of a family of membrane-spanning glycoproteins

Table 4. Comparative Activity of Amsacrine (10) and CI-921 (25) Against P388 Leukemia and a Range of Murine and Human Solid Tumors

	Amsacrine			*CI-921*		
Tumor System[a]	*OD*[b]	*%ILS*[c]	*Survivors*[d]	*OD*	*%ILS*	*Survivors*
P388 leukemia	13.3	78	0/6	13.3	42	0/6
Lewis lung carcinoma	13	38	0/10	20	168	14/17
B16 melanoma	7.5	93	1/10	31	157	1/10
LOX human melanoma	10	26	0/6	25	78	0/6
Glasgow osteogenic sarcoma	19	8.1[e]	0/10	25	16.4[e]	1/10
Colon 38	36	13.7[e]	0/11	20.5	6.5[e]	0/11
Mammary 16c	11.9	2.2[e]	0/10	9.5	4.1[e]	0/10
CD8F1 papillary adenocarcinoma	33	7.4[e]	0/10	30	8.9[e]	0/10

[a] Data from ref. 66. Intraperitoneal tumor administration, with intraperitoneal drug dosing on days 1, 5, and 9.
[b] OD is the optimal dose of drug (in milligrams per kilogram per day).
[c] %ILS is the percentage increase in life span of treated animals (at the optimal dose) compared with tumor-bearing control animals.
[d] Average number of long-term survivors/group size.
[e] Growth delay in days.

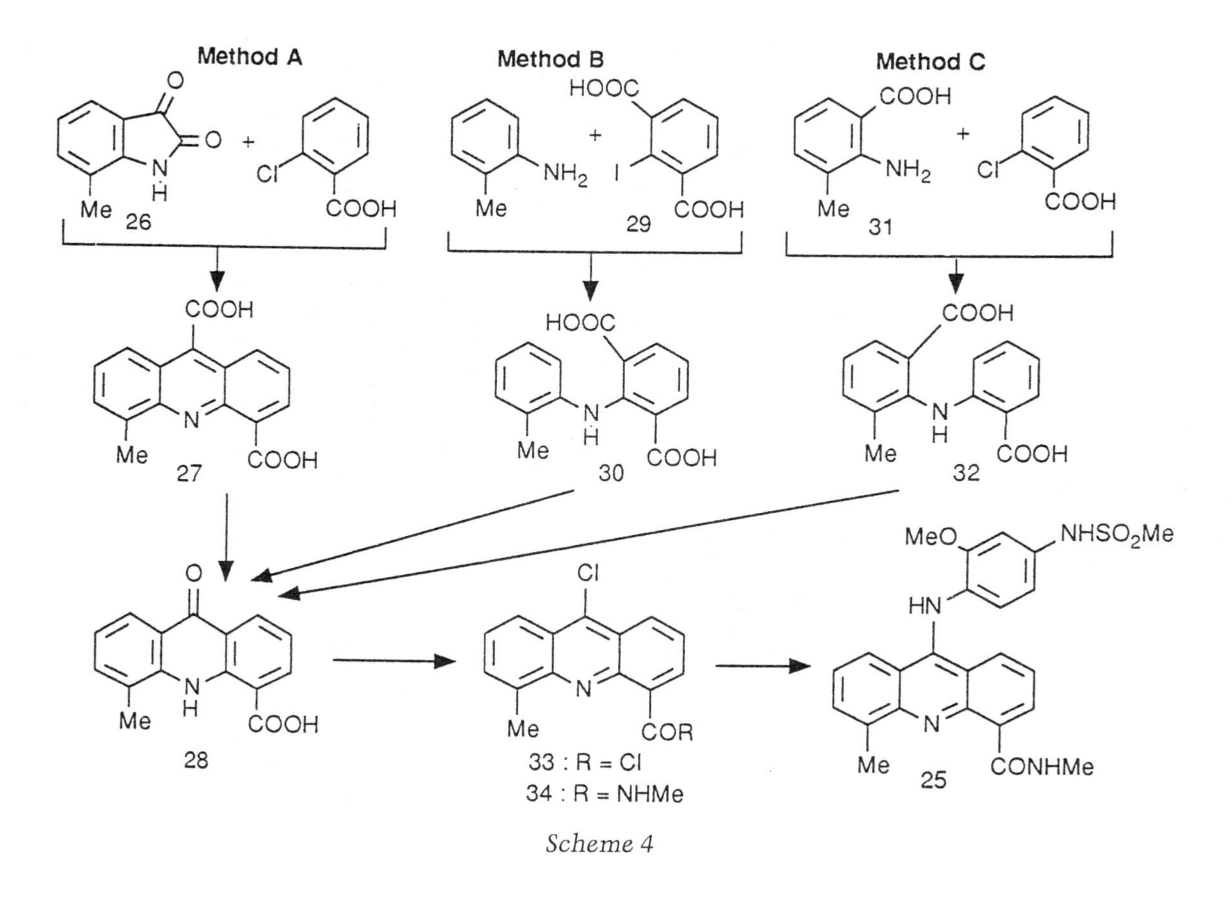
Method A
Method B
Method C
26
27
28
29
30
31
32
33 : R = Cl
34 : R = NHMe
25
COOH
HOOC
NH$_2$
Cl
I
Me
COR
CONHMe
NHSO$_2$Me
MeO
HN

Scheme 4

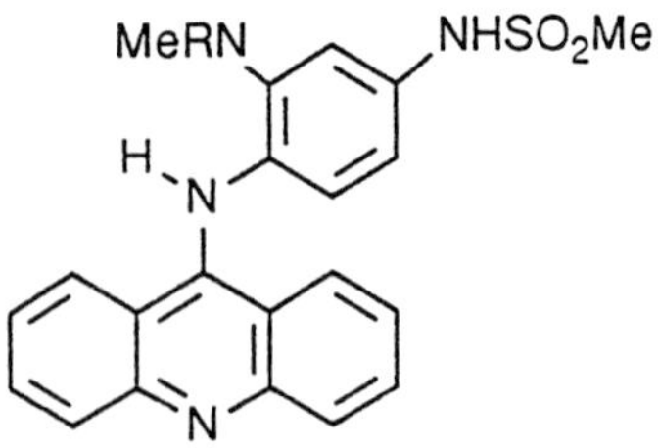

35 : R = H
36 : R = Me

termed P-glycoproteins, which act as an energy-dependent "pump" to accelerate the removal of intracellular drug (76). Both amsacrine and CI-921 show activity against such MDR sublines of P388 cells (77), warranting further study of 9-anilinoacridines in general in this regard. A second form of MDR, evident only with topoisomerase poisons, is altered expression of topoisomerase II isozymes. In both noncycling cells and cells resistant to "classical" topoisomerase II agents, the normally predominant IIα form is replaced by the IIβ isozyme (78). Studies with analogues of amsacrine have shown (77) that redesign of the protein-binding domain (the anilino side chain) can provide analogues (Table 5) with greatly improved relative in vitro and in vivo activity against a resistant subline of P388 leukemia expressing topoisomerase IIβ. Thus, even though amsacrine itself is a potent cytotoxin against wild-type P388, it is 72-fold less effective against the resistant line. Although substitution of the acridine ring of amsacrine alters absolute cytotoxicity, it has little effect on this ratio (77).

However, changes in the anilino ring have dramatic effects, with the 1′-$NHSO_2Ph$ and 1′-NHCOOMe compounds (**37** and **38**, respectively) having virtually equal cytotoxicity against both lines and significant in vivo activity against the resistant line. Several 3′-methylamino-1′-carbamates (e.g., **39**) have also shown significant in vivo activity against the P388 AMSA-resistant line (79).

Summary

The overriding impression left on those of us who have worked with the 9-anilinoacridines is the versatility of this class of compounds. Amsacrine arose

Table 5. Biological Activities of 9-Anilinoacridines Against a P388 Leukemia Line Expressing Topoisomerase IIα (P388/W) and Topoisomerase IIβ (P388/AMSA)

	In Vitro Cytotoxicity		*In Vivo Activity*			
			P388/W		*P388/AMSA*	
Substituent R	*IC_{50}*[a]	*IC_{50} ratio*[b]	*OD*[c]	*%ILS*[d]	*OD*	*%ILS*
1′-$MHSO_2Me$	33	1.3	150	120	150	43
1′-$NHSO_2Me$, 3′-OMe (amsacrine: **10**)	12.5	72	8.9	78	8.9	NA[e]
1′-$NHSO_2C_6H_5$	4.5	1.0	65	100	6.5	61
1′-$NHSO_2C_6H_4pNH_2$	1.1	9.6	3.9	86	3.9	38
1′-NHCOOMe	42	2.2	30	122	20	64

[a] IC_{50} is the concentration of drug (nanomolarity) required to reduce the growth of wild-type P388 (P388/W) cells in culture to 50% of control cultures after 3 days.
[b] IC_{50} is the ratio IC_{50} (P388/AMSA)/IC_{50} (P388/W), the cross-resistance ratio.
[c] OD is the optimal dose of drug in milligrams per kilogram per day.
[d] %ILS is the percentage increase in life span of treated animals (at the optimal dose) compared with tumor-bearing control animals.
[e] No activity (ILS $>$ 20%) at all nontoxic doses.
SOURCE: Data are taken from ref. 80.

37 : R = $NHSO_2Ph$
38 : R = NHCOOMe

39

from a drug development program aimed deliberately at preparing DNA-binding agents. It was the first synthetic DNA-intercalating agent to show useful clinical activity as an anticancer drug, and it has also become a key biological reagent for studying the mechanism of action of the topoisomerase poisons. The analogue CI-921 resulted from a deliberate follow-up program seeking compounds active against solid tumors and having a defined set of physicochemical properties considered to optimize drug distribution. Recent work on the development of analogues tailored to overcome specific forms of drug resistance shows that, even after 20 years, this class of compounds may still provide solutions to the ever more specific problems being probed by drug designers. Additional clinically useful members of the family of 9-anilinoacridines possibly will be found.

Acknowledgments

Core support for this work has been provided over many years by the Auckland Division of the Cancer Society of New Zealand, the Health Research Council of New Zealand, and the Parke-Davis Division of the Warner-Lambert Company.

Literature Cited

1. Purdom, L.; Ambrose, E. J. *Nature (London)* **1958,** *181,* 1586–1588.
2. Hirt, R.; Berchtold, R. *Cancer Chemother. Rep.* **1962,** *18,* 5–7.

3. Yesair, D. W.; Kensler, C. J. In *Antineoplastic and Immunosuppressive Agents*; Sartorelli, A. C.; Johns, D. G., Eds; Springer-Verlag: New York, 1975; Chapter 75.
4. Denny, W. A.; Atwell, G. J.; Baguley, B. C.; Cain, B. F. *J. Med. Chem.* **1979,** *22,* 134–151.
5. Atwell, G. J.; Cain, B. F. *J. Med. Chem.* **1973,** *16,* 673–678.
6. Cain, B. F.; Atwell, G. J.; Seelye, R. N. *J. Med. Chem.* **1969,** *12,* 199–206.
7. Leupin, W.; Chazin, W. J.; Hyberts, S.; Denny, W. A.; Stewart, G. M.; Wuthrich, K. *Biochemistry* **1986,** *25,* 5902–5910.
8. Braithwaite, A. W.; Baguley, B. C. *Biochemistry* **1980,** *19,* 1101–1106.
9. Cain, B. F.; Atwell, G. J.; Seelye, R. N. *J. Med. Chem.* **1971,** *14,* 311–315.
10. Goldin, A.; Serpick, A. A.; Mantel, N. *Cancer Chemother. Rep.* **1966,** *50,* 173–179.
11. Atwell, G. J.; Cain, B. F.; Seelye, R. N. *J. Med. Chem.* **1972,** *15,* 611–615.
12. Cain, B. F.; Atwell, G. J.; Denny, W. A. *J. Med. Chem.* **1975,** *18,* 1110–1117.
13. Cain, B. F.; Seelye, R. N.; Atwell, G. J. *J. Med. Chem.* **1974,** *17,* 992–930.
14. Waring, M. J. *Eur. J. Cancer* **1976,** *12,* 995–1001.
15. Cain, B. F.; Atwell, G. J. *Eur. J. Cancer* **1974,** *10,* 539–549.
16. Dubicki, H.; Parsons, J. L.; Starks, F. W. U.S. Pat. Appl. 25,157, 1981; *Chem. Abstr.* **1981,** *95,* 97615b.
17. Brennan, S. T.; Colbry, N. L.; Leeds, R. L.; Leja, B.; Priebe, S. T.; Reily, M. D.; Showalter, H. D. H.; Uhlendorf, S. E.; Atwell, G. J.; Denny, W. A. *J. Heterocycl. Chem.* **1989,** *26,* 1469–1476.
18. Grove, W. R.; Fortner, C. L.; Wiernik, P. H. *Clin. Pharm.* **1982,** *1,* 320–324.
19. Lawrence, H. J.; Ries, C. A.; Reynolds, B. D.; Lewis, J. P.; Koretz, M. M.; Torti, F. M. *Cancer Treat. Rep.* **1982,** *66,* 1475–1478.
20. Denny, W. A. *Anti-Cancer Drug Design* **1990,** *4,* 241–263.
21. Arlin, Z. *Cancer Treat. Rep.* **1983,** *67,* 967–970.
22. Miller, L. P.; Pyesmany, A. F.; Wolff, L. J.; Rogers, P. C. J.; Siegel, S. E.; Wells, R. J.; Buckley, J. D.; Hammond, G. D. *Cancer* **1991,** *67,* 2235–2240.
23. Denny, W. A.; Atwell, G. J.; Baguley, B. C. *J. Med. Chem.* **1983,** *26,* 1625–1630.
24. Hall, D.; Swann, D. A.; Waters, T. N. M. *J. Chem. Soc., Perkin Trans. 2* **1974,** 1334–1338.

25. Karle, J. M.; Cysyk, R. L.; Karle, I. L. *Acta. Crystallogr., Sect. B.* **1980,** *36,* 3012–3014.
26. Denny, W. A.; Atwell, G. J.; Cain, B. F. *J. Med. Chem.* **1978,** *21,* 5–10.
27. Wadkins, R. M.; Graves, D. E. *Nucleic Acids Res.* **1989,** *17,* 9933-9946.
28. Wilson, W. R.; Baguley, B. C.; Wakelin, L. P. G.; Waring, M. *J. Mol. Pharmacol.* **1981,** *20,* 404–414.
29. Sakore, T. D.; Reddy, B. S.; Sobell, H. M. *J. Mol. Biol.* **1979,** *135,* 763–785.
30. Chen, K. X.; Gresh, N.; Pullman, B. *Nucleic Acids Res.* **1988,** *16,* 3061–3074.
31. Wadkins, R. M.; Graves, D. E. *Biochemistry* **1991,** *30,* 4277–4283.
32. Robbie, M. A.; Baguley, B. C.; Denny, W. A.; Gavin, J. B.; Wilson, W. R. *Cancer Res.* **1988,** *48,* 310–319.
33. Burr-Furlong, N.; Sato, J.; Brown, T.; Chavez, F.; Hurlbert, R. B. *Cancer Res.* **1978,** *38,* 1329–1335.
34. Cain, B. F.; Wilson, W. R.; Baguley, B. C. *Mol. Pharmacol.* **1976,** *12,* 1027–1035.
35. Khan, M. N.; Malspeis, L. *J. Org. Chem.* **1982,** *47,* 2731–2740.
36. Denny, W. A.; Atwell, G. J.; Cain, B. F. *J. Med. Chem.* **1979,** *23,* 1453–1460.
37. Shoemaker, D. D.; Cysyk, R. L.; Padmanabhan, S.; Bhat, H. B.; Malspeis, L. *Drug. Metab. Dispos.* **1982,** *10,* 35–39.
38. Anderson, R. F.; Packer, J. E.; Denny, W. A. *J. Chem. Soc., Perkin Trans. 2* **1988,** 489–496.
39. Jurlina, J. L.; Lindsay, A.; Packer, J. E.; Denny, W. A. *J. Med. Chem.* **1987,** *30,* 473–480.
40. Wong, A.; Cheng, H.-Y.; Crooke, S. T. *Biochem. Pharmacol.* **1986,** *35,* 1071–1078.
41. Shoemaker, D. D.; Cysyk, R. L.; Gormley, P. E.; DeSouza, J. J. V.; Malspeis, L. *Cancer Res.* **1984,** *44,* 1939–1945.
42. Lee, H. H., Palmer, B. D.; Denny, W. A. *J. Org. Chem.* **1988,** *53,* 6042–6047.
43. Robbie, M. A.; Palmer, B. D.; Denny, W. A.; Wilson, W. R. *Biochem. Pharmacol.* **1990,** *39,* 1411–1421.
44. Pryzbylski, M.; Cysyk, R. L.; Shoemaker, D. D.; Adamson, R. H. *Biomed. Mass Spectrom.* **1981,** *8,* 485–488.
45. Paxton, J. W.; Jurlina, J. L. *Pharmacology* **1985,** *31,* 50–56.
46. Paxton, J. W.; Kim, S. N.; Whitfield, L. R. *Cancer Res.* **1990,** *50,* 2692–2697.
47. Jurlina, J. L.; Varcoe, A. R.; Paxton, J. M. *Cancer Chemother. Pharmacol.* **1985,** *14,* 21–25.

48. Liu, L. F. *Annu. Rev. Biochem.* **1989,** *58,* 351–375.
49. Nelson, E. M.; Tewey, K. M.; Liu, L. F. *Proc. Natl. Acad. Sci. U.S.A.* **1984,** *81,* 1361–1364.
50. Rowe, T. C.; Chen, G. L.; Hsiang, Y.-H.; Liu, L. F. *Cancer Res.* **1986,** *46,* 2021–2026.
51. Baguley, B. C.; Denny, W. A.; Atwell, G. J.; Cain, B. F. *J. Med. Chem.* **1981,** *24,* 520–525.
52. Baguley, B. C.; Nash, R. *Eur. J. Cancer* **1981,** *17,* 671–679.
53. Covey, J. M.; Kohn, K. W.; Kerrigan, D.; Tilchen, E. J.; Pommier, Y. *Cancer Res.* **1988,** *48,* 860–865.
54. Heck, M. M. S.; Hittelman, W. N.; Earnshaw, W. C. *Proc. Natl. Acad. Sci. U.S.A.* **1988,** *85,* 1086–1090.
55. Atwell, G. J.; Cain, B. F.; Denny, W. A. *J. Med. Chem.* **1977,** *20,* 1128–1134.
56. Denny, W. A.; Atwell, G. J.; Baguley, B. C.; Rewcastle, G. W. In *QSAR in the Design of Bioactive Compounds*; Kuchar, M., Ed.; J. R. Prous SA: Barcelona, 1984; pp 97–118.
57. Baguley, B. C.; Cain, B. F. *Mol. Pharmacol.* **1982,** *22,* 486–492.
58. Denny, W. A.; Atwell, G. J.; Cain, B. F.; Leo, A.; Panthananickal, A.; Hansch, C. *J. Med. Chem.* **1982,** *25,* 276–316.
59. Hansch, C.; Leo, A. J. *Substituent Constants for Correlation Analysis in Chemistry and Biology*; John Wiley & Sons: New York, 1979.
60. Cain, B. F.; Atwell, G. J.; Denny, W. A. *J. Med. Chem.* **1977,** *20,* 987–996.
61. Denny, W. A.; Atwell, G. J.; Baguley, B. C. *J. Med. Chem.* **1984,** *27,* 363–366.
62. Goldin, A.; Venditti, J. M.; MacDonald, J. S.; Muggia, F. M.; Henney, J. E.; DeVita, V. T. *Eur. J. Cancer* **1981,** *17,* 129–142.
63. Atwell, G. J.; Rewcastle, G. W.; Baguley, B. C.; Denny, W. A. *J. Med. Chem.* **1987,** *30,* 652–658.
64. Baguley, B. C.; Denny, W. A.; Atwell, G. J.; Finlay, G. F.; Rewcastle, G. W.; Twigden, S. J.; Wilson, W. R. *Cancer Res.* **1984,** *44,* 3245–3252.
65. Leopold, W. R.; Corbett, T. H.; Griswold, D. P.; Plowman, J.; Baguley, B. C. *JNCI* **1987,** *79,* 343–349.
66. Denny, W. A. In *The Chemistry of Antitumour Agents* Wilman, D.E.V., Ed.; Blackie: London, 1990; pp 1–29.
67. Rewcastle, G. W.; Denny, W. A. *Synthesis* **1985,** 217–220.
68. Atwell, G. J.; Cain, B. F.; Baguley, B. C.; Finlay, G. J.; Denny, W. A. *J. Med. Chem.* **1984,** *27,* 1481–1485.
69. Hardy, J. R.; Harvey, V. J.; Paxton, J. W.; Evans, P. C.; Smith, S.; Grillo-Lopez, A.; Grove, W.; Baguley, B. C. *Cancer Res.* **1988,** *48,* 6593–6596.

70. Paxton, J. W.; Hardy, J. R.; Evans, P. C.; Harvey, V. J.; Baguley, B. C. *Cancer Chemother. Pharmacol.* **1988,** *22,* 235–240.
71. Kestell, P.; Paxton, J. W.; Evans, P. C.; Young, D.; Jurlina, J. L.; Robertson, I. G. C.; Baguley, B. C. *Cancer Res.* **1990,** *50,* 503–508.
72. Harvey, V. J.; Hardy, J. R.; Evans, P. C.; Paxton, J. W.; Grove, W.; Grillo-Lopez, A.; Baguley, B. C. *Proc. Am. Soc. Clin. Oncol.* **1988,** *5,* 204.
73. Atwell, G. J.; Rewcastle, G. W.; Denny, W. A.; Cain, B. F.; Baguley, B. C. *J. Med. Chem.* **1984,** *27,* 367–371.
74. Atwell, G. J.; Baguley, B. C.; Finlay, G. J.; Rewcastle, G. W.; Denny, W. A. *J. Med. Chem.* **1986,** *29,* 1769–1776.
75. Atwell, G. J.; Rewcastle, G. W.; Baguley, B. C.; Denny, W. A. *J. Med. Chem.* **1987,** *30,* 652–658.
76. Endicott, J. A.; Ling, V. *Annu. Rev. Biochem.* **1989,** *58,* 137–171.
77. Baguley, B. C.; Holdaway, K. M.; Fray, L. M. *JNCI* **1990,** *82,* 398–402.
78. Chung, T. D. Y.; Drake, F. H.; Tan, S. R.; Per, M.; Crooke, S. T.; Mirabelli, C. K. *Proc. Natl. Acad. Sci. U.S.A.* **1989,** *86,* 9431–9435.
79. Rewcastle, G. W.; Baguley, B. C.; Atwell, G. J.; Denny, W. A. *J. Med. Chem.* **1987,** *30,* 1576–1581.
80. Denny, W. A.; Wakelin, L. P. G. *Cancer Res.* **1986,** *46,* 1717–1721.

Index

Index

B

D

E

F

H

I

M

N

Q

R

S

T

Copy editing: P. M. Gordon Associates, Inc.
Indexing: Steven Powell
Production: C. Buzzell-Martin
Cover design: Teddy Vincent Bell
Acquisition: Barbara C. Tansill

Typeset by ATLIS Publishing Services Inc.
Printed and bound by Quinn-Woodbine Incorporated, Woodbine, NJ